PID Control

tapir academic press

ISBN 82-519-1945-2

Layout: Finn Haugen

Cover design: Tapir Academic Press

Printed by Tapir Uttrykk

Binding: Grafisk Produksjonsservice AS

Tapir Academic Press

N–7005 TRONDHEIM

Tel.: + 47 73 59 32 10
Fax: + 47 73 59 32 04
E-mail: forlag@tapir.no
www.tapirforlag.no

Contents

Preface

This book gives an introduction to PID control of dynamic systems. The PID controller (PID = Proportional Integral Derivative) is the dominating (most frequently used) controller function in industry. This book can be used as a text-book in control courses in B.Sc. studies and in M.Sc. studies. It may also serve as a reference for engineers working in the industry.

The book describes the theory, but does not (except in a few cases) describe computer tools for analysis and design. However, lots of supplementary material are available from the homepage of the book on **http://techteach.no**. This material is in the form of documents which describes how analysis, simulation, and design of dynamic systems can be performed in MATLAB, Octave[1], SIMULINK, and LabVIEW. From this homepage there is also a link to KYBSIM (http://techteach.no/kybsim) which is a library of freely available simulators. Many of these simulators are used in this text book.

To benefit from all parts of the book, you must be familiar with systems theory of continuous-time dynamic systems – specifically basic mathematical modeling, differential equations, transfer functions, block diagrams, first and second order systems and frequency response.[2]

The theoretical tools for analysis and design described in this book is for continuous-time feedback control systems. The theoretical tools for analysis and design of discrete-time (sampled) feedback systems are quite similar to tools for continuous-time systems, and they are described in documents available for free on http://techteach.no.

[1] Octave is a free mathematical tool, quite similar to MATLAB, with lots of in-built function categories, like the toolboxes in MATLAB. Octave is available from http://www.octave.org.

[2] These topics are included in the textbook **Dynamic systems – modelling, analysis and simulation** by F. Haugen, Tapir Academic Publisher, 2004. (Information on http://techteach.no.)

A textbook covering advanced control topics building on the present book will be available during 2004. (Information is given on http://techteach.no.)

The book focuses on topics which I have found practically important. I have tried to describe the material in a simple and understandable way. I will appreciate suggestions and comments about both the presentation in the book and the choice of topics (e-mail to finn@techteach.no).

A comment about mathematical notation used in the book: Given a function of time, say $f(t)$. Taking the Laplace transform of $f(t)$ yields, say $F(s)$. Different symbols are used since they are different functions. However, because it is very convenient to do it, I have chosen to use the same symbol for both the time function and the corresponding Laplace transform in this book. So I write $f(s)$ for the Laplace transform of $f(t)$. It is my experience that this style of notation does not cause problems or misunderstandings.

The book is written with the text formatting program Scientific Word. LabVIEW, MATLAB, and SIMULINK are used as computer-based tools for analysis and simulation. Most simulations are performed with LabVIEW.

An exercise book with solutions is available during 2004 (information will be given on http://techteach.no).

A few words about my background: I have a M.Sc. degree (1985) in Engineering cybernetics from the Norwegian Institute of Technology. I have been doing teaching, writing, programming, and consulting since then. I have now a teaching position at the Telemark University College. I also work in my one-man company TechTeach.

I want to thank my family for giving me good working conditions while writing this book.

FinnHaugen

Skien, Norway, August 2004

Chapter 1

Introduction

1.1 The importance of control

Control engineering is a fascinating and important field. In short, control engineering is the methods and techniques used in technical systems having the ability of automatically correcting its own behaviour so that specifications for this behaviour are satisfied. The following process variables are typical objects of control:

- Level or weight (mass)
- Pressure
- Temperature
- Flow
- pH
- Speed
- Position

Due to control engineering, a supply ship will stay at or close to a specified position without anchor; a painting robot paints accurately and smoothly on a car body; The temperature and the composition in a chemical reactor will follow the specifications defined to give an optimal production; A turbine generator produces AC voltage of the specified frequency of 50 Hertz; The pen of an X-Y-plotter draws (follows) a varying voltage signal with great precision; The tool of a rotational cutting machine cuts the work-pieces with high precision; The emission of ammonia from a fertilizer producing factory is kept within limits established by law; The pH value and the composition of Nitrogen, Phosphate and Potassium in the fertilizer which is sent to the market lies between certain quality limits. And many more examples can be given.

Control engineering may be of crucial importance for the following applications:

- **Product quality**: A product will have acceptable quality only if the difference between certain process variables and their setpoint values – this difference is called the *control error* – are kept less than specified values. Proper use of control engineering may be necessary to achieve a sufficiently small control error, see Figure 1.1.

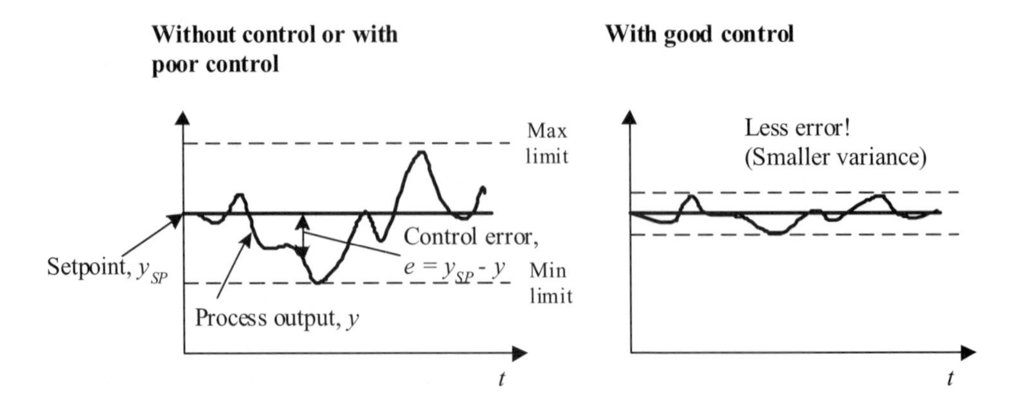

Figure 1.1: Good control reduces the control error

One example: In fertilizers the pH value and the composition of Nitrogen, Phosphate and Potassium are factors which express the quality of the fertilizer (for example, too low pH value is not good for the soil). Therefore the pH value and the compositions must be controlled.

- **Production economy**: The production economy will be deteriorated if part of the products has unacceptable quality so that it can not be sold. Good control may maintain the good product quality, and hence, contribute to good production economy. Further, by good control it may be possible to tighten the limits of the quality so that a higher price may be taken for the product!

- **Security**: To guarantee the security both for humans and equipment, it may be required to keep variables like pressure, temperature, level, and others within certain limits– that is, these variables must be controlled. Some examples:
  - An aircraft with an autopilot (an autopilot is a positional control system).
  - A chemical reactor where pressure and temperature must be controlled.
- **Environmental care**: The amount of poisons to be emitted from a factory is regulated through laws and directions. The application of control engineering may help to keep the limits. Some examples:
  - In a wood chip tank in a paper factory, hydrogen sulfate gas from the cookery is used to preheat the wood chip. If the chip level in the tank is too low, too much (stinking) gas is emitted to the atmosphere, causing pollution. With level control the level is kept close to a desired value (set-point) at which only a small amount of gas is expired.
  - In the so-called washing tower nitric acid is added to the intermediate product to neutralize exhaust gases from the production. This is accomplished by controlling the pH value of the product by means of a pH control system. Hence, the pH control system ensures that the amount of expired ammonia is between specified limits.
  - Automatically controlled spray painting robots avoid humans working in dangerous areas. See Figure 1.2.

Figure 1.2: Spray painting robot (IRB580, ABB)

- **Comfort**:

  - The automatic positional control which is performed by the autopilot of an aircraft to keep a steady course contributes to the comfort of the journey.
  - Automatic control of indoor temperature may give better comfort.

- **Feasibility**: Numerous technical systems could not work or would even not be possible without the use of control engineering. Some examples:
  - An exothermal reactor operating in an unstable (but optimal) operating point
  - Launching a space vessel (the course is stabilized)
  - A dynamic positioning system holds a ship at a given position without an anchor despite the influence of waves, wind and current on the ship. The heart of a dynamic positioning system is the positional control system which controls the thrusters which are capable of moving the ship in all directions. See Figure 1.3.

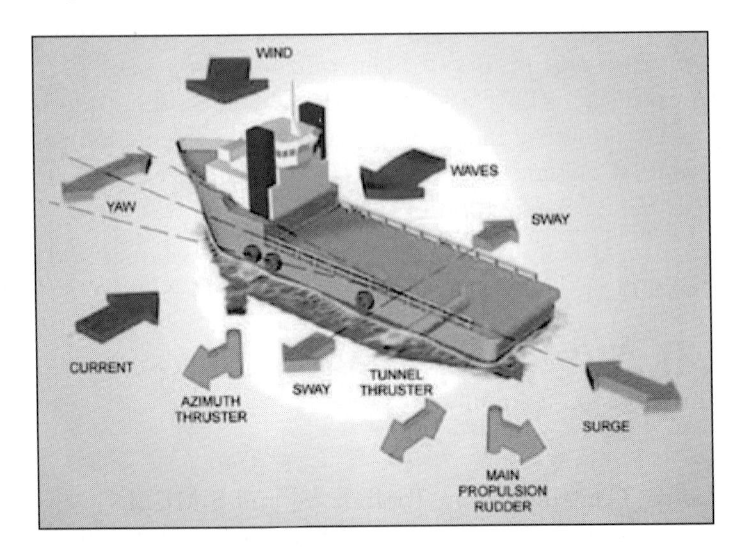

Figure 1.3: A dynamic positioning system holds a ship at a given position without an anchor despite the influence of waves, wind and current on the ship (Kongsberg Simrad, Norway)

- **Automation**: Due to automatic control the operators can perform various tasks in stead of continuously controlling the process, for example perform maintenance or just resting.

1.2 Software tools for analysis and design of control systems

Some typical tasks for software tools for analysis and design of control systems are:

- **Analysis of control systems:**
  - Calculating poles and eigenvalues to observe dynamic properties and stability properties.
  - Calculating frequency response to observe dynamic properties in the term of bandwidth and stability properties.
  - Simulating control systems to observe
    - dynamic properties,
    - control system robustness against noise and parameter variations,
    - implications of nonlinear elements in the control loop, as saturation, hysteresis, etc.
- **Design of control systems:**
  - Calculation of controller parameters on basis of a mathematical model of the control system from specifications to time response, frequency response or stability.
  - Tuning controller parameters by applying an experimental method in a simulator.
  - Trying out various control system structures and control methods on a simulator.

MATLAB[1] with Control System Toolbox [8] and SIMULINK covers the above items. Octave[2], which is a freely available MATLAB-like computer tool for numeric analysis and visualization, includes a set of functions which are similar to the functions of Control Toolbox in MATLAB. LabVIEW[3] with Control Design Toolkit, PID Control Toolkit and Simulation Module also supports the items above. In addition, LabVIEW has powerful tools for developing graphical user interfaces, and has comprehensive I/O-support (Input/Output) to physical processes. (On the

[1] Produced by The MathWorks
[2] http://www.octave.org
[3] Produced by National Instruments

homepage of this book you can find documents and other files which describes using MATLAB, SIMULINK and LabVIEW to such analysis and design, including simulation.)

Computer tools as described above assumes that a mathematical model of the process to be controlled, is available. To develop a precise physics based models for industrial processes is a demanding task. Months of work may be required, except for the most simple processes, as the wood-chip tank described in Example 2.3 (page 19). However, tools are available, as MATLAB's System Identification Toolbox and LabVIEW's System Identification Toolkit, for development of input-output-models in the form of transfer functions or state-space models from experimental data, and these models can be used for analysis and design, as described above. Note that for teaching and training testing (trial) purposes simplified models can be very useful.

There are commercially available simulator for processes, including control systems, based on precise models of processes as heat exchangers, reactors, and columns. The instrumentation diagram of the process constitutes the user interface. Examples of such simulators are Hysys [4] and ASSETT[5].

1.3 A short history of control

Back in 2000 B.C. the Babylonians constructed automatic watering systems based on level control. The old Greeks constructed level control systems for water clocks and oil lamps. The weight control system shown in Figure 1.4 seems to be an automatic bartender.

In the fifteenth and sixteenth century there were made temperature control systems for incubators (heating boxes for eggs), pressure control systems for boilers, and position control systems for wind mills.

In 1788 James Watt constructed a speed control system for a steam engine, see Figure 1.5. Watt's speed control system was based on feedback from measured rotational speed to the opening of the steam valve via a centrifugal controller, which works as follows: The larger the speed, the smaller the valve opening (and steam supply), and vice versa. In this way the speed was held at or near a constant set-point value, despite the disturbances as variations in the steam pressure and changing load torques acting on the engine shaft. Watt's speed control system is regarded as the

[4] Produced by AspenTech

[5] Produced by Kongsberg Simrad

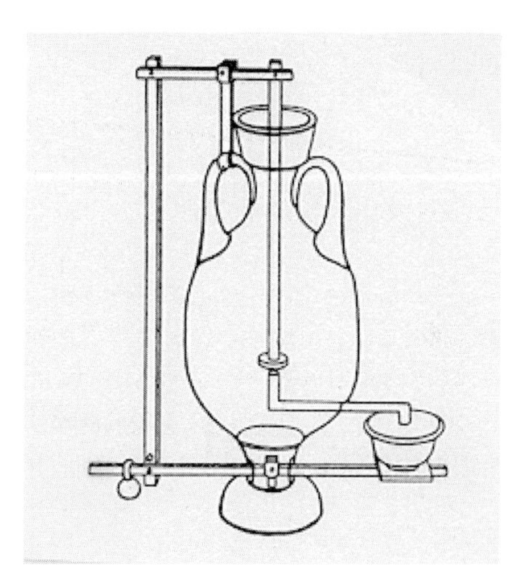

Figure 1.4: A weight control system from the Antics. An automatic bartender? [12]

first industrial application of control engineering.

Watt's control system was not based on any accurate mathematical analysis, but on experiments and trial-and-error. In 1868 James C. Maxwell made a mathematical analysis of the speed control system, and this analysis may be regarded as the stating point of the theoretical methods for analysis and design of control systems.

The field of control engineering and control theory has had an enormous development since 1930. Mechanical and/or pneumatic controllers were developed for the process industry. The first controllers has proportional action only, and later integral and derivative action was implemented. The controller was typically a physical unit mounted on the control valve. There was lack of good methods for tuning the controller parameters. However, this problem was solved by Ziegler and Nichols [20] around 1940. Their controller tuning methods remains among the very best methods available today, and their two methods are described in this book. Their work increased the availability of control engineering in the process industry. Mr. Ziegler was also involved in the first commercial PID controller (Fulscope 100 produced by Taylor Instruments & Co. at the end of the nineteen-thirties).

The big steps, or the new directions, in the control theory have typically been initiated by practical problems which had to be solved. One example is the development of feedback electronic amplifiers with Bell Telephone Lab. in USA in the thirties which led to the *frequency response* methods

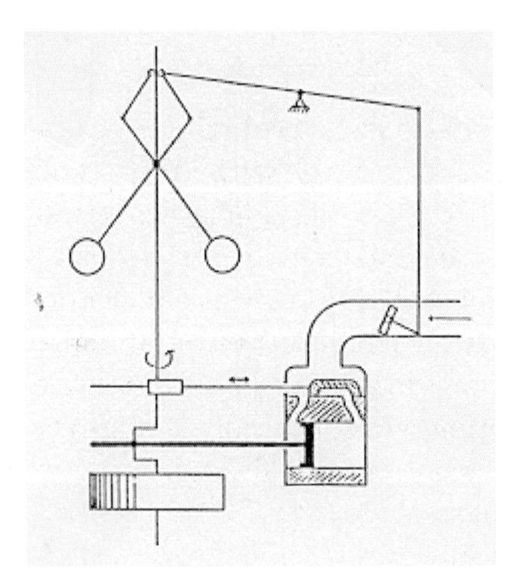

Figure 1.5: Principal diagram of James Watt's speed control system. (Based on [25].)

for analysis and design of feedback amplifiers and feedback control systems. One other example is the development of control systems for radar systems and artillery under The Second World War. The development of the space technology in the Soviet Union and the USA in the fifties and the sixties raised problems which were attempted to be solved by *optimal control* which is formulated by using *state-space methods*. (A state-space model is a set of first order differential equations describing the system.) In an optimal control system there is an optimal balance between the "amount" of control power used and the control error. The optimal solution minimizes a certain optimal criterion.

The development of auto-pilots required *adaptive controllers* as there was a need for control systems which adapted to the varying dynamics of the aeroplane during the flight. The first adaptive controllers were *gain scheduling* controllers, in which the PID parameters are found from a table-lookup in a table or schedule of precalculated PID parameter values. In the 1980's the first generic commercial adaptive PID controllers were introduced. In these adaptive controllers a process model is estimated continuously, and the PID parameters are automatically adjusted from this model.

In the late 1980's and the 1990's there were much interest in *fuzzy control*. Fuzzy control is available in several commercial controllers. The theoretical basis stems from the fuzzy-logic developed by Lotfi Zadeh around 1965. Fuzzy control is particularly suited for processes where the knowledge about how to control it is in the form of an empirically developed set of rules.

From the mid 1980's *model-based predictive control* or MPC has been in the focus of the research of control methods. Several vendors now offer MPC-modules, and MPC has been applied in various industries. MPC is based on a mathematical process model, which can be in the form of a transfer function model or a step-response model or a state-space model. The models used in MPC include the physical limits of the process to be controlled. The MPC algorithm calculates a future sequence of the control variable from an criterion which typically is a criterion containing quadratic terms of the control variable and the control error. From this sequence the first element is used to actually control the process. The MPC algorithm is executed regularly with a fixed time-step. MPC has proven to give good control of difficult processes, as nonlinear multivariable processes with dead-time. We may say that MPC is the next most important control method in the industry today (next to PID control).

Chapter 2

Introduction to feedback control

2.1 Introduction

I this chapter the control problem is defined, and the principle of feedback is introduced as the most important solution to the control problem. Furthermore, standard industrial controller functions based on feedback are described. These are versions of the PID controller. The on/off controller is also introduced. Many practical aspects of the PID controlled are described. The chapter also shows how control systems can be documented in process and instrumentation diagrams – P&I-diagrams or just P&IDs – and block diagrams.

2.2 Terminology. Formulation of the control problem

Control engineering solves a *control problem*. We will soon formulate it, but first we need to define the terminology which will be used.

Figure 2.1 shows a general *block diagram* of the process, which can be of material, mechanical, thermal or electrical type (concrete examples follows soon). Below are definitions of the quantities shown in Figure 2.1.

- **The process** is the physical system which is to be controlled.

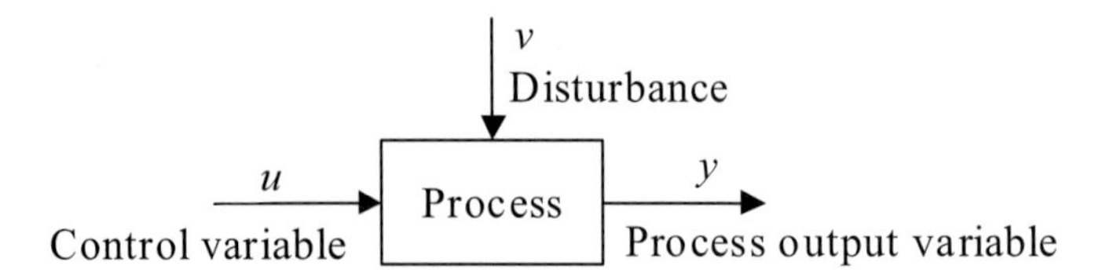

Figure 2.1: Block diagram representation of a process with input and output variables

Included in the process is the actuator, which is the equipment with which (the rest of) the process is controlled.

- **The control variable or the manipulating variable** is the variable which the controller uses to control or manipulate the process. In this book u is used as a general symbol of the control variable. In commercial equipment you may see the symbol MV (manipulating variable).
- **The process output variable** is the variable to be controlled so that it becomes equal to or sufficiently close to the setpoint. In this book y is used as a general symbol of the process output variable. In commercial control equipment PV (process variable or process value) may be used as a symbol.

  The process output variable is not necessarily a physical output from the process! One example: In a heat exchanger where the temperature of the product flow is to be controlled, the temperature – and *not* the product flow – is the process output variable.
- **The disturbance** is a non–controlled input variable to the process which affects the process output variable. This influence is undesirable, and the controller will adjust the control variable to compensate for the influence. In this book v is used as a general symbol for the disturbance.

Above, the control variable, the process output variable and the disturbance were assumed to be scalar values. In general, however, there may be several of these variables, most typically: More than one disturbance.

To formulate the control problem, we need a few more definitions:

- **The setpoint or the reference** is the desired or specified value of the process output variable. The general symbol y_{SP} will be used in this book.

- **The control error** is the difference between the setpoint and the process output variable:

$$e = y_{SP} - y \tag{2.1}$$

Now let us formulate:

The control problem:

Adjust the control variable u so that the control error e is within acceptable limits.

"Within acceptable limits" typically means that *the steady-state or static control error*, e_s, is zero:

$$e_s = \lim_{t \to \infty} e(t) = 0 \tag{2.2}$$

The static control error is the error when all variables have (converged to) constant values.

In practical control systems there are random noise and disturbances acting on various parts of the system, causing the control error to fluctuate randomly around its mean value, see Figure 2.2. The requirement $e_s = 0$

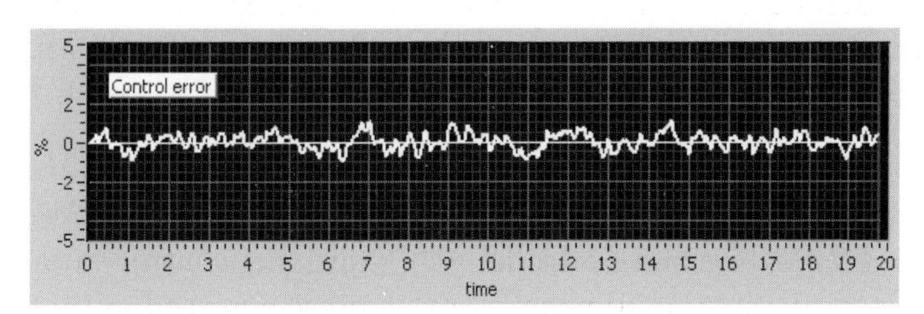

Figure 2.2: In practical control systems the control error fluctuates more or less randomly around its mean value.

then must be interpreted as zero mean value of e. In most of the examples of control systems described in this book, the system is assumed to be noise-free. However, consequences of random measurement noise in control systems is treated in Section 2.7.3.

The present value of the process output variable y defines the *operating point* of the process, for example water level of 3.2 meters or product temperature of 150°C. If y is equal to y_{SP}, we say that the process is in the specified operating point. Usually a specific operating point is steady-state which means that all process variables have constant values. If necessary, the control variables and the disturbances can also be included in the specification of an operating point.

2.3 Solutions to the control problem

2.3.1 Introduction

The control problem is about finding the value of the control variable u so that the control error e becomes sufficiently small. Two ways to try to solve the control problem are as follows:

- Using a *constant control signal*, independent of the present value of the control error.
- Using a *control signal which is continuously adjusted as a function of the control error.*

These two solutions are described in more detail in the following sections.

There is actually a third way to control a process: By continuously calculating the control signal from a mathematical model of the process to be controlled. This control method is however not easy to use in practice since an accurate model expressing the dynamics of the process is not easily available for most processes. The method is called *feedforward control*, and it is described in Section 9.1.

2.3.2 Control using a constant control signal

Using a constant control signal is the simplest way to control a process. Figure 2.3 shows a block diagram of the process controlled by a constant control signal

$$u = u_0 = \text{constant} \tag{2.3}$$

The constant control signal u_0 can be tuned in two ways:

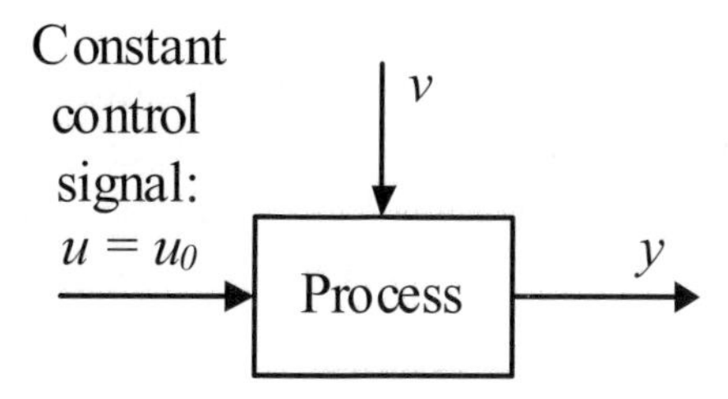

Figure 2.3: Controlling the process with a constant control signal, $u = u_0$.

- **Experimentally**: u_0 is adjusted until we observe that the process output variable y (or its measurement value) is approximately equal to the setpoint y_{SP} in steady-state, and then u_0 is fixed at this value.

- **Calculated from a mathematical process model**: This approach eliminates possibly expensive or time consuming experiments on the physical process, but a mathematical process model is required. The procedure is the same as for finding the nominal control variable used in feedback control, which is described – with a concrete example – in Section 2.6.2.

If there are no changes in the setpoint or in the disturbance, using a constant control variable is an acceptable solution. But if the setpoint or the disturbance varies – a common situation in real control systems – the control error may be too large. A better, but more complicated solution is feedback control which is described in Section 2.3.3.

Example 2.1 ***Controlling a process with constant control signal***

Figure 2.4 shows typical responses for a simulated process controlled by a constant control signal.[1] The control variable u has a constant value, $u_0 = 50$. (The unit of this value is not important here, but it may be percent.) Initially, the setpoint is $y_{SP} = 50$ and the disturbance is $v_0 = 0$. The setpoint y_{SP} is changed as a step from 50 to 70 (so the amplitude is 20) at $t = 5$, and the disturbance v is changed as a step from 0 to -20 (amplitude -20) at $t = 15$. Figure 2.4 shows a *steady-state control error different from zero* (more specifically 20) after the step in y_{SP}, and the error increases to 40 after the step in v.

[End of Example 2.1]

[1] The simulated process is a first order system with time delay.

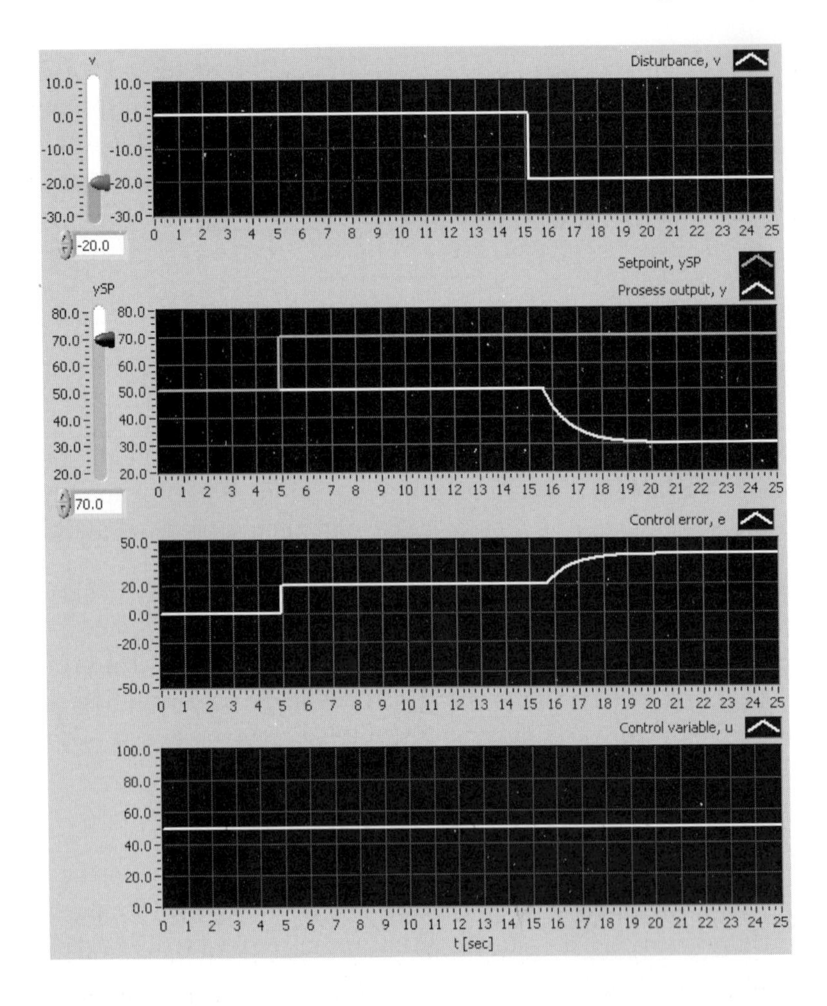

Figure 2.4: Responses in a simulated process controlled by a constant control signal. The control system is excited by a setpoint step and a disturbance step.

The solution of using a constant control variable is sometimes denoted open loop control, since it can be regarded an alternative to closed loop control which is described in Section 2.3.3). The solution can also be regarded as static feedforward, cf. Chapter 9.1.

2.3.3 Control using error-based control signal (feedback control)

The problem with control with constant control signal, cf. Section 2.3.2, is that there is no adjustment of the control variable if there are changes in the setpoint or in the disturbance. Consequently the control error can be different from zero and maybe too large. We should expect better control, that is, smaller control error, if the control signal is calculated continuously *as a function of the control error*. Since the error $e = y_{SP} - y$, y must be *measured*. Figure 2.5 illustrates this solution, which is error-driven control or *feedback control* since there is a connection from the process output variable y back to the control variable (the process input) u. The loop which consists of process, sensor and controller is called the *control loop*.

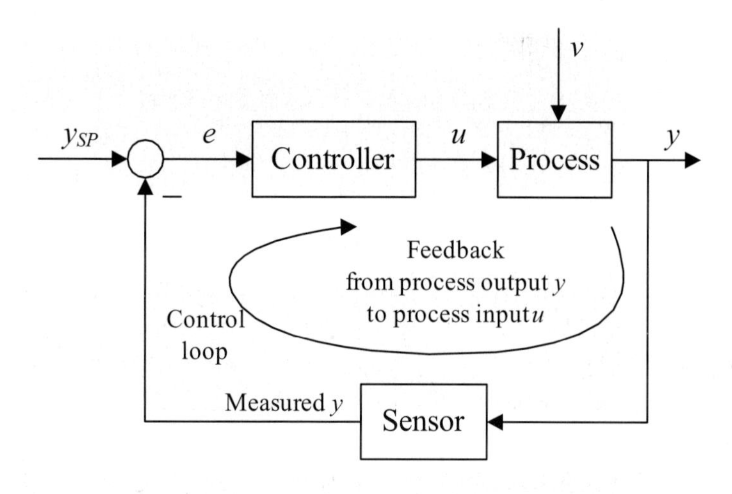

Figure 2.5: Feedback control

The calculation of u takes place in *the controller*. The term controller here means *controller function*, which usually is implemented in a computer program in the control equipment. We will use the term controller for both the control function and the physical equipment in which the control function is implemented.

Example 2.2 ***Controlling a process with error-based control signal***

Figure 2.6 shows typical responses in a feedback control system for a simulated process (the process is the same as used in the simulation shown in Figure 2.4).[2] The control variable u has initially value 50, the setpoint has value 50, and the disturbance has value 0. The control system is excited by a step in the setpoint y_{SP} from 50 to 70 (amplitude 20) at $t = 5$, and with a step in the disturbance v from 0 to -20 (amplitude -20) at $t = 15$. Figure 2.6 shows responses in the control variable u and in the process output variable y. We see that the control variable changes value when the control error e changes value (from zero), which takes place after

[2]The simulated process is a first order system with time delay. The controller is a PI-controller.

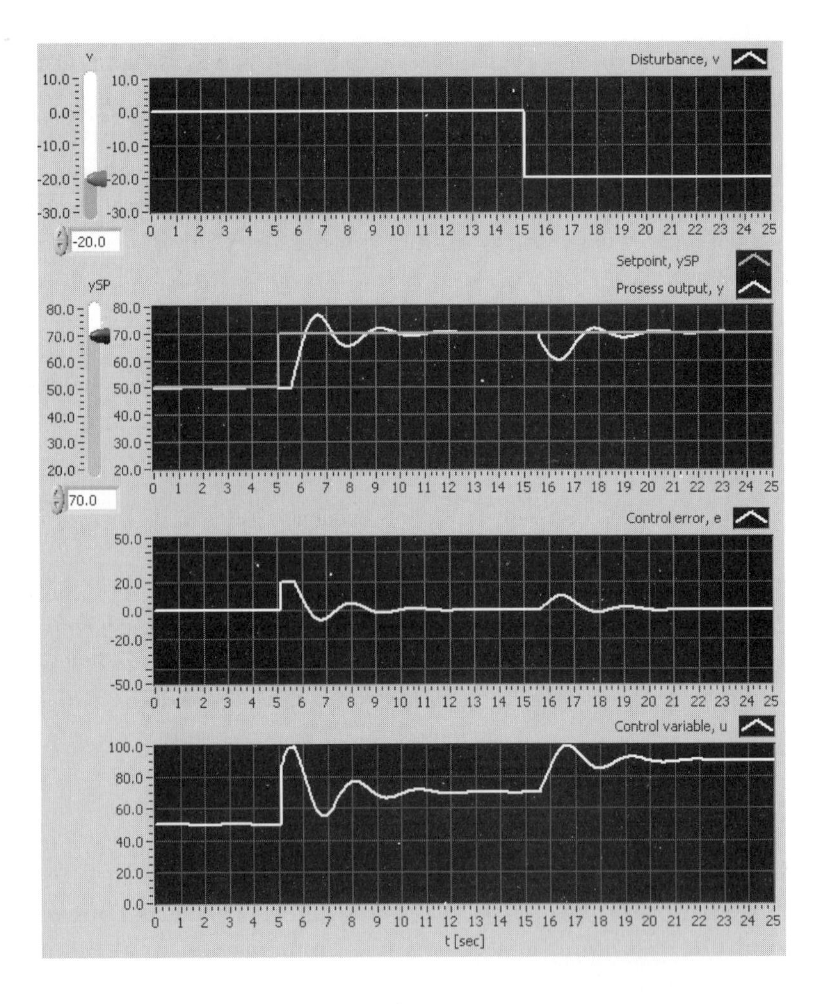

Figure 2.6: Typical responses in a feedback control system, where the control variable is continuously calculated as a function of the control error. The control system is excited by a setpoint step and a disturbance step.

the steps in y_{SP} and v. In this simulated control system the static control error, e_s, becomes zero both after the steps in y_{SP} and after the step in v. This is a *large improvement* compared to using a constant control variable, cf. Example 2.1 and Figure 2.4.

[End of Example 2.2]

A control system which is capable of getting static control error for any constant setpoint value and any constant disturbance value is said to give *perfect static control.*

2.4 Examples of control systems. Documentation with P&I diagram and block diagram

Sections 2.3.2 and 2.3.3 indicated that the control error becomes smaller with feedback control (error-based control) than with a constant control variable. Therefore, feedback control is the main control principle. In the following several examples of feedback control systems are described. These are control system for a wood-chip tank, a heated liquid tank, a motor, and a shower. The control systems, except the latter, are described or documented in two ways:

- **Process and instrumentation diagram or P&I diagrams** which is a common way to document control systems in the industry. This diagram contains easily recognizable drawings and symbols of the process to be controlled, together with symbols for the controllers and the sensors and the signals in the control system. Appendix A gives a small overview over some of the most frequently used symbols in P&I diagrams. There are international standards for instrumentation diagram, but you must expect that company standards are used.
- **Block diagram**, which are useful in principal and conceptual descriptions of control systems.

Example 2.3 ***Level control of a wood-chip tank***

Figure 2.7 shows a P&I diagram and a block diagram of a level control system for a wood-chip tank with feed screw and conveyor belt (which moves with constant speed).[3] Chip is consumed via a outlet screw in the bottom of the tank. This outflow is a disturbance to the control system. The chip level h shall be controlled to be equal or approximately equal to a given level setpoint h_{SP}.[4] LT (Level Transmitter) represents the level sensor. (The levels sensor is based on ultrasound: The level is calculated from the reflection time for a sound signal emitted from a transmitter to a

[3] Such a tank is in the beginning of the process string in the paper mass factory Södra Cell Tofte in Norway.

[4] A few words about the need for a level control system for this chip tank: Hydrogene sulphate gas from the cookery is used to preheat the wood chip. If the chip level in the tank is too low, too much (stinking) gas is emitted to the athmosphere, causing pollution. With level control the level is kept close to a desired value (set-point) at which only a small amount of gas is expired. The level must not be too high, either, to avoid overflow and reduced preheating temperature increase.

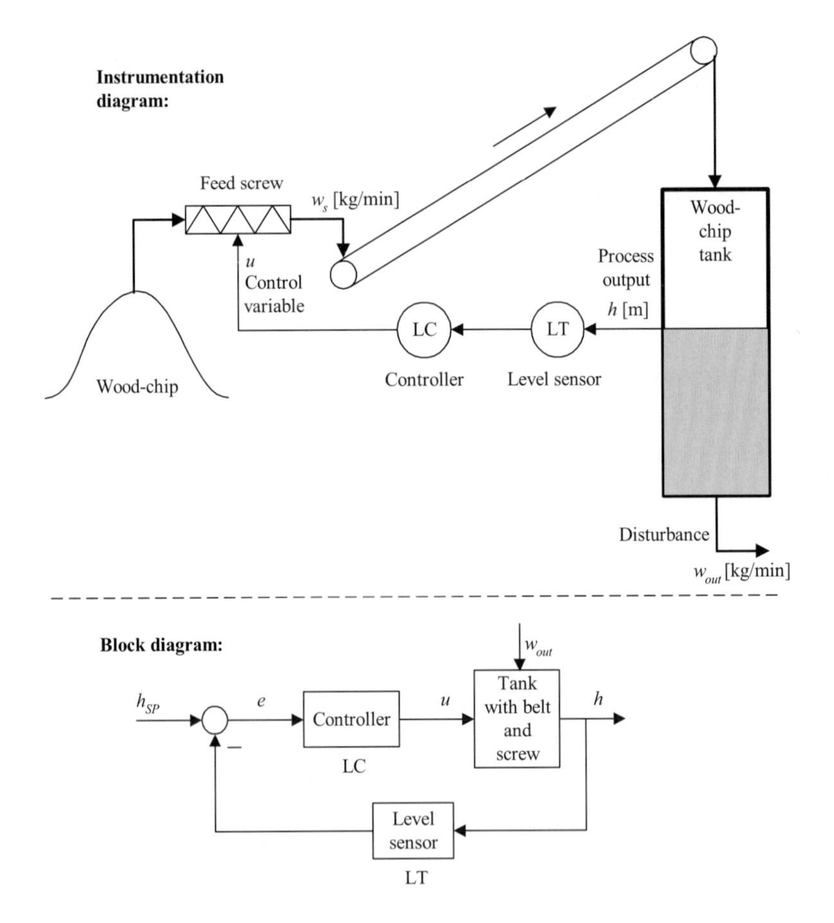

Figure 2.7: P&I diagram and block diagram of a level control system for a wood-chip tank

receiver.) LC is the Level Controller. The setpoint is usually not shown explicitly in an P&I diagram (it is included in the LC block). The controller controls the chip level by manipulating the (rotational speed of the) feed screw.

[End of Example 2.3]

Example 2.4 *Temperature control of heated liquid tank*

Figure 2.8 shows a P&I diagram and a block diagram of a temperature control system for a heated liquid tank with continuous inlet and outlet flow. This process can represent a heat exchanger or a heated reactor in a process line. The temperature T is to be controlled. The temperature setpoint is T_{SP}. Important disturbances of the temperature control system are the inflow temperature T_{in} and the environmental temperature T_e. The controller controls the temperature by manipulating the power via the control signal u to the power amplifier.

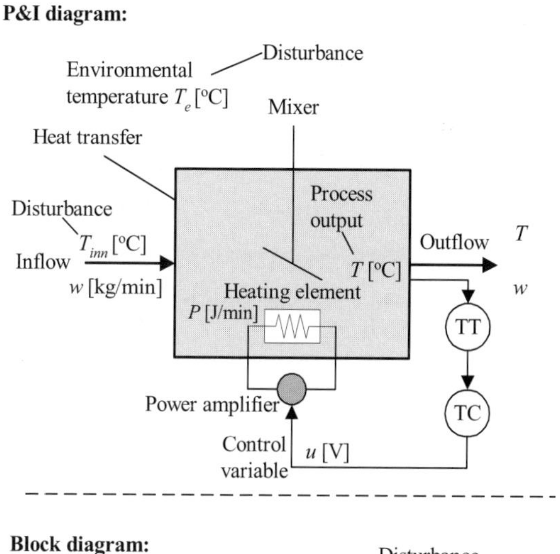

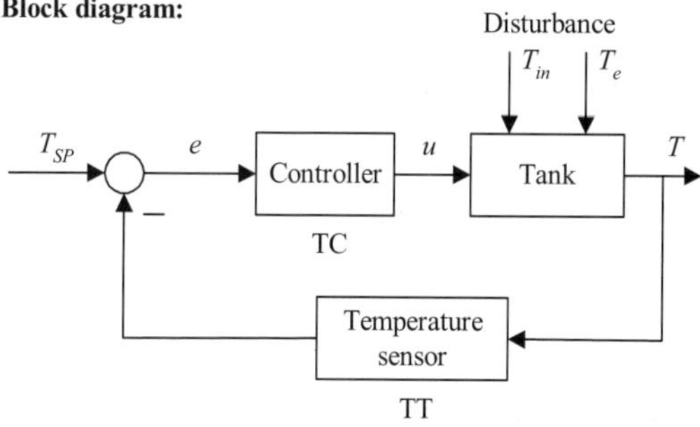

Figure 2.8: P&I diagram and a block diagram for a temperature control system for a heated liquid tank

[End of Example 2.4]

Example 2.5 ***Speed control of motor***

Figure 2.9 shows a P&I diagram and a block diagram of a speed control system for a motor (electrical or hydraulic). The load can be a tool or a conveyor belt. The rotational speed n is to be controlled. The speed setpoint is n_{SP}. The motor is influenced by a load torque T_L, which is a disturbance on speed control system. The controller controls the rotational speed by manipulating the power supplied to the motor.

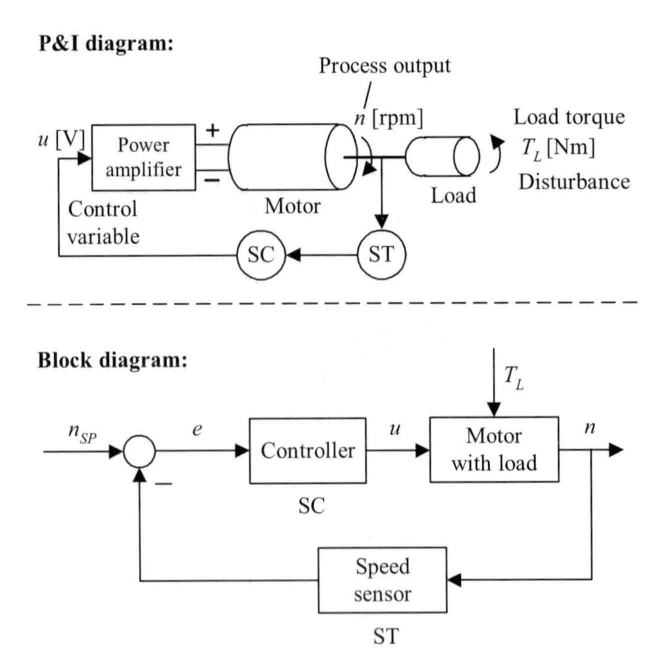

Figure 2.9: P&I diagram and a block diagram of a speed control system for a motor. SC is speed controller. ST (Speed Transmitter) is speed transmitter.

[End of Example 2.5]

Example 2.6 ***Control of shower water temperature***

When we adjust the position of the crane in the shower on the basis of the hand measured temperature to obtain a pleasant water temperature, we actually implement feedback temperature control. The hand is the sensor, the brain is the controller, and the nerve signal which via the (other) hand controls the crane position, is the control variable. Figure 2.10 shows the process and a block diagram of the temperature control system.

[End of Example 2.6]

2.5 Function blocks in the control loop

We will now take a closer look at the function blocks in a control loop. The level control system for the wood-chip tank, cf. Example 2.3, will be used as a concrete example. Figure 2.11 shows a detailed block diagram of

The process:

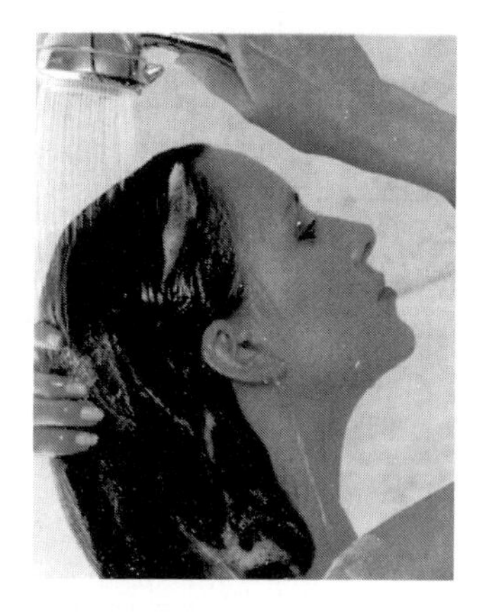

Block diagram:

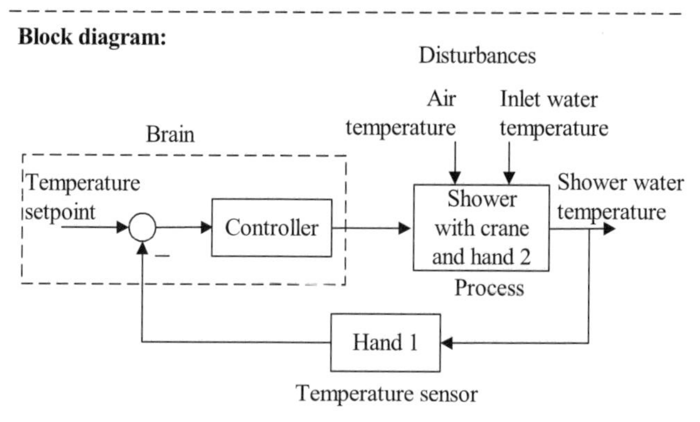

Figure 2.10: Feedback control of shower water temperature

the level control system. The block diagram contains a switch between *automatic mode* and *manual mode* of the controller:

- **Automatic mode:** The controller calculates the control signal using the control function (typically a PID control function).
- **Manual mode**: A human operator may adjust the control variable directly on the equipment, with the control (PID) function being inactive. The process is controlled by the manually adjusted control variable signal u_0 – *the nominal control value* . Typically, u_0 can not be adjusted in automatic mode. Its value is however included in the control signal in switching from manual to automatic mode to ensure bumpless transfer between the control modes.

  The operator sets the controller into manual mode during controller tuning or maintenance (you can then just not turn off the controller, otherwise e.g. the reactor would stop).

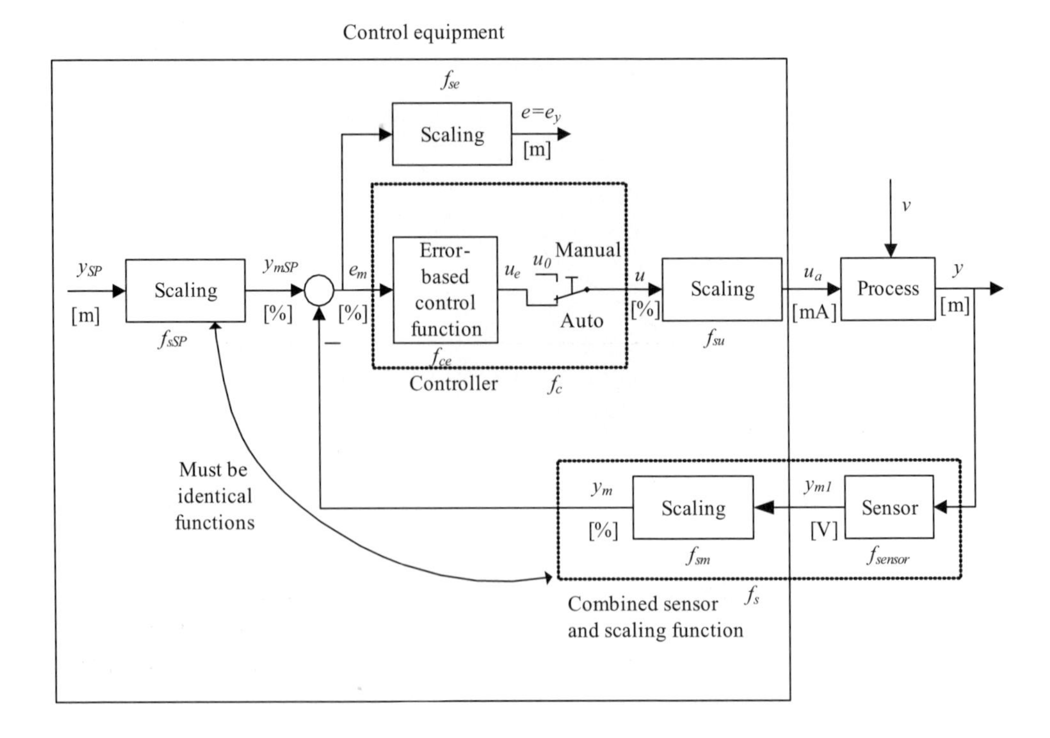

Figure 2.11: Detailed block diagram of the level control system

On commercial control equipment the operator typically can switch the controller between automatic and manual mode e.g. via a physical button or a menu on a screen.

The block diagram in Figure 2.11 contains scaling blocks for calculating the signals in proper units. Let us look at some of the blocks.

- **Sensor or measurement function**: The relation between the process output variable y and the physical sensor signal y_{m_1} can be expressed by the *sensor or measurement function* f_{sensor}:

$$y_{m_1} = f_{sensor}(y) \tag{2.4}$$

The sensor function is in most cases linear and can then be written on the form:

$$y_{m_1} = \underbrace{K_{m_1}(y - y_0) + y_{m_{1_0}}}_{f_{sensor}(y)} \tag{2.5}$$

where K_{m_1} is the measurement gain and y_0 is the value of the process signal y which gives the measurement value $y_{m_{1_0}}$. Figure 2.12 illustrates (2.5).

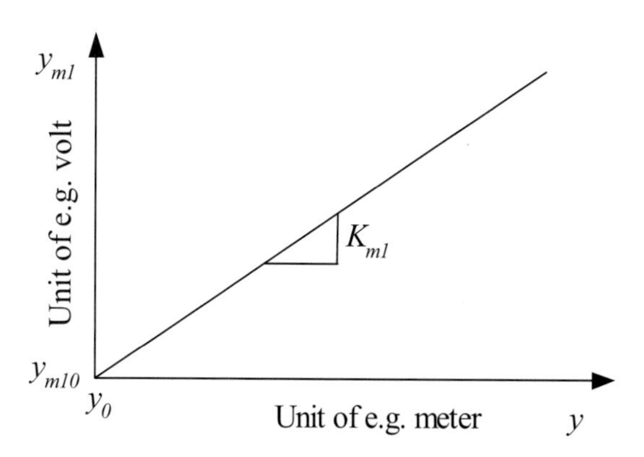

Figure 2.12: The sensor or measurement function $y_{m_1} = K_{m_1}(y - y_0) + y_{m_{1_0}}$

Usually the physical measurement signal y_m is a voltage signal or a current signal. For industrial applications a number of standard signal ranges of measurement signals are defined. Common standard ranges are 0 — 5V, −10 — +10V and 4 — 20mA.

Example (the wood-chip tank): Assume that the measurement signal range is 0 — 5V, and that this range corresponds to 0 — 15m (linearly). Thus,

$$y_0 = 0\text{m},\ y_{m_{1_0}} = 0\text{V},\ K_{m_1} = \frac{5 - 0\text{m}}{15 - 0\text{V}} \approx 0.33\text{V/m} \tag{2.6}$$

- **Scaling of measurement signal**: It is quite usual that the measurement signal is in unit % and that the %-value is used in the control function. The measurement signal y_{m1} in unit V is scaled to a corresponding signal y_m in unit % by the scaling function f_{sy}. With a linear f_{sy},

$$y_m = \underbrace{K_{sm}(y_{m_1} - y_{m_{1_0}}) + y_{m_0}}_{f_{sm}} \tag{2.7}$$

Commercial control equipment contains functions for scaling. During configuration of the controller, the user typically gives information about the minimum and the maximum values of the physical sensor signal, e.g. 4mA and 20mA, and the minimum and the maximum values of the scaled measurement signal, e.g. 0% and 100%. From this information the controller automatically configures the scaling function (2.7).

Example (wood-chip tank): Assume that the 0 — 5V measurement signal range corresponds to the range 0 — 100%. Then,

$$y_{m1_0} = 0\text{V};\ y_{m0} = 0\%;\ K_{sy} = \frac{100 - 0\%}{5 - 0\text{V}} = 20\%/\text{V} \tag{2.8}$$

- **Combined scaling and sensor/measurement function**: The function f_s in Figure 2.11 is the combined function of (2.4) and (2.7). If both these functions are linear, the combined function is on the form

$$y_m = f_s(y) = K_m(y - y_0) + y_{m0} \tag{2.9}$$

Example (wood-chip tank): The combined scaling and measurement function from level in meter to level measurement signal in percent can be found from (2.4) and (2.7). However, in this case it can more easily be set up from the information that 0 – 15m corresponds to 0 – 100%. Thus,

$$y_0 = 0\text{m},\ y_{m0} = 0\%,\ K_m = \frac{100 - 0\%}{15 - 0\text{m}} = 6.67\%/\text{m} \tag{2.10}$$

- **Setpoint scaling**: The setpoint and the measurement signal must of course be represented with the same unit, otherwise subtracting the measurement signal from the setpoint is meaningless. The setpoint must be scaled using a scaling function which is equal to the combined scaling function for the measurement signal, (2.9). Thus,

$$y_{mSP} = \underbrace{K_m(y_{SP} - \overbrace{y_{SP_0}}^{=y_{m0}}) + y_{mSP_0}}_{f_{sSP} = f_s} \tag{2.11}$$

- **The controller** calculates the control variable according to the control function (control functions are described in Section 2.6).

- **Scaling the control variable**: If the controller calculates the control variable u in % (which is quite usual), then a scaling of the %-value to the physical unit used by the actuator, is necessary. A linear scaling function is

$$u_a = \underbrace{K_{su}(u - u_{p0}) + u_{a0}}_{f_{su}} \tag{2.12}$$

Example (wood-chip tank): Assume that the control variable in the range 0 – 100% is to be scaled to the range 4 – 20mA. In this case

$$u_{p0} = 0\%,\ u_{a0} = 4\text{mA},\ K_{su} = \frac{20 - 4\text{mA}}{100 - 0\%} = 0.16\text{mA}/\% \tag{2.13}$$

- **Scaling the control error**: To scale the control error e_m from measurement unit (typically %) to the unit used of the process output variable y, the following scaling function can be used:

$$e_y = \underbrace{K_{se}}_{f_{sy}} = \frac{1}{K_m} e_m \tag{2.14}$$

where K_m is the same as in (2.9).

Example (wood-chip tank): Scaling of the control error in % to meter is realized by

$$K_{se} = \frac{15 - 0\text{m}}{100 - 0\%} = 0.15\text{m}/\% \tag{2.15}$$

2.6 Controller functions

2.6.1 Introduction

Figure 2.11 shows where the control function – usually denoted simply "controller" – is placed in a control loop. The most common control functions are the following:

- On/off controller
- P controller (proportional)
- PD controller (proportional-derivative)
- PI controller (proportional-integral)
- PID controller (proportional-integral-derivative)

The P-, PD- and PI controllers can be derived from the PID controller. The PI- and the PID controller is by far the most commonly used in the industry since they gives best control: Zero static control error is achieved. The derivative term creates problems due to amplification of high frequent measurement noise, so that the control variable would become noisy, and therefore the D-term is not used in many practical cases. Thus, the PI controller is probably most frequently used.

The On/off controller is particularly easy to implement using an electrical or a mechanical on/off-element (as in a thermostat) or by simple expressions in a control program. However, the On/off controller gives a somewhat imprecise control because there will be sustained oscillations in the control loop. The On/off controller may be used for tuning of a PID controller, cf. Section 4.5.

The control functions which are described in detail in the following, are all on the following form:

$$u = u_0 + u_e \tag{2.16}$$

where u_0 is the nominal (manually adjusted) control variable and u_e is a function of the control error e, cf. Figure 2.11. (2.16) is illustrated in Figure 2.13. u_0 is the control signal required to keep the process in or close

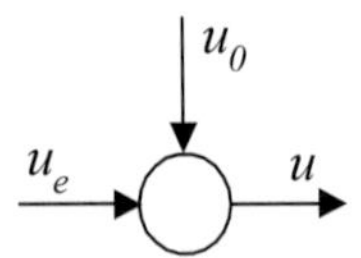

Figure 2.13: The control variable (or signal) is calculated as $u = u_0 + u_e$.

to the nominal (specified) operation point when the controller is in manual mode. u_0 can be adjusted by the operator while the controller is in manual mode, but is typically fixed while the controller is in automatic mode. u_0 may be used by the control function as an initial value of the control variable u when the controller is switched from manual to automatic mode. The term u_e in (2.16) represents the feedback term or the error based term, which gives compensation for changes in the setpoint or in the disturbances.

Tuning u_0 is described in Section 2.6.2. Various control functions producing (calculating) the feedback term u_e are described in Sections 2.6.3 – 2.6.7.

2.6.2 Tuning the nominal control signal

The nominal control signal u_0 can be tuned in two ways:

- **Experimentally**: u_0 is adjusted until we observe that the process output variable y (or, more correctly: its measurement value) is approximately equal to the setpoint y_{SP} in steady-state, and then u_0 is fixed at this value.

  Example: Figure 2.7 shows a feedback control system for a wood-chip tank. The nominal control signal is adjusted until it is observed that the level is constant and approximately equal to the level setpoint.

- **Calculated from a mathematical process model:** This approach eliminates possibly expensive experiments on the physical process, but it is required that a mathematical model exists. The procedure is as follows: *u_0 is calculated as the steady-state or static solution u of the equations which constitute the static process model in which the setpoint y_{SP} is substituted for the process output variable y.* The static process model is derived from a dynamic model in the form of differential equations by setting all time-derivatives equal to zero and neglecting any time-dependencies, as time delays.

Example 2.7 ***Nominal control signal for wood-chip tank***

A level control system for a wood-chip tank is described in Example 2.3 (page 19). We will now calculate the nominal control signal u_0 for the following operation point: The level h is equal to the level setpoint h_{SP}, and the chip outflow w_{out} has a constant (static) value of w_{out_s}.

We need a mathematical model. Assume the following: h [m] is the level. A [m^2] is the cross sectional area. ρ [kg/m^3] is the chip density. ρAh [kg] is the mass of chip in the tank. w_{in} [kg/min] is the chip inflow from the belt. w_s [kg/min] is the chip inflow to the belt from the screw. w_{out} [kg/min] is the chip outflow from the outlet in the bottom of the tank. K_s [(kg/min)/%] is the screw gain. τ [min] is the time delay or transport delay on the conveyor belt, which runs with constant speed. u [%] is the control variable.

Mass balance of the chip in the tank yields

$$\frac{d\left[\rho Ah(t)\right]}{dt} \equiv \rho A\dot{h}(t) = w_{in}(t) - w_{out}(t) \tag{2.17}$$

$$= w_s(t-\tau) - w_{out}(t) \tag{2.18}$$

$$= K_s u(t-\tau) - w_{out}(t) \tag{2.19}$$

or, rearranged,

$$\dot{h}(t) = \frac{1}{\rho A}\left[K_s u(t-\tau) - w_{out}(t)\right] \tag{2.20}$$

A *static* process model can be found from the dynamic the model (2.20) by setting the time-derivative equal to zero and neglecting the time delay:

$$\rho A\dot{h}_s(t) = 0 = \underbrace{K_s u_0}_{w_{in_s}} - w_{out_s} \tag{2.21}$$

or[5]

$$u_0 = \frac{w_{out_s}}{K_s} \quad [\%] \tag{2.22}$$

This control signal will keep the chip level at a constant value, but at which value? Actually, at any level, since (2.22) does not contain information of the level – it just ensures that the inflow and the outflow are equal. For this process – and for many other tank processes - it is insufficient to use a fixed control signal to control the level. The control variable must contain an error-based term or feedback term to obtain control of the level, as explained in the following sections.

[End of Example 2.7]

2.6.3 On/off controller

The On/off controller calculates the control variable u according to (2.16), which is repeated here:

$$u = u_0 + u_e \tag{2.23}$$

The nominal control signal u_0 can be calculated as explained in Chapter 2.6.2. With the On-off controller the error-based or feedback term u_e is calculated as a function of the control error e as follows:

$$u_e = \left\{ \begin{array}{c} A \text{ for } e \geq 0 \\ -A \text{ for } e < 0 \end{array} \right\} \tag{2.24}$$

where A is the *amplitude*. The control error e has same unit as for the setpoint and the process measurement in the expression $e = y_{SP} - y$, typically %, or V, mA, m, ^{o}C or some other. Figure 2.14 illustrates u_e and u.

If your controller does not implement an On/off controller, you can get one by using a P controller with a very large (ideally: infinite) value of the controller gain K_p. The P controller is described in detail in Section 2.6.5.

Maybe you recognize the On/off controller from the room temperature control at home? The thermostat is actually a combined temperature sensor and an On/off controller.

Unfortunately, with an On/off controller all signals in the control loop will oscillate continuously (unless some saturation limit is reached). These oscillations comes *automatically*. This can be explained as follows: Assume that the control error e is positive. Then the control signal u equals

[5] I guess you could have come up with this relation without deriving the full dynamic model...

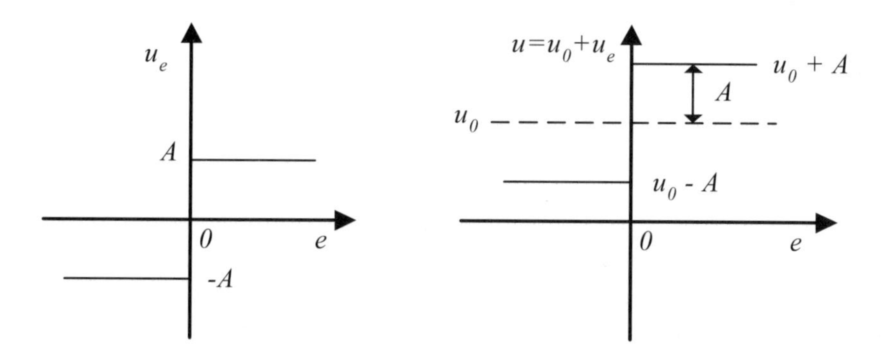

Figure 2.14: The total control signal u and the error-based term u_e in the On/off-controller

$u_0 + A$, which causes the process output variable y and consequently the process measurement y_m to increase. When y_m has increased so much that it becomes greater that the setpoint, the error changes sign, and the control signal becomes $u_0 - A$, which causes the process output variable to decrease, and then the error eventually is positive and the control variable becomes $u_0 + A$, and so on. Thus, there are oscillations.

The oscillations in u are in the form of a square wave. If the actuator is a mechanical device, e.g. a feed screw or a valve, the stepwise movements may cause wear. But if the actuator is an electronic device, e.g. an electronically controlled heating element, there will not be any wear problems. The oscillations in the process output variable y becomes sinusoidal for most processes, but for processes having only integrator dynamics (as a tank), the oscillations are triangular, as we will see in Example 2.8 (below).

Example 2.8 ***On/off-control of chip level of a wood-chip tank***

Figure 2.15 shows the front panel of a simulator for the wood-chip tank.[6] (The front panel also shows parameters of a PID controller, but this controller is not used in this example.) Here is some information about the simulation: The amplitude A is 20%. The initial level is 10 m. The setpoint is initially 10 m and is increased to 12 m at approximately $t = 60$ min. The chip outflow w_{out} (the disturbance) is initially 1500 kg/min and is increased to 1800 kg/min at approx. 120 min. The nominal control

[6] The simulator, which is implemented in LabVIEW, is based on a numeric solution of the differential equation (2.20) which expresses the mass balance of the tank.

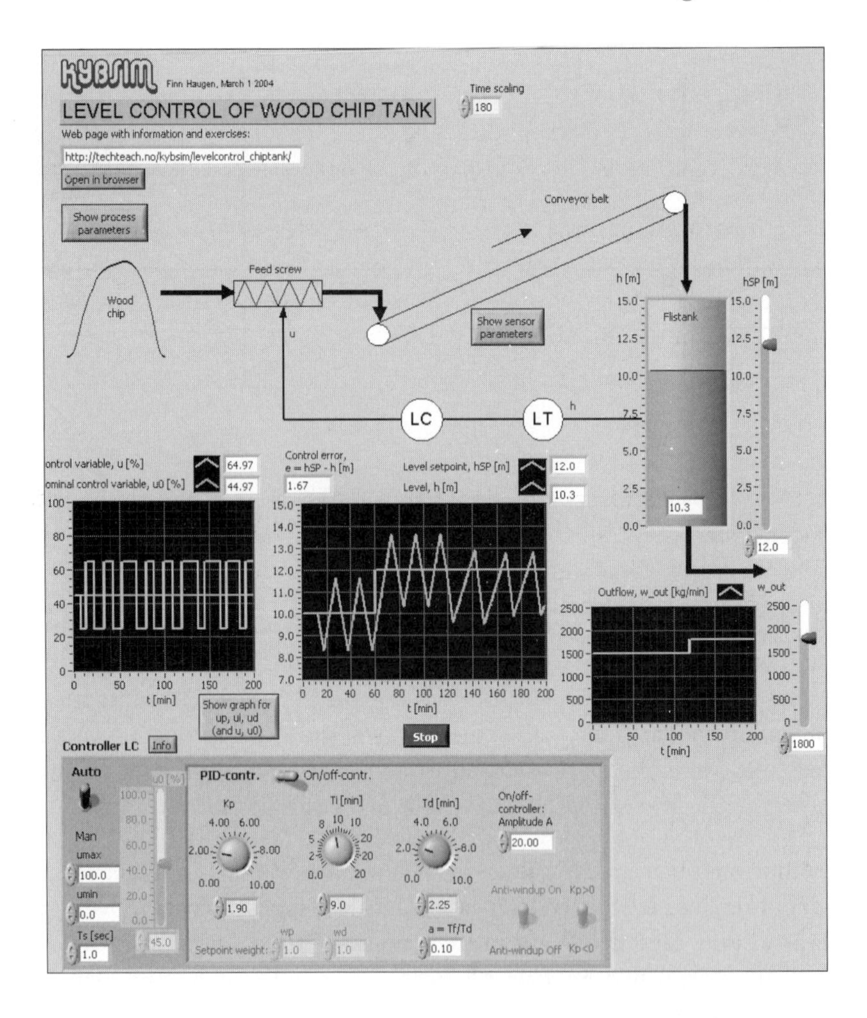

Figure 2.15: Example 2.8: Level control of the wood-chip tank with an On/off-controller. (The front panel shows also PID-parameters, but they are irrelevant in this simulation.)

signal u_0 is 45%, which is calculates according to (2.22) where $w_{out_s} = 1500$ kg/min and $K_s = 33.36$ (kg/min)/%.

The simulation shows the following:

- The control variable oscillates as a square wave. It is symmetric about the nominal control value u_0 (45%). The oscillations become asymmetric if u_0 is no longer correctly tuned (after $t = 120$ min.).
- The oscillations in the level are triangular (not sinusoidal as they may be for many other processes), which is due to the integrator dynamic of the tank (the time-integral of a piecewise constant inflow

is a piecewise ramp).

- The level oscillates about the setpoint with a mean value equal to the setpoint as long as the nominal control signal u_0 has correct value.
- The level oscillates about the setpoint with a mean value which is *not* equal to the setpoint and with an asymmetric form when u_0 has an incorrect value, which is the case after the disturbance was changed from 1500 til 1800 kg/min.

[End of Example 2.8]

In the following sections you will see that the controller can perform far better (without oscillations and, for some of the controllers, with zero static control error) if the error-based term u_e in the control variable (2.16) is calculated using a "softer" and more dynamic function than the abrupt On/off function.

2.6.4 Overview: The PID controller

In the following sections, the P, PI and PID controllers are described. Typically, if you need the only a P- or PI control function, these are achieved as special cases of the PID controller. Therefore, although the PID controller is described in detail later, in Section 2.6.7, it is proper to present the (ideal) PID controller now:

$$u = u_0 + \underbrace{K_p e}_{u_p} + \underbrace{\frac{K_p}{T_i}\int_0^t e\,d\tau}_{u_i} + \underbrace{K_p T_d \frac{de}{dt}}_{u_d} \tag{2.25}$$

The controller parameters are as follows: K_p is the proportional gain. T_i [s] or [min] is the integral time. T_d [s] or [min] is the derivative time. Furthermore, u_0 is the nominal value of the control variable. u_p is the P-term. u_i is the I-term. u_d is the D-term.

From the PID controller (2.25), the P controller and the PI controller can be found from the PID controller as follows:

- A P controller is achieved by setting $T_i = \infty$ (or to a very large value) and $T_d = 0$.[7]

[7] In some commercial controllers you can set T_i to 0 (zero), which is a code expressing that the integral term is de-activated.

- A PI controller is achieved by setting $T_d = 0$.

2.6.5 P controller

The *P controller* (proportional) calculates the control variable according to (2.16) as follows:

$$u = u_0 + \underbrace{K_p e}_{u_p} \tag{2.26}$$

where u_p is the *P-term*. The nominal control variable term u_0 can be found experimentally or by model-based calculations as explained in Section 2.6.2. K_p is the *proportional gain*. Figure 2.16 illustrates the controller function (2.26).

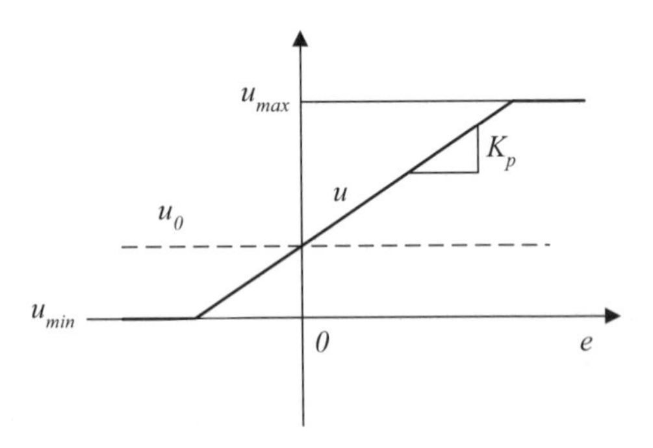

Figure 2.16: P controller as given by (2.26)

In some commercial controllers the *proportional band* P_B, is used in stead of the proportional gain. The proportional band is given by

$$P_B = \frac{100\%}{K_p} \tag{2.27}$$

where K_p is the gain, which here is assumed to be dimensionless. (It will be dimensionless if the control error e and the control variable u have the same unit, typically percent). It is typical that P_B has a value in the range of $10\% \leq P_B \leq 500\%$, which corresponds to K_p being in the range of $0.2 \leq K_p \leq 10$. It is important to note that P_B is inverse proportional to K_p. Thus, a small P_B corresponds to a large K_p, and vice versa.

What does *proportional band* actually mean? One explanation is that P_B is the size of the control error interval Δe (or the size of the measurement signal interval) which gives a control signal interval Δu equal to 100%: From (2.26) we see that $\Delta e = \Delta u/K_p = 100\%/K_p = P_B$.

How does the P controller *work*? Let us look at the wood-chip tank, cf. Example 2.3 (page 19). The P controller changes the control signal proportionally to the error. Assume that the level is less than the setpoint. Then, the control error e is positive, and consequently the controller calculates a control signal change K_pe which is positive, which again increases the chip inflow, so that that the level increases and the error is reduced.

Although the P controller increases the control signal if the control error increases, it will in practice not achieve zero error: Assume that the nominal control value u_0 *does not have a correct value*, so that e is different from zero (the process not in the specified operation point). In this situation, the P controller can not bring e to zero, since if it could, e would be zero, and e was assumed to be different from zero. In other words: *As long as u_0 does not have the correct value (and this is, strictly, always the case), the static control error is different from zero with a P controller.*

The static control error which exists with a P controller, can be reduced by increasing the controller gain K_p, since an increased K_p gives more control variable adjustment, K_pe, for a given error e, and this again gives less error. The drawback of increasing K_p is that the control loop gets *reduced stability*, and if K_p becomes too large, the control loop becomes unstable. The stability of control loops is discussed further in Section 2.11 and in Chapter 6.4.

Example 2.9 ***P control of the level of a wood-chip tank***

Figure 2.17 shows simulated responses for the level control system for wood-chip tank described in Example 2.3 (page 19). The front panel of the simulator is as shown in Figure 2.15 (page 32). The controller gain, tuned with the Ziegler-Nichols' closed loop tuning method, cf. Section 4, is

$$K_p = 1.55 \tag{2.28}$$

The initial level is 10m. The setpoint is initially 10m and is increased to 12m at approx. $t = 10$min. The chip outflow w_{out} (which is a disturbance to the control system) is initially 1500kg/min and is increased to 1800kg/min after approx. 60min. The nominal control signal u_0 is 45%, which is calculated from (2.22) where $w_{out_s} = 1500$kg/min and $K_s = 33.36$ (kg/min)/%. The simulation shows the following:

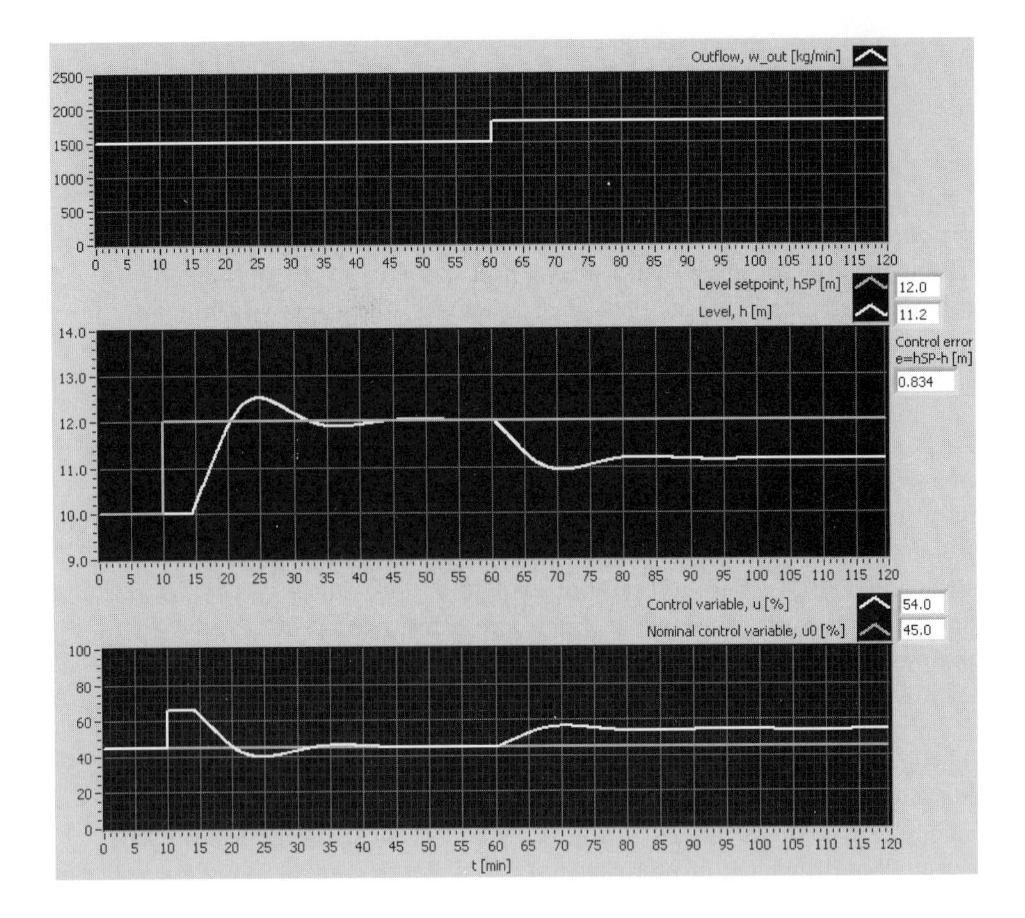

Figure 2.17: Example 2.9: Level control of the wood-chip tank with a P controller. (The front panel of the simulator is as shown in Figure 2.15 on page 2.15.)

- Tracking properties: The steady-state error e_s is zero only as long as the nominal control variable value u_0 has a correct value, which is the case for $t < 60$min.
- Compensation properties: e_s is *different from zero* due to the step in the disturbance w_{out} which comes at $t = 60$min. In Figure 2.15 e_s has value 0.83 at $t = 120$min which is the time when the simulation is stopped. However, if the simulations were run longer (the responses has not converged completely at $t = 120$s) we would have seen that the error actually settles at

$$e_s = 0.87\text{m} \tag{2.29}$$

Finally, what happens if we *increase* K_p? We should expect that e_s is reduced since increasing K_p in the compensation term $u_p = K_p e$ forces e to become smaller. Figure 2.18 shows simulated responses with $K_p = 2.6$. The *steady-state control error will be reduced* from 0.87m to 0.52m, and *the stability of the control loop is reduced.*

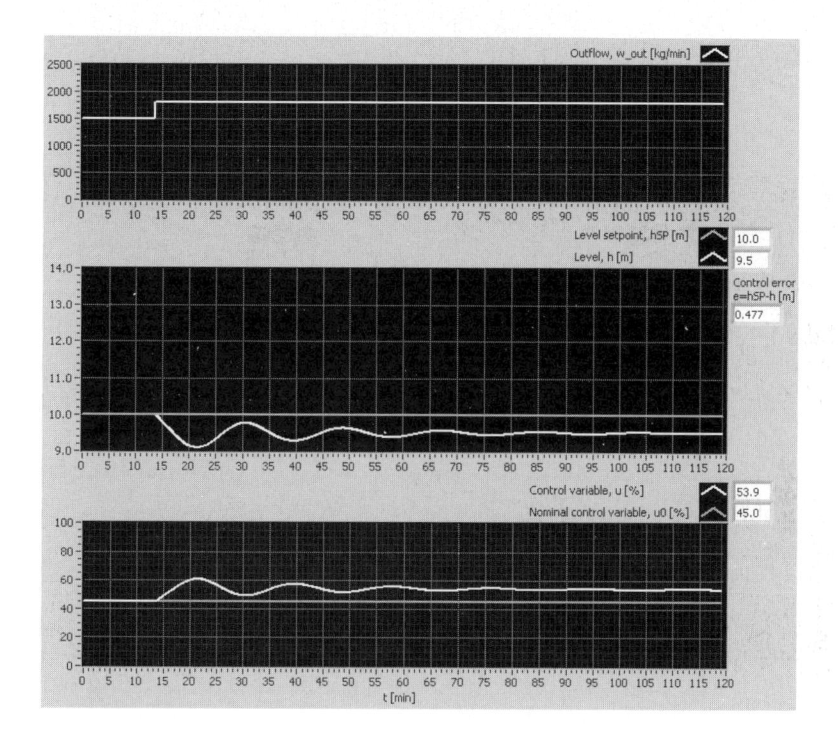

Figure 2.18: Example 2.9: Level control of the wood-chip tank with a P controller with an increased gain of $K_p = 2.6$. (The front panel of the simulator is as shown in Figure 2.15.)

[End of Example 2.9]

2.6.6 PI controller

Example 2.9 demonstrated a problem with the P controller: The steady-state control error e_s becomes different from zero when the nominal control signal u_0 does not have correct value. In practice, u_0 does never have a completely correct value since there are always unknown disturbances acting on the process to be controlled. If the P controller is substituted by a PI controller, $e_s = 0$ can be achieved, for any value of u_0!

In the PI controller (proportional + integral) the control variable is calculated as

$$u = u_0 + \overbrace{\underbrace{K_p e}_{u_p} + \underbrace{\frac{K_p}{T_i} \int_0^t e \, d\tau}_{u_i}}^{u_e} \tag{2.30}$$

Here, u_i is the integral term, K_p is the proportional gain. T_i [s] or [min] is the integral time (also denoted the reset time). In some commercial controllers the fraction K_p/T_i is represented by the *integral gain* K_i:

$$K_i = \frac{K_p}{T_i} \tag{2.31}$$

In some controllers the value of $1/T_i$ is used in stead of the value of T_i. The unit of $1/T_i$ is *repeats per minute*. For example, 5 repeats per minute means that T_i is equal to $1/5 = 0.2\text{min}$. The background of the term repeats per minute is as follows: Assume that the control error e is constant, say E. The P-term has value $u_p = K_p E$. During a time interval of 1 minute the I-term equals $\frac{K_p}{T_i} \int_0^1 E \, d\tau = K_p E \cdot 1[\text{min}]/T_i = u_p \cdot 1/T_i$. Thus, the I-term has repeated the P-term $1/T_i$ times.

How does the PI controller work? The integral term is essential. u_i as calculated as *the time integral* of the control error e from as initial time, say $t = 0$, to the present point of time, thus the integral is being calculated continuously. Let us think about the level control of the wood-chip tank: Assume that e initially is greater than zero (the level is then less than the level setpoint). As long as e is positive, u_i and therefore the total control variable u will get steadily increasing value, since the time integral of a positive value increases with time. The increasing u gives an increasing wood-chip inflow. Consequently, the chip level in the tank increases. Due to this level increase the control error eventually becomes less positive. The increase of the integral term (the inflow) continues until the error has become zero. The conclusion is that the *integral term ensures zero steady-state control error.* The zero steady-state control error is achieved even if the nominal control signal u_0 has an incorrect value (even if the value is zero).

The potential of achieving zero static control error is why the PI controller, besides the PID controller which also contains an integral term, is the most commonly used control function in the industry.

Example 2.10 ***PI control of chip level of a wood-chip tank***

Figure 2.19 shows simulated responses for the level control system for the wood-chip tank described in Example 2.3 (page 19). The front panel of the

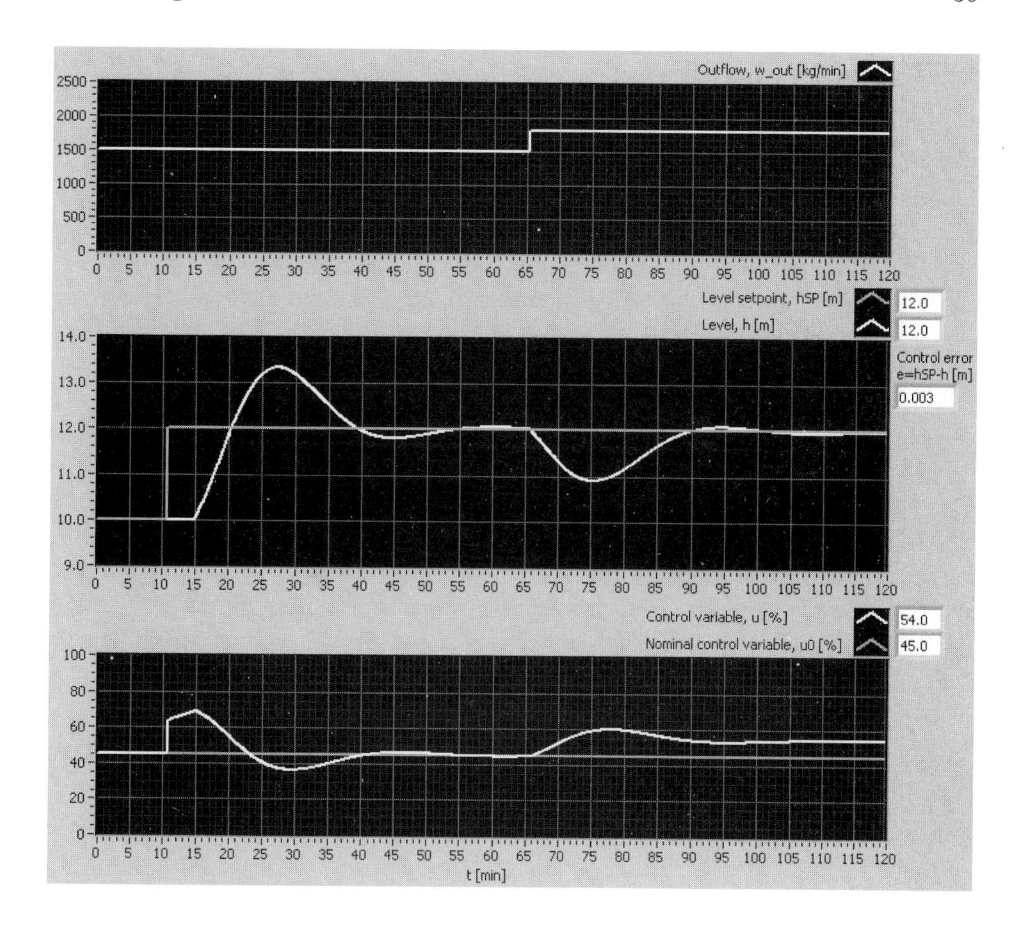

Figure 2.19: Example 2.10: Level control of the wood-chip tank with a P-controller. (The front panel of the simulator is as shown in Figure 2.15 (page 2.3).)

simulator is as shown in Figure 2.15. The controller parameters have values

$$K_p = 1.40;\ T_i = 900\text{s} = 15.0\text{min} \tag{2.32}$$

(found using the Ziegler-Nichols' closed loop-method, cf. Section 4.4). The simulation shows that the steady-state control error is zero both due to a step in the level setpoint and due to a step in the disturbance (chip outflow).

[End of Example 2.10]

2.6.7 PID controller

We may be quite content with the PI controller since it gives zero steady-state control error. But in some cases it would be desirable to have *faster control* than with the PI controller. This can be achieved by including a term in the control variable that is *proportional to the time derivative* or the rate of change of the error e. Then we have the *PID controller* (proportional + integral + derivative)

$$u = u_0 + \overbrace{\underbrace{K_p e}_{u_p} + \underbrace{\frac{K_p}{T_i}\int_0^t e\, d\tau}_{u_i} + \underbrace{K_p T_d \frac{de}{dt}}_{u_d}}^{u_e} \tag{2.33}$$

u_d is the derivative term. K_p is the proportional gain. T_i [s] or [min] is the integral time . T_d [s] or [min] is the derivative time. In some commercial controllers the product $K_p T_d$ is represented by the *derivative gain* K_d:

$$K_d = K_p T_d \tag{2.34}$$

The derivative term of the PID controller works as follows: Assume that the control error e is increasing. Then the time derivative de/dt is positive, and the derivative contributes with a positive value to the total control signal u. This will in general give *faster control.*

All commercial controllers implements a PID controller. But none of them implements (2.33)! It is namely an *ideal PID controller*, and its D-term must be modified, otherwise the controller will not work properly in practical applications (we will return to this practical aspect later in this section).

Example 2.11 ***PID control of chip level of a wood-chip tank***

Figure 2.20 shows simulated responses for the level control system for wood-chip tank described in Example 2.3. The PID parameters have values

$$K_p = 1.86;\ T_i = 540\text{s} = 9.0\text{min};\ T_d = 135\text{s} = 2.25\text{min} \tag{2.35}$$

(found using the Ziegler-Nichols' closed loop-method, cf. Section 4.4). The simulation shows the following:

- The steady-state or static control error is zero both after a step in the level setpoint and after a step in the disturbance (chip outflow).

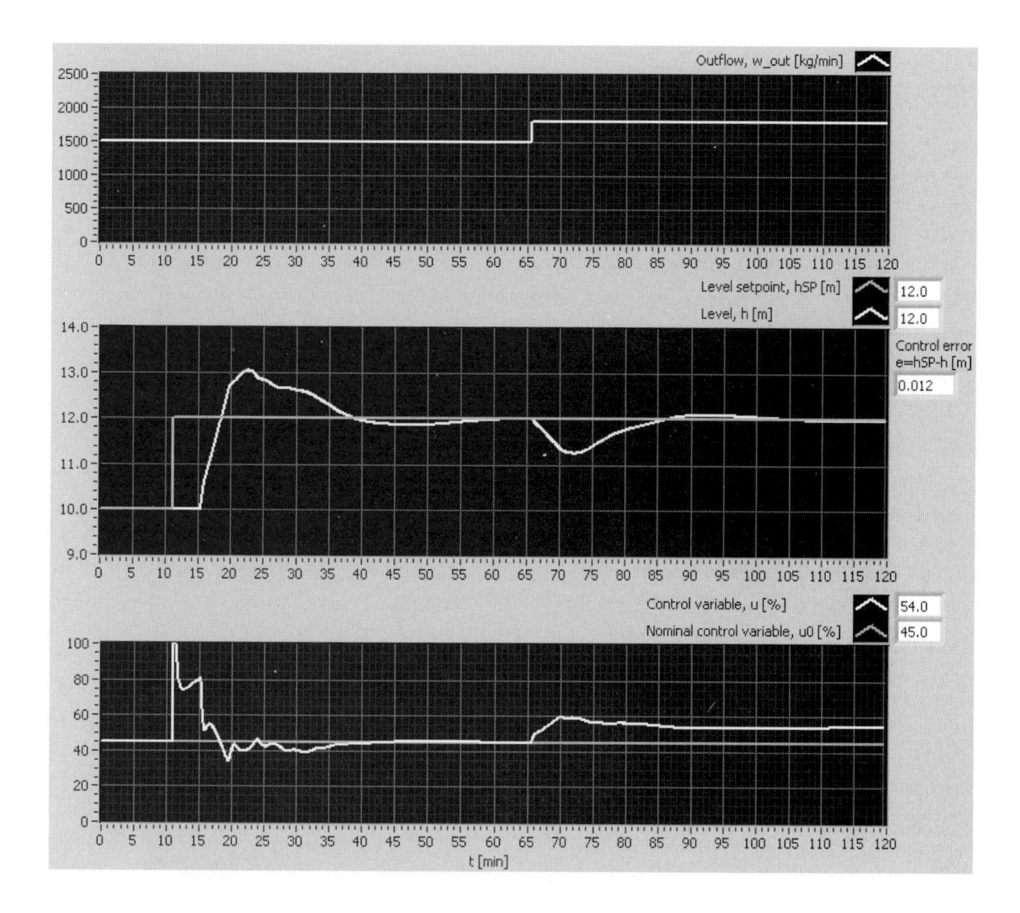

Figure 2.20: Example 2.11: Level control of the wood-chip tank with a P-controller. (The front panel of the simulator is as shown in Figure 2.15.)

- The setpoint tracking is faster than with a PI controller, cf. Example 2.10. Also, the compensation of the step change of the disturbance (outflow) is faster than with a PI controller.

[End of Example 2.11]

Behaviour of the PID controller terms

Do you want to see how each of the terms of a PID controller function works? Figure 2.21 shows the time-responses due to a step in the outflow w_{out} (the disturbance) from the initial value of 1500kg/min to 1800kg/min. The controller parameters have values as in Example 2.11. The nominal control signal u_0 has value 45% which gives an inflow that is equal to the initial value of w_{out}. We can observe the following:

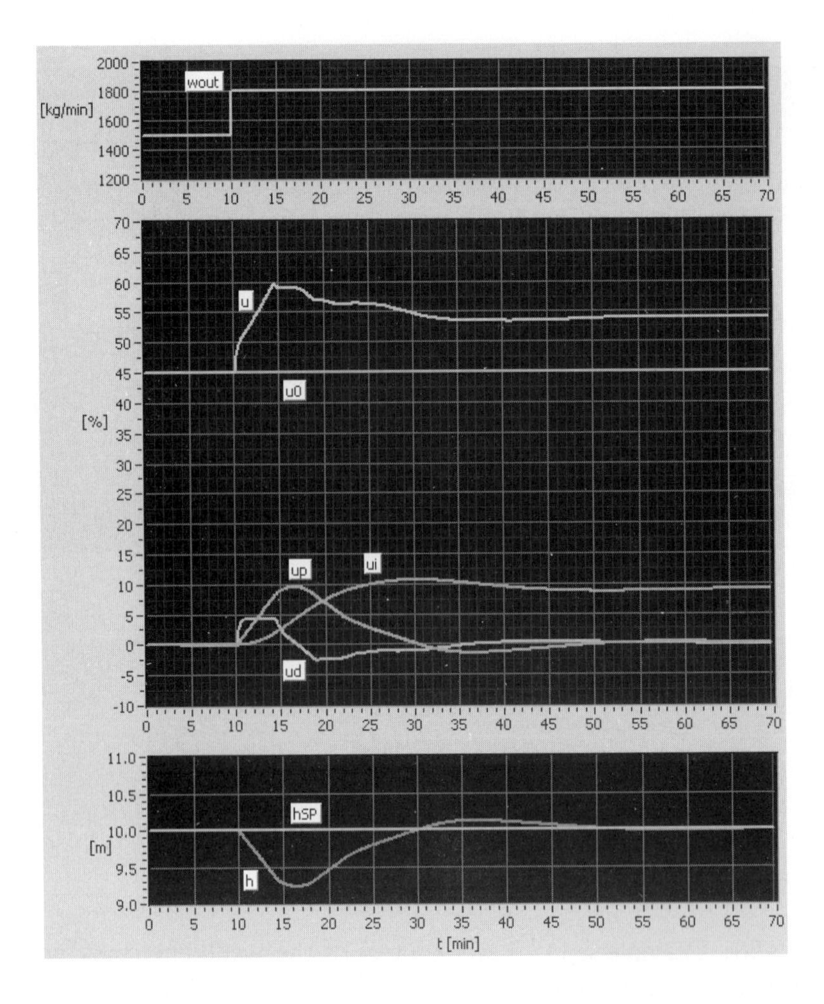

Figure 2.21: Behaviour of the various terms of the PID-controller terms after a step in the outflow w_{out} (disturbance)

- The D-term u_d reacts abruptly and its value converges towards zero (the time-derivative of a constant error is constant).

- The I-term is relatively sluggish. Its value changes as long as the control error is different from zero. u_i goes to a (new) constant value after the step in w_{out}. The change in u_i constitutes the compensation for the incorrect value in u_0 after the change in w_{out} (the disturbance).

- The P-term u_p is quicker than the I-term, but more sluggish than the D-term, and its value goes to zero since $K_p e$ goes to zero when e goes to zero.

Measurement noise and lowpass filter in the D-term

There is a potential problem using the PID controller: It may give a very unsteady high frequent control signal due to noise in the process measurement – and such noise is always present, more or less. See Figure 2.22. The measurement noise can stem from electronic noise sources or

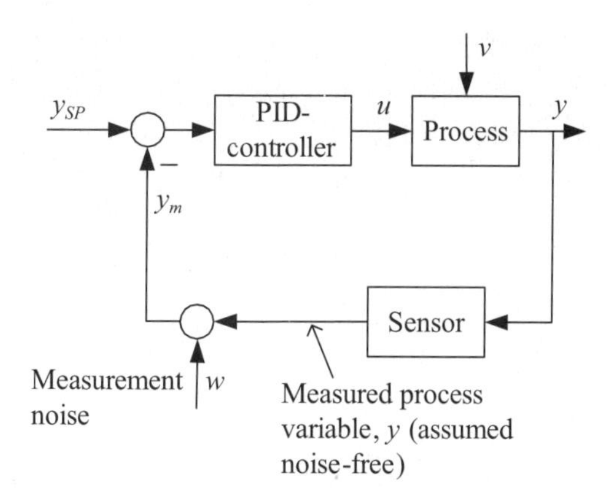

Figure 2.22: Measurement noise in the control loop

from the measurement principle, e.g. ultrasound based level measurement of a liquid surface.

The unsteady control signal is due to the differentiation of the control error e in the D-term. The control error consist of the following terms:

$$e = y_{SP} - y_m = y_{SP} - (y + w) \tag{2.36}$$

where y_{SP} is the setpoint, y_m is the process measurement, y is the (noise-free) process variable, and w is the measurement noise. The D-term becomes

$$u_d = K_p T_d \frac{d(y_{SP} - y_m)}{dt} \tag{2.37}$$

$$= K_p T_d \frac{d\left[y_{SP} - (y + w)\right]}{dt} \tag{2.38}$$

$$= K_p T_d \underbrace{\frac{d(y_{SP} - y)}{dt}}_{\frac{de}{dt}} - K_p T_d \frac{dw}{dt} \tag{2.39}$$

The term dw/dt is the time-derivative of the noise and it is a term in the control variable. If the noise w is high frequent, its time-derivative (rate of change) may get very large values, and the control variable u may be very unsteady. We can see this by assuming that w is sinusoidal:

$$w(t) = W \sin(\omega t) \tag{2.40}$$

From this we get

$$\frac{dw}{dt} = \underbrace{\omega W}_{A_w} \cos(\omega t) \tag{2.41}$$

If the frequency ω is large, the amplitude $A_w = \omega W$ of the time-derivative dw/dt may be large, and consequently the term $-K_p T_d dw/dt$ in the D-term u_d may get a large amplitude.

How can we reduce the problem of the time-differentiation of the measurement noise? If we can not reduce or remove the noise itself, we can *lowpass filter* the control error used in the D-term before it is differentiated. This is a standard solution used in commercial controllers. Let us use the symbol e_f for the filtered error. The modified PID controller is then

$$u = u_0 + \underbrace{K_p e}_{u_p} + \underbrace{\frac{K_p}{T_i} \int_0^t e \, d\tau}_{u_i} + \underbrace{K_p T_d \frac{de_f}{dt}}_{u_d} \tag{2.42}$$

The filter is typically a first order lowpass filter. It is convenient to represent the filter with its Laplace transfer function. The relation between e_f and e is then[8]

$$e_f(s) = \frac{1}{T_f s + 1} e(s) \tag{2.43}$$

where T_f is the filter time constant which usually is expressed as a fraction of the derivative time T_d:

$$T_f = a T_d \tag{2.44}$$

a is a constant which typically is chosen between 0.05 and 0.2. If no special requirements exist, we can set $a = 0.1$.

Figure 2.23 shows simulations of a control system (not the wood-chip tank this time). The setpoint y_{SP} and the process measurement y_m are shown in one diagram, and the control variable u is shown in the other diagram. The controller is a PID controller where K_p and T_i have constant values. The setpoint is constant. The measurement contains random measurement noise w (uniformly distributed between $\pm 0.2\%$). The simulation shows three situations:

- Between $t = 120$ and 140s: No D-term, that is, the controller is a PI controller ($T_d = 0$ in the PID controller). The simulation shows

[8] Although it is not mathematically correct, it is convenient to use the same symbol for the time function and the Laplace transform of the function.

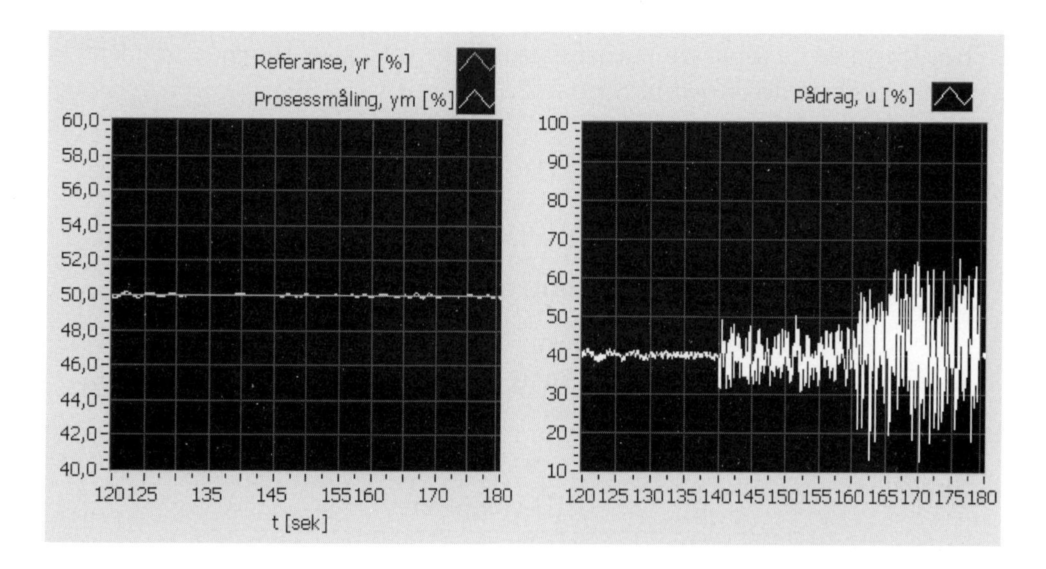

Figure 2.23: Simulation of a PID-control system with measurement noise for three different situations, cf. the text

naturally enough some noise in the control signal u. The noise propagates to the control variable mainly via the P-term, but also somewhat via the I-term.

- Between $t = 140$ and 160s: Ordinary PID controller with lowpass filter with $a = 0.1$. The noise gives a larger response in the control variable than with the PI controller due to the noise sensitivity in the D-term. This demonstrates that *the PID controller gives more noisy control signal than PI controller.*

- Between $t = 160$ and 180s: PID controller with an (approximately) ideal D-term, that is, the lowpass filter in the D-term is (approximately) removed. The response of the noise in the control signal is very noisy . This demonstrates that *the lowpass filter in the D-term is important for attenuating the response of the measurement noise in the control variable.*

If the measurement noise has a mean value m_w different from zero, there will be a steady-state control error different from zero, since m_w will appear as an addition to the setpoint. The PID controller ensures that the process output variable y will track this false setpoint (containing the m_w term).

Above, the solution to the measurement noise was to lowpass filter the control error used in the D-term. If this does not give enough filtering, we can try to use a separate measurement filter acting on the measurement signal. This is described in Section 2.7.3.

Block diagram of the PID controller

Figure 2.24 shows a block diagram of the PID controller given by (2.42).

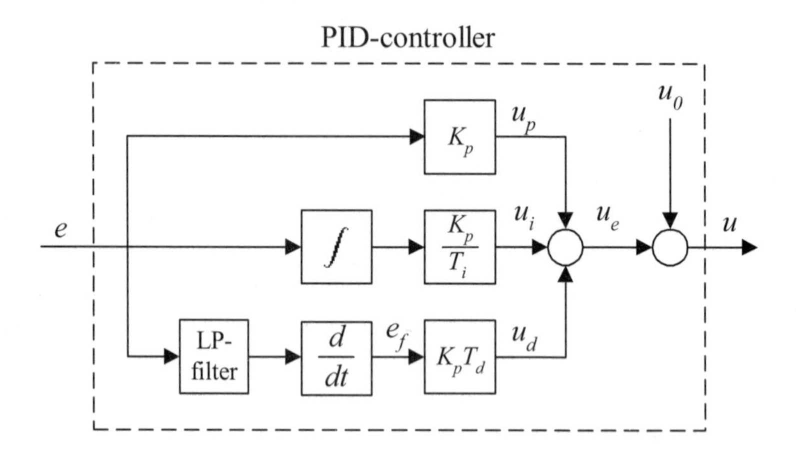

Figure 2.24: Block diagram of the PID-controller (2.42)

Transfer function of the PID controller

In some situations it is useful to represent the PID controller (2.42) by its transfer function. This is the case in frequency response analysis control systems, analytical calculation of time-responses using the Laplace transform, and simulations when it is sufficient to use a compact, linear controller model.

It is quite easy to find the controller transfer function $H_c(s)$ from input e to output u by taking the Laplace transform of (2.42) combined with (2.43) and neglecting u_0 (the controller transfer function is independent of u_0):

$$
\begin{aligned}
u(s) &= K_p e(s) + \frac{K_p}{T_i}\frac{1}{s}e(s) + K_p T_d s e_f(s) && (2.45) \\
&= \underbrace{\left[K_p + \frac{K_p}{T_i s} + \frac{K_p T_d s}{T_f s + 1}\right]}_{H_c(s)} e(s) && (2.46) \\
&= \underbrace{\frac{K_p T_i (T_f + T_d) s^2 + K_p (T_i + T_f) s + K_p}{T_i T_f s^2 + T_i s} e(s)}_{H_c(s)} && (2.47)
\end{aligned}
$$

PID controller on serial form

The PID controller given by (2.33) is said to be on *parallel form* since a block diagram of the controller function shows the P, I and D term in parallel paths, cf. Figure 2.24. There is also a *serial form*. In most cases it is not an important difference between the parallel form and the serial form. The the serial form consists of a PD controller (which is a PID controller with the I-term removed) in series with a PI controller, and with the gain of the PD and the PI controllers combined in the common gain K_p:

$$
\begin{aligned}
u(s) &= \underbrace{K_{p_i}\left(1 + \frac{1}{T_i s}\right)}_{\text{PI}} \cdot \underbrace{K_{p_d}\frac{T_d s + 1}{T_f s + 1}}_{\text{PD}} e(s) && (2.48) \\
&= K_p \frac{(T_i s + 1)(T_d s + 1)}{T_i s (T_f s + 1)} e(s) && (2.49) \\
&= \underbrace{\frac{K_p T_i T_d s^2 + K_p (T_i + T_d) s + K_p}{T_i T_f s^2 + T_i s} e(s)}_{H_s(s)} && (2.50)
\end{aligned}
$$

A few comments:

- The serial form is more practical than the parallel form in frequency response based controller design, cf. Chapter 8, due to the factorized form of (2.49).
- The parallel form is more general since it can have complex zeros in its transfer function (the serial form can only have real zeros).
- The parallel form is somewhat easier to express in the time domain as a differential/integral equation) and to realize as a practical discrete-time algorithm.

- According to [24] the serial form is more frequently used in modern commercial controllers – probably because the serial form behaves similar to the first industrial PID controllers which were pneumatic and was used in the 1930s.[9]

Transformation from serial to parallel form

You can perform a transformation from serial form to parallel form (the reverse transformation is less used). One reason for performing such a transformation is that you used tuning methods which assumes the serial form, while the controller you use, actually implements the parallel form. The transformation can be executed as follows[24] (it is based on comparing coefficients between the ideal PID controller functions, that is, with T_f set to 0): Given the parameters K_{p_s}, T_{i_s} and T_{d_s} of the serial form PID controller. The corresponding parameters, K_{p_p}, T_{i_p}, T_{d_p} and T_{f_p} of a parallel PID controller having approximately the same behaviour as the serial form, is achieved with the following transformations:

$$K_{p_p} = K_{p_s}\left(1 + \frac{T_{d_s}}{T_{i_s}}\right) \tag{2.51}$$

$$T_{i_p} = T_{i_s}\left(1 + \frac{T_{d_s}}{T_{i_s}}\right) \tag{2.52}$$

$$T_{d_p} = T_{d_s}\frac{1}{1 + \frac{T_{d_s}}{T_{i_s}}} \tag{2.53}$$

In addition, the time constant of the lowpass filter in the derivative term can be calculated by

$$T_{f_p} = aT_{d_p} \tag{2.54}$$

where a typically is 0.1. For P and PI controllers the serial and the parallel forms are identical (since T_{d_s} is 0).

From (2.51)-(2.53) we see that the transformations are functions of the ratio T_{d_s}/T_{i_s}. The less T_{d_s}/T_{i_s}, the less importance of the transformations. In the Ziegler-Nichols' tuning methods, cf. Chapter 4,

$$\frac{T_{d_s}}{T_{i_s}} = \frac{1}{4} \tag{2.55}$$

If this relation is used in (2.51)–(2.53),

$$K_{p_p} = 1.25K_{p_s} \tag{2.56}$$

[9] The famous Ziegler-Nichols' methods for controller tuning, cf. Chapter 4, were published in 1942 and they must have been based on pneumatic controllers approximately implementing the serial form.

$$T_{i_p} = 1.25 T_{i_s} \tag{2.57}$$

$$T_{dp} = 0.8 T_{d_s} \tag{2.58}$$

In this case the parameter transformations do not change the PID parameters much, and you can quite safely assume that the two PID controllers behave approximately equally, which implies that you not need to care about which PID form which is actually implemented in the controller. But if you feel a little uncertain about the different implementations, you should still consider to use the transformations. You can use simulations to check if the parallel and serial form causes any substantial difference in the behaviour of the control systems.

Example 2.12 ***Parallel and serial form of the PID controller***

In this example control systems for processes having the following transfer function model are simulated:

$$y(s) = \frac{K_u}{(T_1 s + 1)(T_2 s + 1)} e^{-\tau s} u(s) \tag{2.59}$$

$$+ \frac{K_v}{(T_1 s + 1)(T_2 s + 1)} e^{-\tau s} v(s) \tag{2.60}$$

(The process model is thus a second order system with time delay.) The process parameter are

$$K_u = 1;\ K_v = 2;\ T_1 = 1\text{s};\ T_2 = 1\text{s};\ \tau = 0.5\text{s}; \tag{2.61}$$

For comparison two control system are simulated simultaneously: One with a parallel PID controller and one with a serial PID controller. The process to be controlled, and the setpoint and the disturbance are identical for both control systems. The two control systems are simulated in two scenarios:

1. *Without using PID parameter transformation*: The following PID parameters (found using the Ziegler-Nichols' closed loop method) are used for both the parallel PID controller and the serial PID controller:

   $$K_p = 3.6;\ T_i = 2.0\text{s};\ T_d = 0.5\text{s}; \tag{2.62}$$

   Figure 2.25 shows the simulated responses due to a setpoint step (at $t = 4$s) and a disturbance step (at $t = 20$s). The simulations shows that the responses in the two control systems are somewhat but not dramatically different. The stability is satisfactory in both systems.

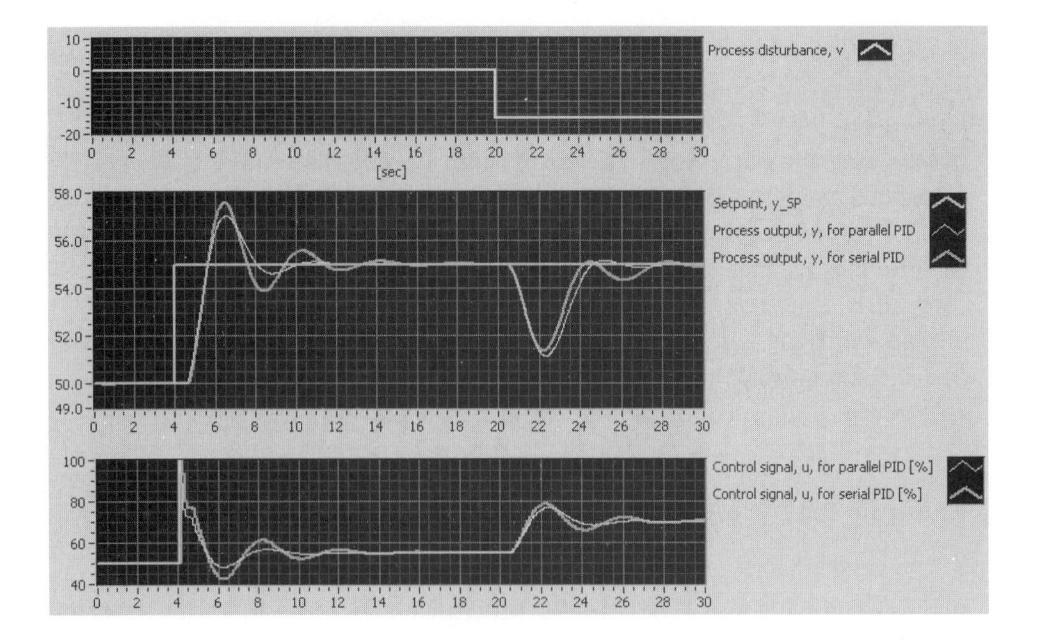

Figure 2.25: Example 2.12: Simulated responses due to a setpoint step and a disturbance step. PID parameter transformation is *not* used.

2. *Using PID parameter transformation*: The PID parameters (2.62) are still used for the serial PID controller, but for the parallel PID controller the following parameters are used:

$$K_p = 4.5;\ T_i = 2.5\text{s};\ T_d = 0.42\text{s}; \tag{2.63}$$

These parameter values are found by transforming the serial form parameters (2.62) to parallel form parameters using (2.51)-(2.53). The parallel PID controller and the serial PID controller should then have the same behaviour. Figure 2.26 shows the simulated responses due to a setpoint step and a disturbance step. The responses in the two control systems are now almost identical. The difference is probably due to simulation technicalities, and possibly due to the fact that the parameter transformations are not ideal since they do not take the derivative filter time constant T_f into account.

[End of Example 2.12]

This above example indicates that it is probably not important to distinguish between the serial PID controller and the parallel PID controller. This conclusion is true if the Ziegler-Nichols' closed loop tuning method is used, in which $T_d/T_i = 1/4$. If a different (larger) ratio between

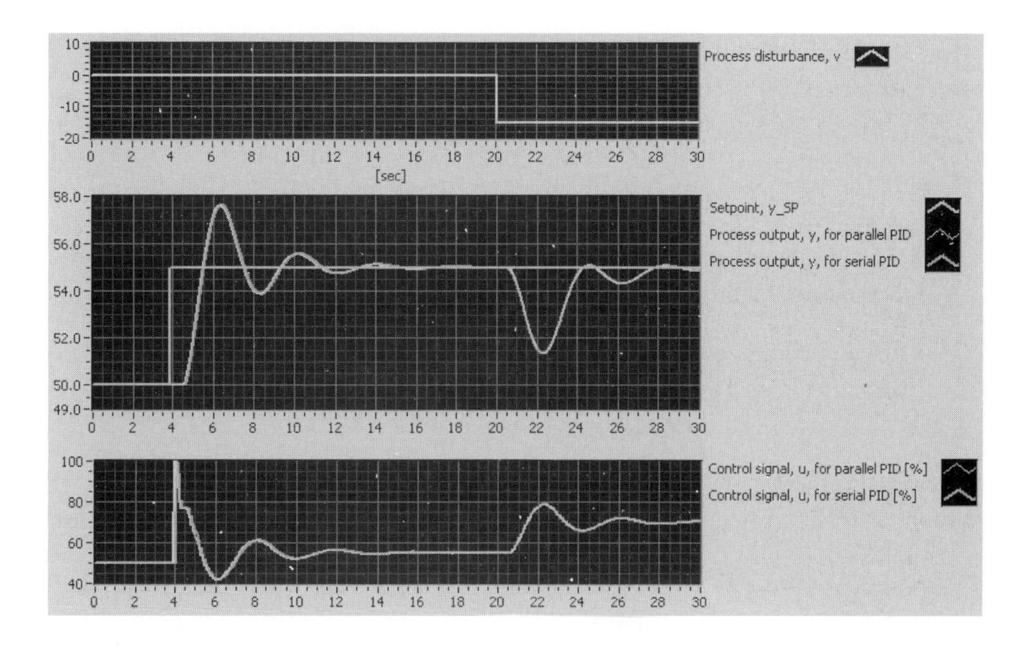

Figure 2.26: Example 2.12: Simulated responses due to a setpoint step and a disturbance step. PID parameter transformation is used.

T_i and T_d is used, the difference between the controllers may be larger and hence the PID parameter transformations more important.

2.6.8 Positive or negative controller gain?

On commercial controllers can you choose whether the controller gain K_p of the PID controller has *positive or negative value*. Let us write the PID controller (2.42) as

$$u = u_0 + \underbrace{K_{sign} \cdot K_{p_1}}_{K_p} \left(e + \frac{1}{T_i} \int_0^t e \, d\tau + T_d \frac{de_f}{dt} \right) \tag{2.64}$$

The controller gain, which is $K_p = K_{sign} K_{p_1}$ where K_{p_1} is always positive, will have positive sign with $K_{sign} = 1$ and negative sign with $K_{sign} = -1$. On commercial controllers the user typically sets the value of K_{p_1}, while the sign, here K_{sign}, is set via a parameter field or a button. The default choice is positive gain ($K_{sign} = 1$).

The consequence of choosing *wrong* sign of K_p is dramatic: The control loop becomes *unstable*. Instability implies that the variables in the control loop shows a steadily increasing amplitude, until some saturation occurs.

How do you know which controller gain sign to use? It is the sign of the *process gain* K which determines the controller gain sign, as follows:

- If the process gain K is positive, K_p shall be positive, that is $K_{sign} = 1$. The controller is in this case said to have *reverse action*, since an increase of the process output variable gives a reduction of the control signal.
- If the process gain K is negative, K_p shall be negative, that is $K_{sign} = -1$. The controller is in this case said to have *direct action*.

Above, the term "process" includes all subsystems in the control loop except the controller. Consequently, the "process" also includes the sensor.

The point is that the total gain of the loop shall be positive whatever the sign of the process is (in this context we disregard the negative gain of the subtraction between the setpoint and the process measurement). This is achieved by requiring that $K_p \cdot K_{sign}$ is positive, cf. Figure 2.27.[10]

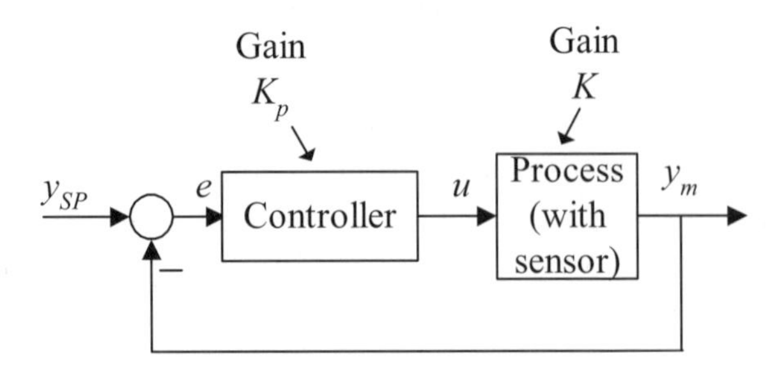

Figure 2.27: To have a stable control loop the product $K_p \cdot K_s$ must be positive.

What is the process gain, K? Simply stated, the gain of a system is the ratio of the output signal (or the rate of change of the output signal if the system has integral dynamics) to the input signal of the system. So, if the output has a positive response to a positive input, the gain is positive. Here are a few examples:

- Figure 2.28 shows a level control system for a liquid tank where the control variable *controls the outflow* of the tank. An increase of the control signal reduces the level and the level measurement (it is

[10] Note that $K_pK_s > 0$ does not ensure stability since the loop will be ustable if K_pK_s has too large positive value. But the loop is certainly unstable if $K_pK_s < 0$.

assumed that the measurement signal decreases as the level decreases). Consequently, the process (including sensor) has a negative process gain.

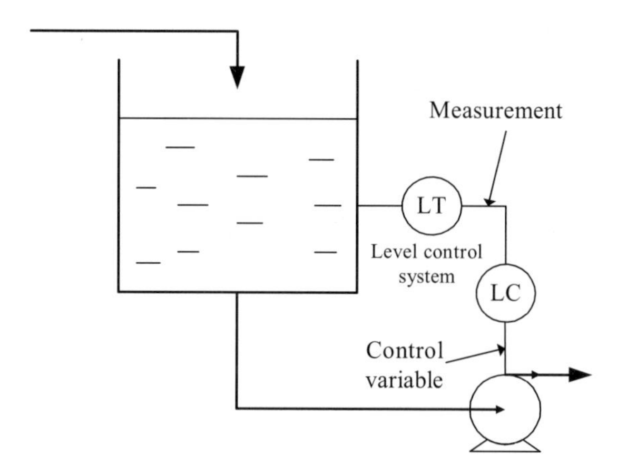

Figure 2.28: An example of a process (with sensor) with negative gain. An increase of the control signal will reduce the level.

- A heat exchanger with temperature control where the control signal *controls the supply of cooling media (e.g. cold water)* has a negative gain since an increase of the control signals increases the cooling and hence decreases the temperature (and the temperature measurement).

The process gain can of course be found from the process model. Here are a few examples:

- The transfer function model (first order with time delay)

$$y_m(s) = \frac{3}{s+1}e^{-2s}u(s) \tag{2.65}$$

  has positive gain, namely 3.

- The transfer function model (integrator with time delay)

$$y_m(s) = \frac{-2}{s}e^{-s}u(s) \tag{2.66}$$

  has negative gain, namely -2.

2.7 Practical problems: Control kicks, windup, and noise

This Section describes several important practical problems which can exist in real control loops, and how to solve these problems.

2.7.1 Reducing P- and D-kick caused by setpoint changes

Introduction

Abrupt changes of the setpoint y_{SP}, for example step changes, may cause unfortunate kicks in the control variable. The problem is connected to the P-term and the D-term of the controller function (2.42). These kicks are denoted proportional kick or P-kick and derivative kick or D-kick, respectively. Such kicks may cause mechanical actuators to move abruptly, resulting in excessive wear.

One solution to the above mentioned problem is to modify the P-term and/or the D-term of the PID controller. Another solution is to accept only smooth setpoint changes! We will study these solutions in detail in the following sections.

For an easy reference the PID controller function with setpoint weights in the D-term and the P-term is repeated here:

$$u = u_0 + \underbrace{K_p e_p}_{u_p} + \underbrace{\frac{K_p}{T_i}\int_0^t e\,d\tau}_{u_i} + \underbrace{K_p T_d \frac{de_{d_f}}{dt}}_{u_d} \tag{2.67}$$

where e_{d_f} is given by

$$e_{d_f}(s) = \frac{1}{T_f s + 1} e_d(s) \tag{2.68}$$

and

$$e_p = w_p y_{SP} - y \tag{2.69}$$

$$e_d = w_d y_{SP} - y \tag{2.70}$$

where w_p and w_d are setpoint weights in the P-term and the D-term, respectively.

Note: Do not try reducing the setpoint weight in the I-term since it will cause the static control error to become different from zero! This is because the integrand becomes zero in steady-state in a stable control system, and if the integrand of the u_i-term is not equal to the difference $y_{SP} - y$, but in stead say $w_i y_{SP} - y$, then of course $y_{SP} - y = e$ can not be equal to zero in steady-state. So, once more: Do not use reduced setpoint weighting in the I-term!

Reduction of D-kick

The derivative term of the PID controller (2.42) is

$$u_d = K_p T_d \frac{de_{d_f}}{dt} = K_p T_d \frac{d\left(w_p y_{SP} - y\right)_f}{dt} \tag{2.71}$$

where index f is for "filtered". Assume initially that $w_d = 1$, that is, no reduced setpoint weight. Due to the time derivative, an abrupt change of the setpoint y_{SP} gives an abrupt change of u_d and a corresponding change of the total control variable u in which u_d is an additive term, cf. (2.67). For example, a stepwise change of the setpoint gives an impulse in u_d since the time derivative of a step is an impulse.

To avoid such changes of u_d, the setpoint y_{SP} can be given a reduced weight in the D-term by giving w_d a value less than 1. In the case of reduced weight, it is common to set $w_d = 0$, causing the setpoint to be removed completely from the derivative term. In many commercial controllers $w_d = 0$ is a fixed factory setting.

One drawback with reduced setpoint weight is more sluggish response to varying setpoint signals. This can be unfortunate in at least the following cases

- Secondary controllers in cascade control systems, cf. Chapter 9.2.
- Servo systems (control systems for motors).

One question is: Will reduced weighting of the setpoint in the D-term influence the ability of the controller to compensate for disturbances? The answer is *no* because the compensation for disturbances takes place after the disturbance has caused a response in the process output variable y and measurement y_m, and the appearance of y_m in the D-term is independent of the setpoint weight.

Example 2.13 ***Reduced setpoint weight in the D-term***

In this example a control system for a process having the following transfer function model is simulated:

$$y(s) = \frac{K_u}{(T_1 s + 1)(T_2 s + 1)} e^{-\tau s} u(s) \tag{2.72}$$

$$+ \frac{K_v}{(T_1 s + 1)(T_2 s + 1)} e^{-\tau s} v(s) \tag{2.73}$$

(The process is thus a second order system with time delay.) u is the control variable, and v is the process disturbance. The process parameter are

$$K_u = 1;\ K_v = 1;\ T_1 = 2\text{s};\ T_2 = 1\text{s};\ \tau = 0.5\text{s}; \tag{2.74}$$

The PID parameters are

$$K_p = 3.6;\ T_i = 2.0\text{s};\ T_d = 0.5\text{s}; \tag{2.75}$$

(tuned with the Ziegler-Nichols' closed loop method). Two cases are simulated:

- *Full setpoint weight, $w_d = 1$:* Figure 2.29 shows simulated responses in the control system due to a setpoint and a disturbance step.

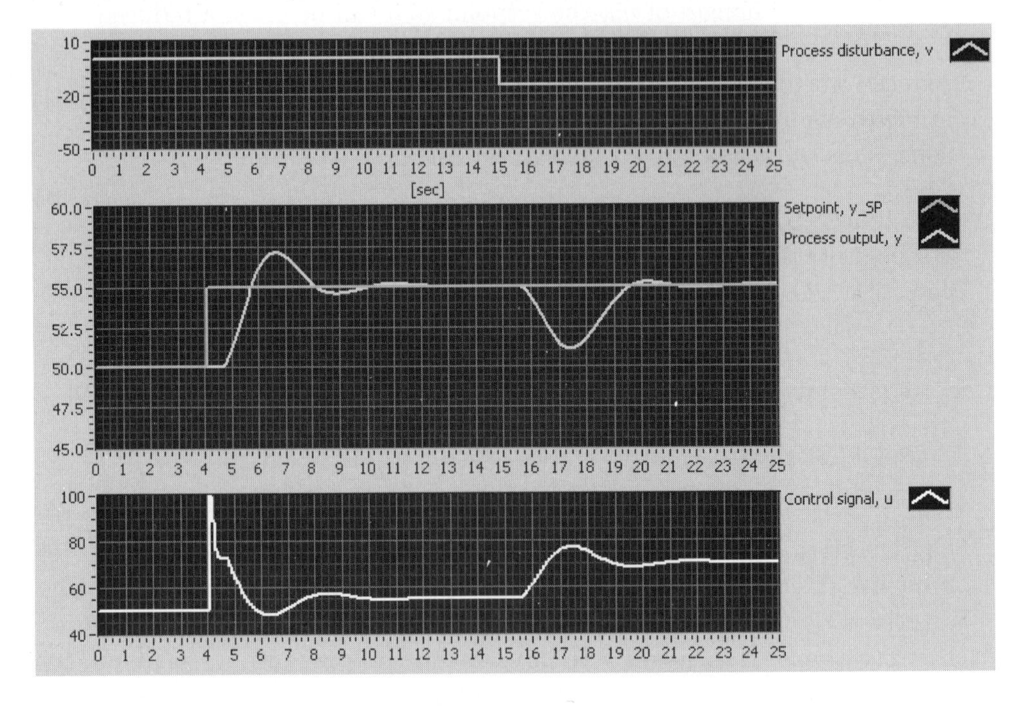

Figure 2.29: Example 2.13: Simulated responses in the control system. There is No reduction of setpoint weight, thus $w_d = 1$.

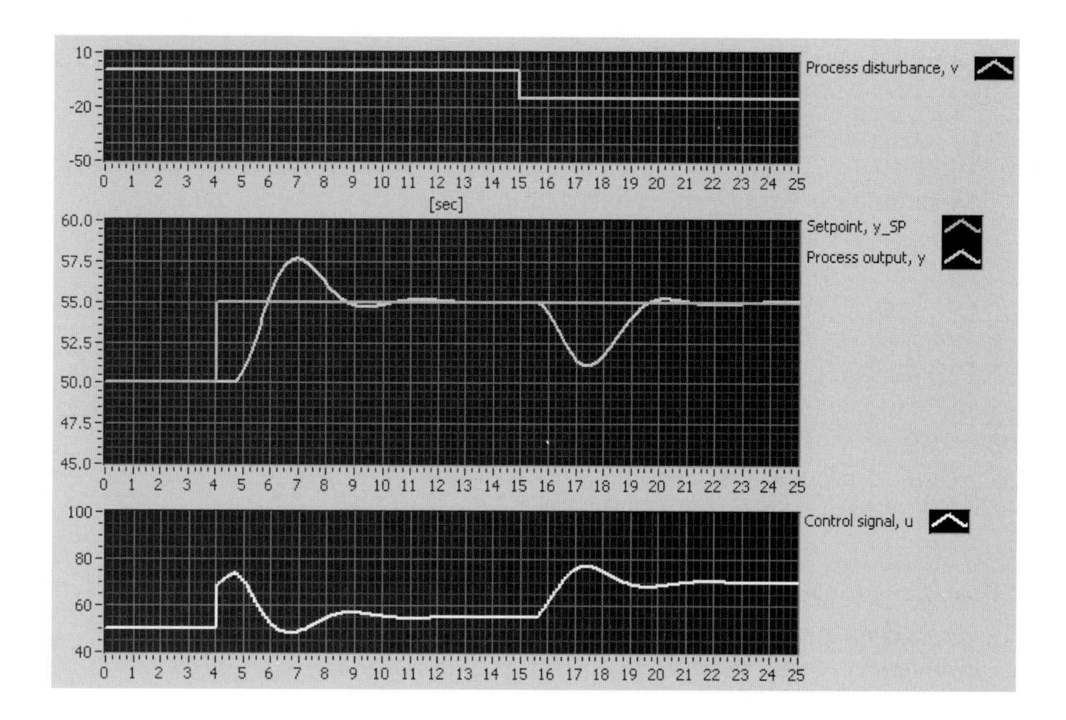

Figure 2.30: Example 2.13: Simulated responses in the control system. The setpoint is removed from the D-term with $w_d = 0$.

- *No setpoint weight,* $w_d = 0$: Figure 2.30 shows simulated responses.

Comparing the responses in Figures 2.29 and 2.30, it is clear that the control signal reacts smoother with $w_d = 0$ than with $w_d = 1$ after the setpoint step. However, there is no difference in the control signals after the disturbance step, as expected.

[End of Example 2.13]

Reduction of P-kick

The proportional term (P-term) in the PID controller (2.67) is

$$u_p = K_p e = K_p(w_p y_{SP} - y) \tag{2.76}$$

Let us assume initially that $w_p = 1$, which means there is no reduction of the setpoint weight in the P-term. If the setpoint is changed, say it is changed as a step, the P-term u_p and therefore the total control variable where u_p appears as an additive term, is changed like a step, too. This can be unfortunate for mechanical actuators, cf. the discussion in Chapter 2.7.1.

By setting w_p less than 1, the setpoint has reduced weight in the P-term, which implies less abrupt changes in u_p caused by setpoint changes. It is however common in commercial controllers to have $w_p = 1$ (which means no reduced weight), but if w_p is reduced, then $w_p = 0.3$ is suggested [24].

Note: If you are to use the Ziegler-Nichols' closed loop method to tune PID parameters, the control system will not react at all to excitations via the setpoint if $w_p = 0$. So, you should not set $w_p = 0$ during the tuning.

Smoothing the control signal by ramping the setpoint

In the previous sections you have seen that abrupt setpoint changes implies abrupt control signal changes. As explained, one way to reduce problem is to use a reduced setpoint weight in the D-term and/or in the P-term (the D-term is the most critical case due to the time differentiation). An alternative solution is to avoid sudden changes, e.g. step changes, in the setpoint. The change of the setpoint from one value to another may follow a ramp in stead of a step, see Figure 2.31. Commercial

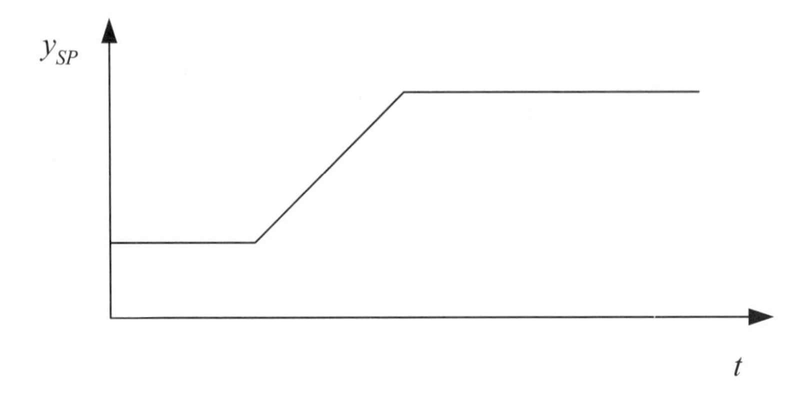

Figure 2.31: The change of the setpoint from one value to another may follow a ramp in stead of a step to avoid kicks in the control signal.

controllers typically supports setpoint ramping.

Example 2.14 ***Setpoint ramping***

Figure 2.32 shows a simulation of the same system which was simulated in Example 2.13, but now there is no reduced setpoint weight. In stead, the setpoint is changed as a ramp. Clearly the control signal varies much smoother compared to the response shown in Figure 2.29 where the setpoint was changed as step. The response after the disturbance step is of course the same since the disturbance is independent of the setpoint.

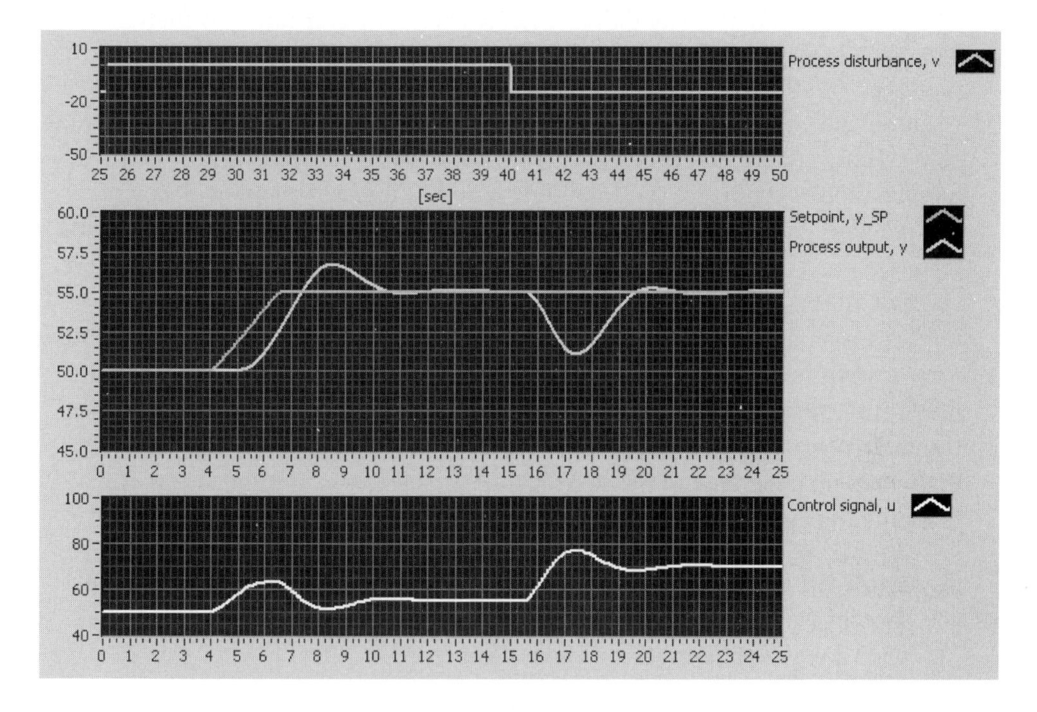

Figure 2.32: Example 2.14: Ramping the setpoint gives smoother control signal.

[End of Example 2.14]

2.7.2 Integrator anti wind-up

All actuators have saturation limits, i.e. a maximum limit and a minimum limit. For example, a power amplifier (for a heater or a motor) can not deliver an infinitely amount of power, and a valve can not have an infinitely large opening and can not be more closed than closed(!). Under normal process operation the control variable should not reach the saturation limits.

But everything is not normal all the time. Assume that in a time interval a large process disturbance acts on the process so that the process output variable is reduced. The control error then becomes large and positive, and the control variable will increase (because of the integral term of the PID controller) until the control signal limits at its maximum value, $u_{\max}$. Assume that the disturbance is so large that $u_{\max}$ is not large enough to compensate for the large disturbance. Because of this, the control error keeps large, and the integral of the control error continues to increase, which means that the calculated integral term u_i continues to increase. This is *integrator wind-up.*

Assume that the process disturbance after a while goes back to its normal value. This causes the process output variable to increase since the disturbance is reduced (the load is removed), and the error will now change sign (it becomes negative). Consequently the integral term starts to integrate downwards (its value is continuously being reduced), so that the calculated u_i is reduced, which is ok, since the smaller disturbance requires a smaller control signal. However, the problem is that is may take *a long time* until the large value of the calculated u_i is reduced (via the down-integration) to a normal (reasonable) value. During this long time the control variable is larger than what is required to compensate for the disturbance, causing the process output variable to be larger than the setpoint during this time.

To sum it up: A large and long-lasting process disturbance which forces the control variable (via the controller) to one of its saturation limits, implies a long-lasting error different from zero.

A practical PID controller must be able to cope with the possibility of integrator wind-up, that is, it must have some *anti wind-up* mechanism. You can assume that anti wind-up is implemented in commercial controllers. The principle of an anti wind-up mechanism is simple: Since the problem is that the integral term increases continuously during actuator saturation, the solution is to halt the integration when the control signal reaches either its maximum or minimum limit. An analogy of anti wind-up is to mount an overflow outlet in a liquid tank, see Figure 2.33. A tank is dynamically an integrator, so it represents here the I-term of the controller.

Note that you can not implement integrator anti-windup by just limiting the control signal, u, calculated by the PID controller. It is crucial to halt the integration of the control error.

There are several ways to implement anti wind-up in a continuous-time PID controller [24], but these are not described here. It is more likely that, if you are to implement integrator anti wind-up in a controller, it will be on a discrete-time controller, cf. Chapter 5.

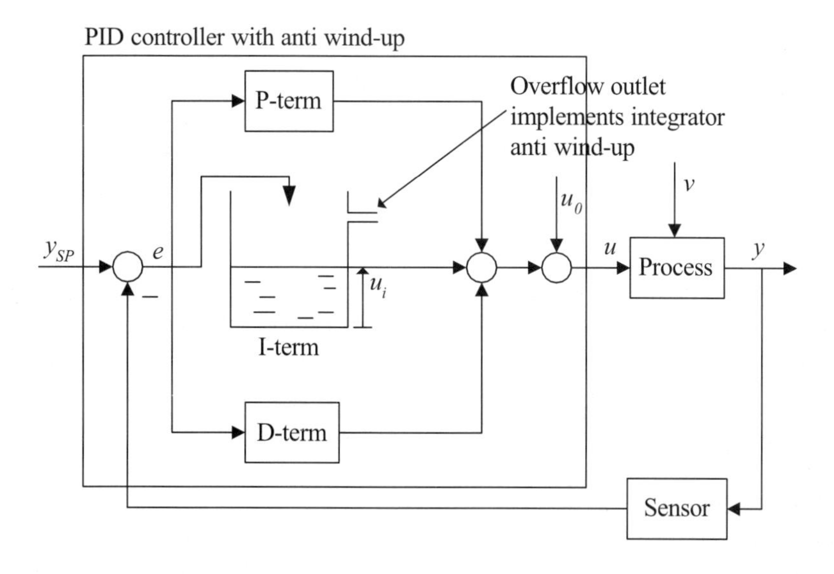

Figure 2.33: An analogy of anti wind-up: The overflow outlet limits the *integral* of the inflow (which is the volume, or the level) in a liquid tank

Example 2.15 ***Integral anti wind-up in a temperature control system***

Figure 2.34 shows the front panel of a simulator for a temperature control system for a liquid tank with continuously mass flow. The disturbance is here the inlet temperature T_{in}, which is is changed as a step from 40°C to 10°C at approx. 210min and back to 40°C at approx. 300min. The temperature setpoint T_{SP} is 70°C (constant). The parameters of the PID controller are $K_p = 6.7$, $T_i = 252\text{s} = 42\text{min}$ and $T_d = 63\text{s} = 10.5$ min (found using Ziegler-Nichols' closed loop method). The maximum value of the control variable is 100% and the minimum value is 0%. When T_{in} is reduced to 10°C, the actuator (heating element) goes into saturation (100%), trying to compensate for the (cold) disturbance. It can be shown that the control variable u should have a value of 122.5% (which corresponds to more heat power than what is available) to be able to compensate for $T_{in} = 10$°C.

Figure 2.34 shows the simulated responses in the control system *without* using integrator anti wind-up, and Figure 2.35 shows the responses *with* integrator anti wind-up. In the case of no anti wind-up, it was observed (but this is not indicated in Figure 2.34) that the integral term u_i in the PID controller reached a maximum value of approximately 2200%! The simulations clearly show that it is beneficial to use integrator anti wind-up (as the temperature returns much quicker to the setpoint after the

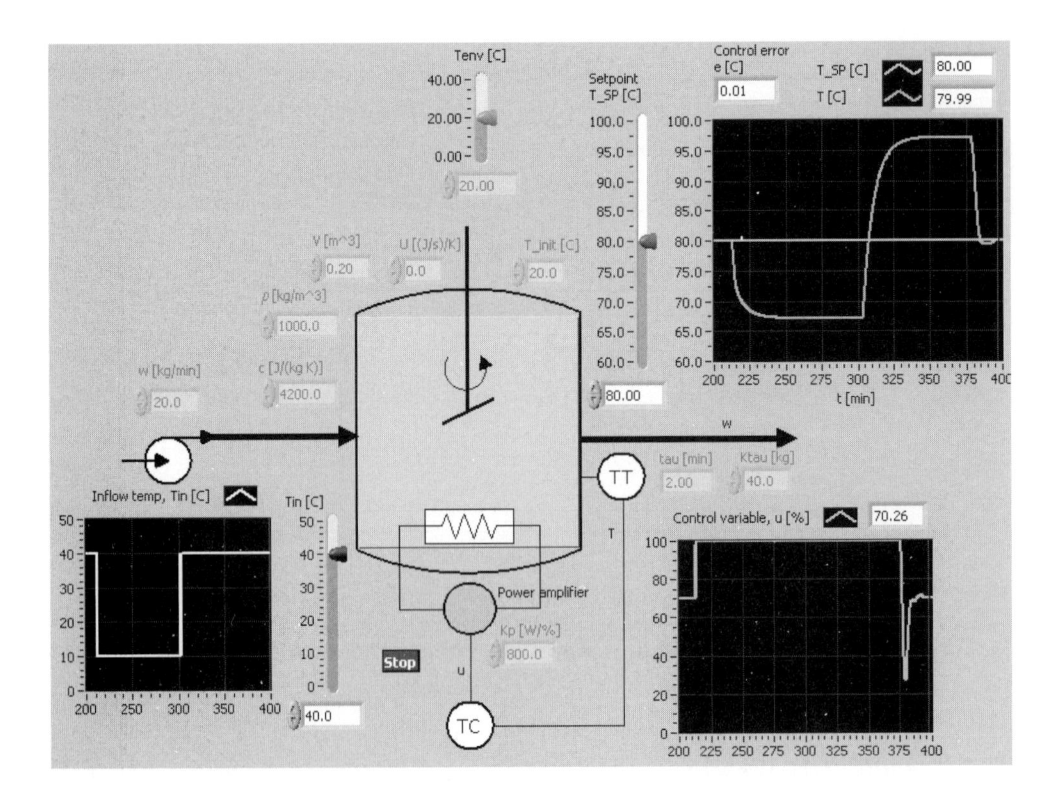

Figure 2.34: Example 2.15: Temperature control *without* anti wind-up

disturbance has changed back to its normal value).

[End of Example 2.15]

2.7.3 Measurement noise. Signal variance

Introduction

You have already seen the problems concerning measurement noise in a control loop, cf. Section2.6.7. Figure 2.22 shows where the measurement noise enters the control loop. Section2.6.7 describes a necessary modification of the derivative term in a PID controller: A lowpass filter is inserted before (in series with) the D-term to attenuate the noise before it is time-differentiated to avoid too large noise-generated responses in the control signal. But what if the lowpass filter in the D-term does not give sufficient noise filtering? Then an additional filter should be included in the feedback path, acting on the measurement signal. This filter can be

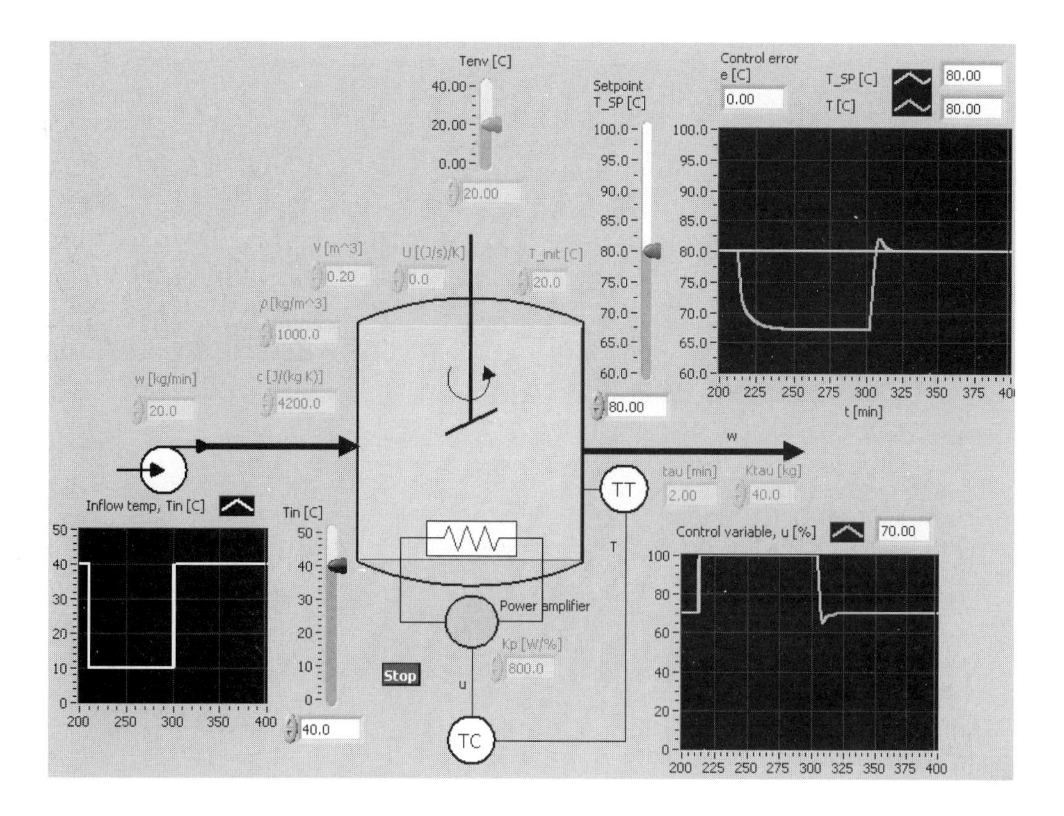

Figure 2.35: Example 2.15: Temperature control *with* anti wind-up

- either a linear dynamic lowpass filter,
- or a deadband filter.

These solutions are described in more detail below.

Calculating the variance

Measurement noise is typically a random signal. The noise propagates through the control system via the controller, causing variations in all variables in the control system. Figure 2.36 shows typical examples of a noisy process measurement and the control variable and in a simulated control system. The variances shown in the figure are calculated as explained below from the 50 most recent samples.

To express the variation of a process variable, the statistical *variance* can be calculated, alternatively the standard deviation which is the square root of the variance. The larger variance, the larger the variations. The

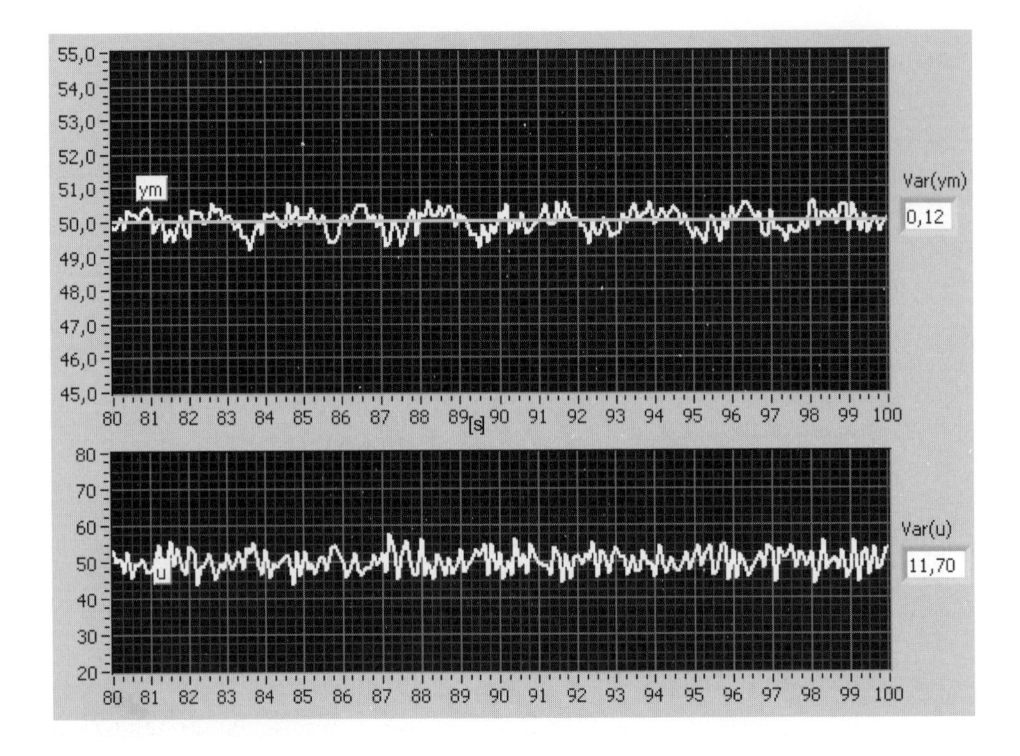

Figure 2.36: Typical examples of a noisy process measurement and the control variable

variance is the mean square deviation about the mean value[11]

$$\mathrm{Var}(y_m) = \frac{1}{N-1}\sum_{k=1}^{N}\left[y_m(t_k) - m_{y_m}\right]^2 \tag{2.77}$$

where N is the number of samples and m_{y_m} is the mean value of y_m, which may be calculated by

$$m_{y_m} = \frac{1}{N}\sum_{k=1}^{N} y_m(t_k) \tag{2.78}$$

The numerical value of the variance is usually not particularly useful in itself, but it is useful when comparing signals.

In Example 2.16 variances will be used to express the improvements by using a lowpass filter on the process measurement signal.

[11] To obtain a so-called nonbiased estimate of the variance, you must divide by $N-1$, not by N.

Using a dynamic lowpass filter

Figure 2.37 shows a control loop having a lowpass filter acting on the measurement signal. The filter can be a discrete-time filter implemented in

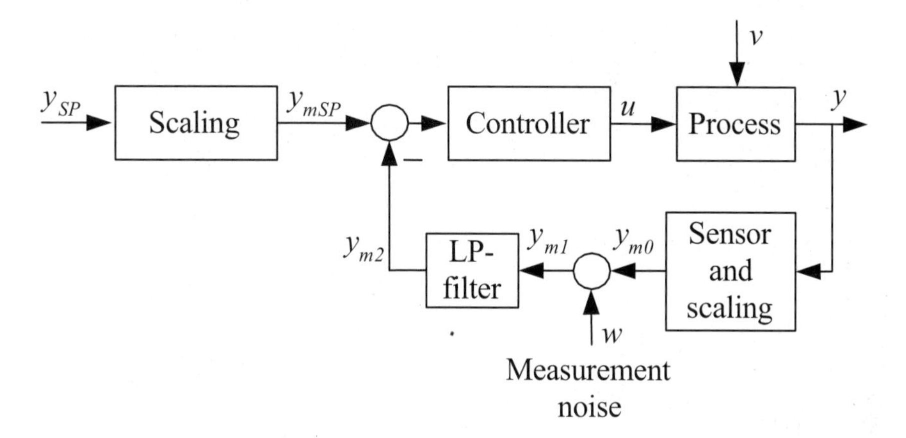

Figure 2.37: Control loop having a lowpass filter acting on the measurement signal. (LP = lowpass.)

the control equipment. It is common that control equipment have inbuilt lowpass filter functions. Alternatively, the filter can be a continuous-time lowpass filter implemented using electronic components external to the control equipment. For example a first order filter can be implemented as an RC-circuit.

Let us assume that the measurement lowpass filter is a first order filter. Such a filter has the following transfer function from filter input x_{in} to filter output x_{out}:

$$\frac{x_{out}(s)}{x_{in}(s)} = H(s) = \frac{1}{\frac{s}{\omega_b} + 1} = \frac{1}{\frac{s}{2\pi f_b} + 1} \tag{2.79}$$

The bandwidth of the lowpass filter is $\omega_b = 2\pi f_b$ where ω_b has unit rad/s and f_b has unit Hz. The bandwidth must be given a value which is smaller than the frequency of the substantial noise frequency components so that these components fall within the stopband of the filter. The bandwidth may be tuned experimentally. Figure 2.38 shows a typical amplitude gain function of first order lowpass filter. One example of a noise frequency component is shown in the figure (it is in the stopband of the filter). The bandwidth is typically defined as the frequency where amplitude gain is $1/\sqrt{2} = 0.71 \approx -3\text{dB}$.

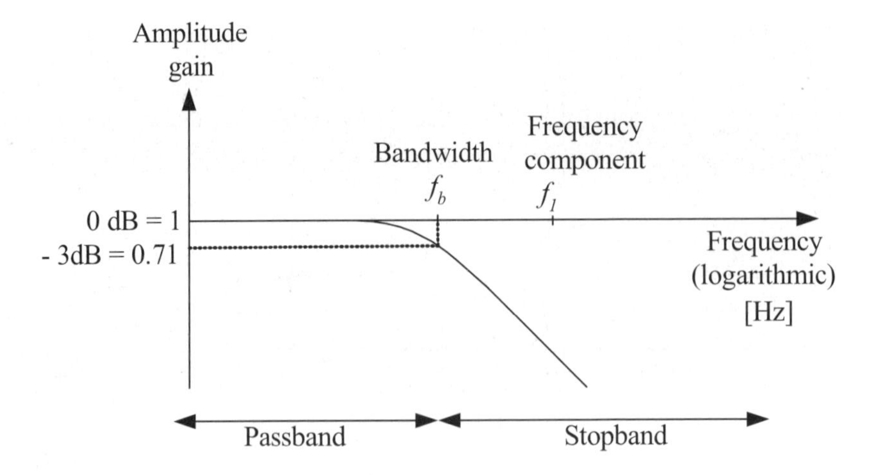

Figure 2.38: Typical amplitude gain function of a lowpass filter. One example of a noise frequency component is shown.

Example 2.16 ***Measurement noise filter in a control loop***

In this example a control system for a process having the following transfer function model is simulated:

$$y(s) = \frac{K_u}{(T_1 s + 1)(T_2 s + 1)} e^{-\tau s} u(s) \tag{2.80}$$

$$+ \frac{K_v}{(T_1 s + 1)(T_2 s + 1)} e^{-\tau s} v(s) \tag{2.81}$$

(The process is thus a second order system with time delay.) u is the control variable, and v is the process disturbance. The process parameter are

$$K_u = 1;\ K_v = 1;\ T_1 = 1\text{s};\ T_2 = 0.5\text{s};\ \tau = 0.3\text{s}; \tag{2.82}$$

The PID parameters are

$$K_p = 2.8;\ T_i = 1.2\text{s};\ T_d = 0.3\text{s}; \tag{2.83}$$

(tuned with the Ziegler-Nichols' closed loop method). Figure 2.39 shows simulated responses in the control system. The measurement noise is a random signal uniformly distributed[12] between -1 and $+1$. The lowpass filter acts on the process measurement, cf. Figure 2.37. It is switched into the loop at $t = 10$s. From Figure 2.37 we see that the filter removes noise from the process measurement and that the control variable (therefore) is less noisy. The filter is a first order lowpass filter with bandwidth 1.5Hz.

[12] which means that there is equal probability for any value between -1 and $+1$.

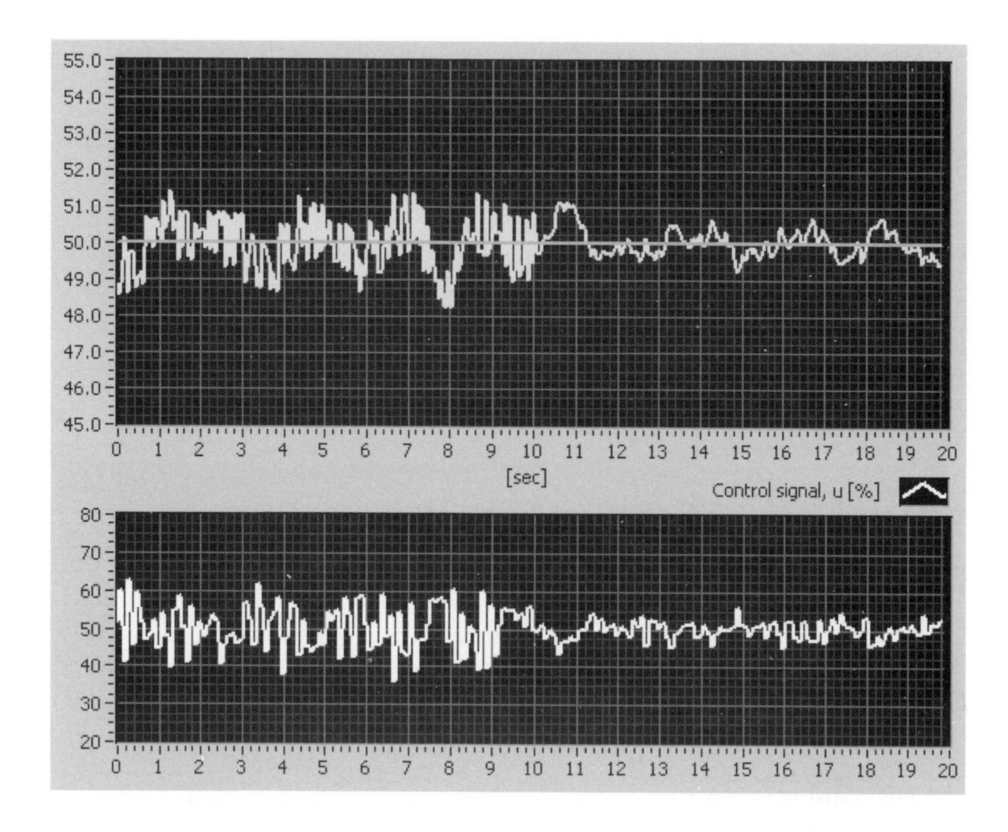

Figure 2.39: Example 2.16: Simulated responses of a control system. A lowpass filter acting on the process measurement signal is switched into the control loop at $t = 10$s.

Table 2.1 shows the variances of the control signal u and the process measurement (after the filter) without and with lowpass measurement filter. The variances are calculated from the 50 most recent signal samples. It is clear from the variances that the filter reduces the influence of the noise in the loop.

Without Filter	**With Filter**
$\text{Var}(u) = 36.4$	$\text{Var}(u) = 5.2$
$\text{Var}(y_m) = 0.33$	$\text{Var}(y_m) = 0.13$

Table 2.1: Variances of control signal u and measurement signal y_m without and with lowpass filter

[End of Example 2.16]

Including a filter in the control loop *changes the dynamic properties* of the loop! Actually, it can cause stability problems in the control loop. In most cases, the less bandwidth (i.e., more sluggish filter), the more reduction of the stability of the loop.

Example 2.17 ***Poor stability because of measurement filter***

Figure 2.40 shows simulated responses for the same control system simulated in Example 2.16. Before $t = 10$s there is no lowpass filter in the loop, while after $t = 10$s a first order lowpass filter with bandwidth 0.2Hz is switched into the loop. The filter causes the control system to have very poor stability.

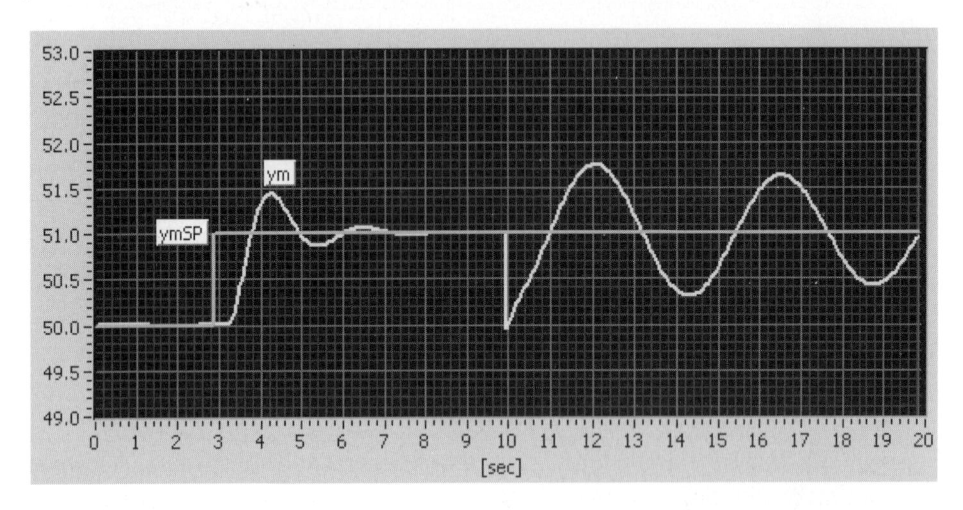

Figure 2.40: Example 2.17: A first order lowpass filter is switched into the control loop at $t = 10$s, causing the control system to have poor stability.

[End of Example 2.17]

If a measurement filter results in poor stability of the control loop – how can that problem be avoided? By *tuning (or re-tuning) the controller with the filter in the control loop.*

It is tempting to select a very small bandwidth of the measurement lowpass filter to achieve strong attenuation of the measurement noise. But in addition to attenuating noise, also frequency components in the ideal (noise-free) process output signal is attenuated. In other words: Important process information may be removed from the measurement signal. This in turn may cause the controller to calculate the control signal on basis of an erroneous control error value. One way to solve this problem, is to introduce a similar filter in series with the setpoint, as shown in Figure . This solution is equivalent to placing one filter in series with (or before) the PID controller in Figure 2.41. A setpoint filter implies that the setpoint which the controller observes, becomes more sluggish since high frequency components are attenuated.

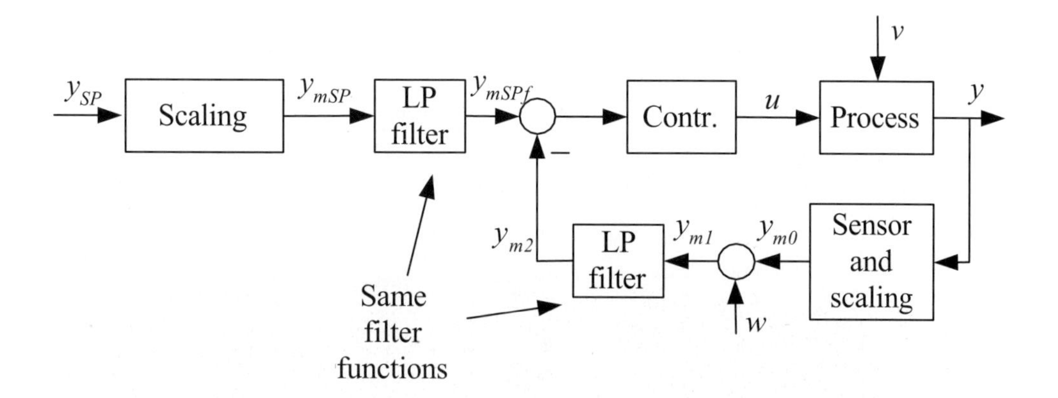

Figure 2.41: Lowpass filter acting on the setpoint

Using a deadband filter

If you know the maximum amplitude of the measurement noise, the noise can be removed from the noisy measurement signal by letting the signal pass through a deadband filter, see Figure 2.42. The output value of the

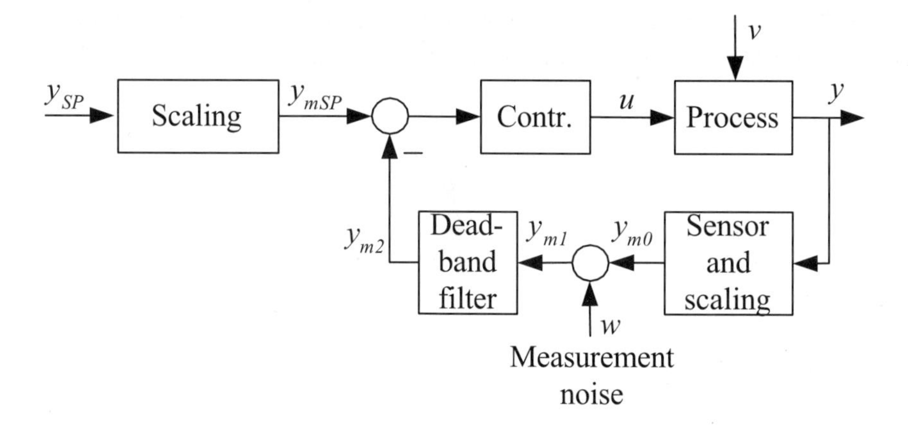

Figure 2.42: Deadband filter acting on the process measurement signal

deadband filter changes value only if the change of the input signal is larger than the deadband.

Example 2.18 ***Deadband measurement filter in the control loop***

Figure 2.43 shows a simulation of a control system with deadband measurement filter. The process and the PID controller are as described in Example 2.16. In the simulation the measurement noise is a random signal

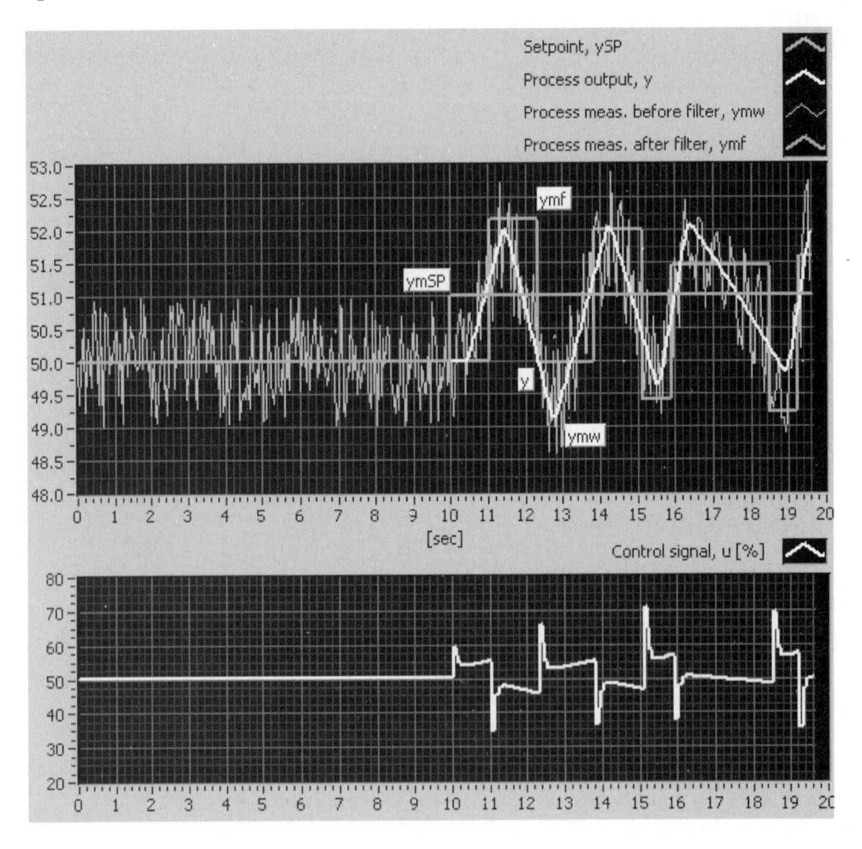

Figure 2.43: Example 2.18: Simulation of a control system with deadband measurement filter

uniformly distributed between −1% and +1%. The deadband of the filter is 2%. The simulation shows the following:

- Up to time $t = 10$s the setpoint is constant (50%). The measurement signal which is the output of the deadband filter is constant since the amplitude of the noise is smaller than the deadband. And since the measurement signal is constant, the control signal generated by the controller is constant – which is good!
- At $t = 10$s the setpoint is changed as a step (from 50% to 51%) which implies that the measurement signal due to the overshoot in

the step response changes value beyond the deadband of 2%. Thereafter the deadband filter acts similar to an on/off-element in the loop, and there are sustained oscillations in the loop – not good!

[End of Example 2.18]

You have in this Section seen two ways of filtering measurement noise:

- *Using a dynamic lowpass filter*: The dynamic filter can be easily tuned via the bandwidth. The filter influences the dynamics and hence the stability of the loop. The controller should be tuned with the filter in the loop.
- *Using a deadband filter*: This filter may give a constant measurement signal, as long as the input signal to the deadband filter does not change more than the deadband of the filter. Once the deadband is exceeded, the deadband filter may behave almost like an on/off controller, causing oscillations in the control loop.

From the above results it seems that the dynamic filter is a safer way than deadband filter to handle measurement noise.

2.8 Performance index of control systems

Assume that different control systems are to be compared, or that different controller parameter for one control system are to be compared. There are several ways to express the performance of the control systems, e.g. bandwidth and stability margins. These measures are based on a mathematical model of the control system, and they are described in Chapter 6.

Alternatively, we can use performance indices which are functions of the observed control error e. These indices does not require a mathematical model. Probably the most frequently used index is the IAE – Integral of Absolute value of control Error [15]:

$$\mathrm{IAE} = \int_0^\infty |e| \, dt \tag{2.84}$$

The less IAE value, the better performance. The IAE value is finite only if e converges towards zero in steady-state, which in practice requires the controller to have have integral action, as in a PI controller and in a PID controller. In discrete time the IAE value can be regarded as the sum of the absolute values of the sample values of the control error, since

$$\mathrm{IAE} = \int_0^\infty |e|\,dt \approx \sum_{k=0}^{\infty} h\,|e(t_k)| = h\sum_{k=0}^{\infty} |e(t_k)| \tag{2.85}$$

where h is the time step (time interval between each discrete point of time). h is $t_k - t_{k-1}$. k is a time index: $t_k = hk$.

We can in practical applications calculate the IAE value only over a finite time interval, say from $t = 0$ to t_k. We can derive a recursive algorithm of the IAE as follows:

$$\mathrm{IAE}(t_k) = \int_0^{t_k} |e|\,dt \tag{2.86}$$

$$= \int_0^{t_{k-1}} |e|\,dt + \int_{t_{k-1}}^{t_k} |e|\,dt \tag{2.87}$$

$$\approx IAE(t_{k-1}) + h|e(t_k)| \tag{2.88}$$

The expression $h|e(t_k)|$ in (2.88) is an Euler backward (rectangular) approximation to the latter integral in (2.87).

Example 2.19 ***IAE for level control of chip tank***

The level control system for the wood-chip tank described in Example 2.3 (page 19). Figure 2.19 shows simulated responses with a PI controller, and Figure 2.20 shows responses with a PID controller. In both cases the IAE index is computed for the control error after the step in the outflow w_{out} (disturbance), which is from $t = 65$s to 120s. The results are as follows:

- PI controller: IAE = 6296
- PID controller: IAE = 3767

We see that the PID controller has better IAE performance than the PI controller. This is in accordance with the better compensating performance of the PID controller that we can easily see in the simulations.

[End of Example 2.19]

2.9 Selecting P, PI, PD, or PID?

In general, the PID controller is the first choice since it gives zero static control error and relatively quick control. Here are a few guidelines for choosing other controller functions than PID:

- If there is much process measurement noise the derivative term should be dropped. What remains is a PI controller.

- If the process is of first order or is a pure integrator and in addition has a time delay, the stability margins of the control system may be small if the controller has a derivative term. This means that the control system stability may become poor after small parameter changes in the process. If you want to be on the safe side, the derivative term may be left out, and a PI controller remains.

- If the process has fast dynamics compared to other processes it is connected to, the derivative term may be dropped since the control system will be quick enough with a PI or a P controller. On example is quick flow control loops which works as inner or secondary loops within more sluggish loops for temperature control or level control, as in cascade control, cf. Chapter 9.2. The increase of control speed due to the derivative term is usually not important since the inner loop in any case will be faster than the outer, primary loop, due to the fast dynamics of the inner process.

- The P controller may be a sufficiently good controller for processes containing a pure integrator, as motors where position is to be controlled, and liquid tanks where level is to be controlled when the disturbance (e.g. load force or load torque on the motor or tank inflow or outflow) is zero or small. In these cases, the integral action in the "inbuilt" integrator in the process ensures zero static control error at constant setpoint. However, a further analysis, as described in Chapter 6, should be done to see if the simple P controller is sufficient to obtain the specifications of maximum control error and quickness (bandwidth).

- The PD controller may be applied in electrical servomechanisms where the steady-state control error due to disturbances (as load torque) is sufficiently small. The D-term may increase the quickness (bandwidth) of the control loop. The D-term may cause stability problems in hydraulic servomechanisms because of the hydraulic resonance.

2.10 Reduction of control error by process changes

Earlier in this chapter we have seen how to use the control variable u to control the process so that the control error becomes sufficiently small. However, the control error, $e = y_{SP} - y$, depends not only of only the control variable, but also of the disturbance and the process itself. This implies that it may be possible to reduce the control error by change the disturbance and/or the process. This is explained in more detail below.

1. **Reducing control error by reducing or isolating the disturbance(s).** In most processes it difficult or impossible realize this point. This is because the disturbance often is closely related to the function of the process.

   Here are a few examples:

   - *Example 1*: In a level control system for the wood-chip tank, cf. Example 2.3, the chip outflow (disturbance) can not be reduced since it is the feed to the cookery downstream in the process line.
   - *Example 2*: In a temperature control system for liquid the tank, cf. Example 2.4, it is hardly possible to change the ambient temperature (disturbance). But it may be possible to change the inlet temperature (also a disturbance).
   - *Example 3*: In a motor speed control system, cf. Example 2.5, it is not realistic to be able to change the load torque (disturbance) since it is probably closely related to the function of the motor, as in a grinding machine or a conveyor belt.

2. **Reducing control error by changing the process construction**. Of course it may be impractical to change the construction of a process which is already built, but a process under planning can be changed more easily.

   A few examples:

   - *Example 1*: In a level control system for a liquid tank a wider tank will reduce the level variations (but not mass variations).
   - *Example 2*: In the level control system for the wood-chip tank, cf. Example 2.3, reducing the transport delay on the conveyor belt may give quicker control and hence smaller control error.

Imagine the inlet screw rotational speed *and* the band speed were controlled simultaneously and proportionally...[13]

- *Example 3*: In a temperature control system, better tank isolation will reduce the effects of the ambient temperature, see Figure 2.44. And an increase of the tank volume would give

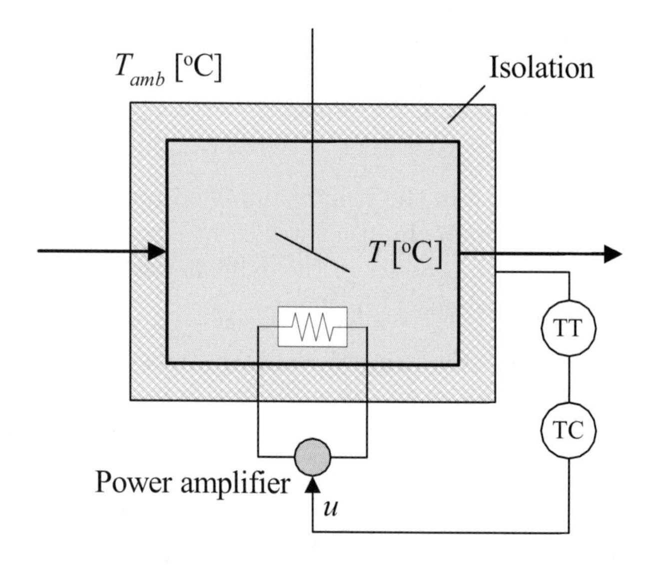

Figure 2.44: Better isolation of the tank will reduce the influence of the ambient temperature T_{amb} on the tank temperature T.

better attenuation (dynamically, but not statically) of temperature disturbances in the tank.

- *Example 4*: In a speed control system a larger motor or using a gear reduces the effects of the load torque on the motor speed.

2.11 Control loop stability

It is important to be aware that there may be stability problems in a control loop. It is an basic requirement to a control loop that it is stable. Simply stated this means that the response in any signal in control loop converges towards a finite value after a limited change (with respect to the amplitude) of the setpoint or the disturbance or any other input signal to the loop. For example, the control loop constituting the control system for the wood-chip tank in Example 2.11 is stable.

All methods for tuning controller parameters have as the main aim that

[13] Then the transport delay was eliminated.

the control loop is stable. The PID parameters used in the level control system in Example 2.11 where tuned using the Ziegler-Nichols' closed loop method, cf. Chapter 4.4. However, there is always a possibility that a feedback control system which is originally stable, *may become unstable* due to parameter changes in the loop. Instability implies that signals in the loop starts to increase in amplitude until some saturation limit is reached (for example, a valve have a limited opening).

Instability can be explained in two ways:

- The signal one place in the loop is amplified too much through the subsystems in the loop. In other words, *the loop gain is too high.* The loop gain is the product of the gains in each of the subsystems (controller, process, sensor) in the loop.
- There is *too much time delay* through the subsystems in the loop.

Chapter 6.4 describes ways to analyze control loop stability theoretically.

Example 2.20 ***Instability in the wood-chip tank level control system***

We will see that the level control system for the wood-chip tank becomes unstable if the controller gain K_p in the PID controller becomes too large, and if the transport delay related to the conveyor belt becomes too large. Figure 2.15 (page 32) shows the front panel of the simulator. The control system is initially stable for the following PID parameters:

$$K_p = 1.9;\ T_i = 540\text{s} = 9.0\text{min};\ T_d = 135\text{s} = 2.25\text{min} \tag{2.89}$$

(found by the Ziegler-Nichols' closed loop-method, cf. Section4). The time delay is 250s = 4.17min.

Figure 2.45 shows responses for $K_p = 6$, which is considerably larger than the optimal (Ziegler-Nichols) value of 1.9. The setpoint and the disturbance (chip outflow) are constant. The control error has initially the very small value of 0.0004. We see from the simulation that the control system is unstable. One explanation of the instability is that the loop gain is too large, due to the large K_p-value. The amplitude of the oscillations are limited due to the limits of the control variable (the maximum value is 100%, and the minimum value is 0%).

Figure 2.46 shows responses with the original PID values, but for an increased value of the time delay τ, namely 600s = 10min (the nominal

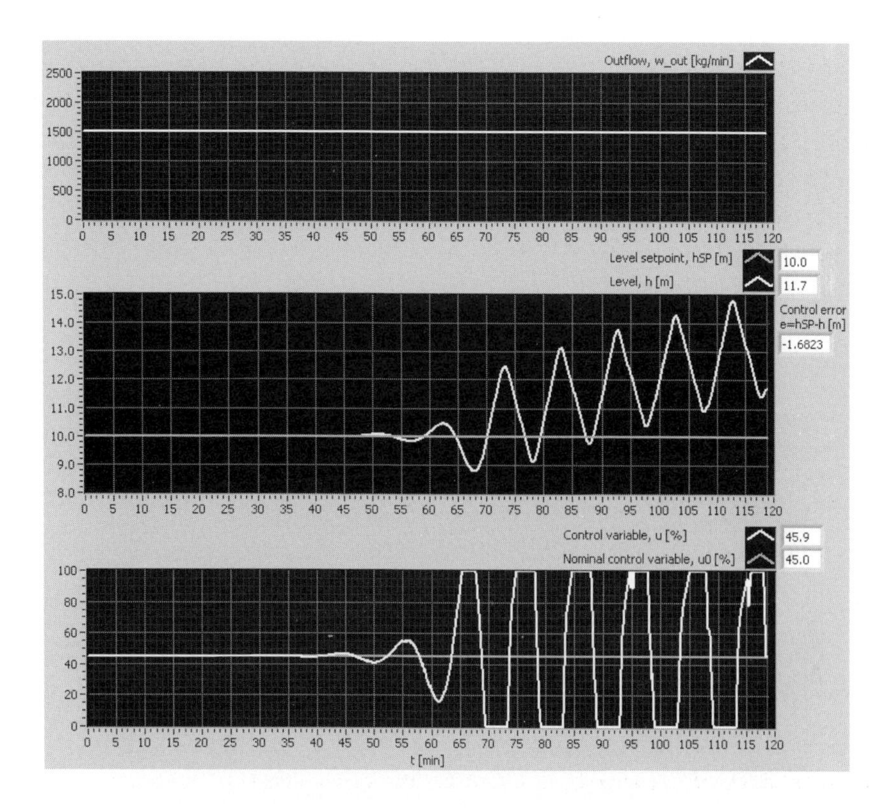

Figure 2.45: Example 2.20: Level control of the wood-chip tank with a (too) large K_p-value of 6, which causes the control system to become unstable. (The front panel of the simulator is as shown in Figure 2.15.)

value is 250sec= 4.17min). The simulation shows that the control system is unstable, which is due to the (too) large time delay in the loop.

[End of Example 2.20]

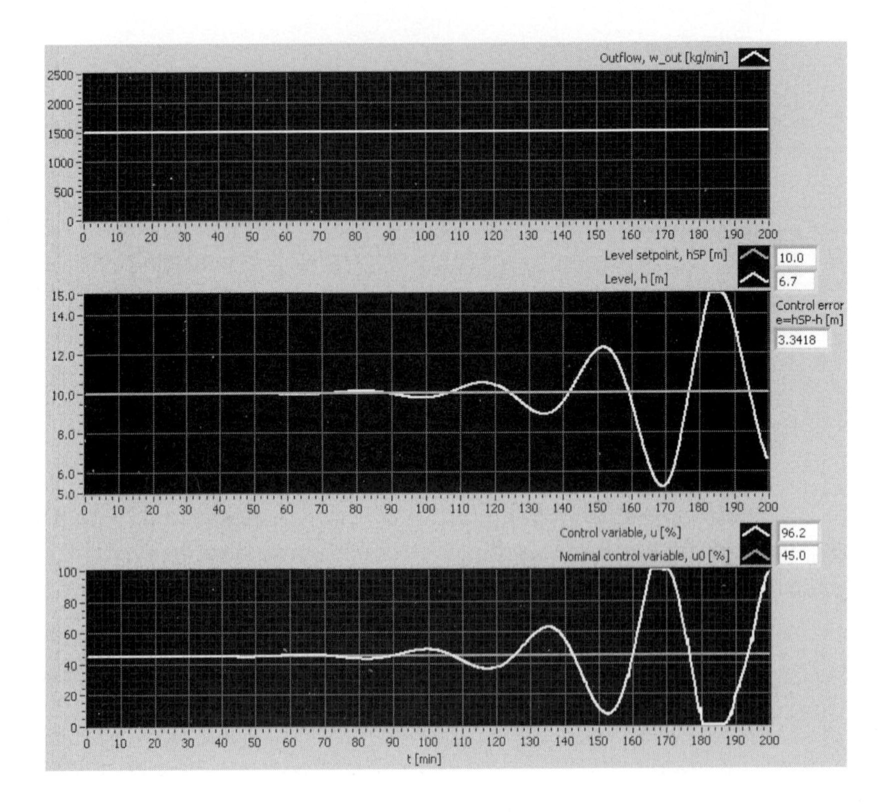

Figure 2.46: Example 2.20: Level control of the wood-chip tank with a (too) large τ-value of 9, which causes the control system to become unstable. (The front panel of the simulator is as shown in Figure 2.15.)

Chapter 3

Control equipment

This chapter gives an overview over various kinds of commercial control equipment.

3.1 Process controllers

A *process controller* is a single controller unit which can be used to control one process output variable. Figure 3.1 shows an example of a process controller (ABB's ECA600). Today new process controllers are implemented digitally with a micro processor. The PID controller function is in the form of a program which runs cyclically, e.g. each 0.1s, which then is called the time step of the controller. Earlier, controllers were build using analog electronics, and even earlier, pneumatic and mechanical components were used. Such controllers are still in use today in some factories since they work well and are safe in dangerous (explosive) areas.

Here is a list of typical characteristics of process controllers:

- The front panel of the controller has vertical bar *indicators and/or numeric displays* showing, see Figure 3.1,
  - the process measurement signal (a common symbol is PV – Process Value),
  - the setpoint (symbol SP),
  - the control variable (MV – control variable).

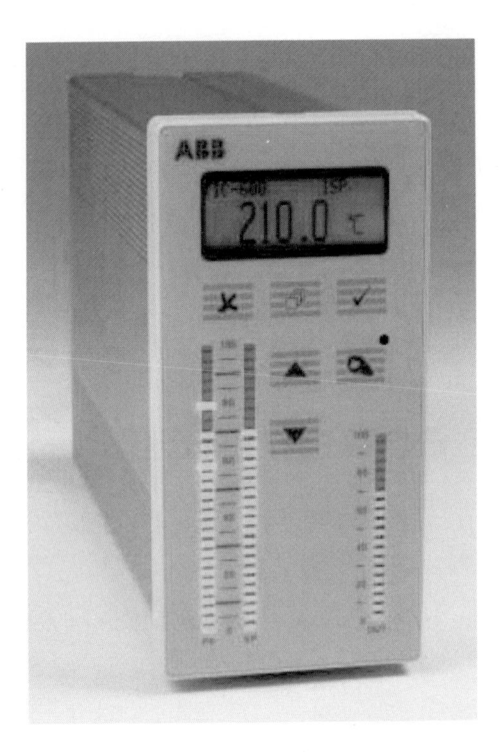

Figure 3.1: A process controller (ABB ECA600)

- The controller has *analog input* (AI) (for measurement signals) and *analog output* (AO) (for the control variable). The input signal is a voltage signal (in volts) or a current signal (in amperes). The output signal is a current signal. In general, current signals are preferred before voltage signals, because:
  - A current signal is more robust against electrical noise.
  - With long conductors the voltage drop along the conductor may become relatively large due to the internal resistance in the conductor.
  - A current signal of 0 mA is abnormal and indicates a break of the conductor.

The standard current range is 4–20 mA (also 0–20 mA is used). There are several standard voltage ranges, as 1–5V[1] and 0–10V. The physical measurement signal in A or V is usually transformed to a percent value using a linear function. For example, the transformation from the a signal y_A in the range 4-20 mA to a signal

[1] 4–20 mA may be transformed to 1–5 V using a resistor of 250 Ω.

$y_{\%}$ in the range 0-100% is realized using the following formula:

$$y_{\%} = K\left(y_A - y_{A0}\right) + y_{\%0} \tag{3.1}$$

where $y_{\%0} = 0\%$, $y_{A0} = 4\text{mA}$ and $K = 100\%/16\text{mA} = 6,25\%/\text{mA}$. The most important reasons to use 4 mA and not 0 mA as the lower current value, is that the chance that the actual measurement signal is drowned in noise is reduced and that the base signal of 4 mA can be used as an energy source for the sensor or other equipment.

- The controller may have *pulse output*, which may used to implement analog output using a binary actuator, typically a relay. The technique is called *PWM - Pulse Width Modulation.* The PWM-signal is a sequence of binary signals or on/off-signals used to control the binary actuator. See Figure 3.2. The PWM-signal is kept

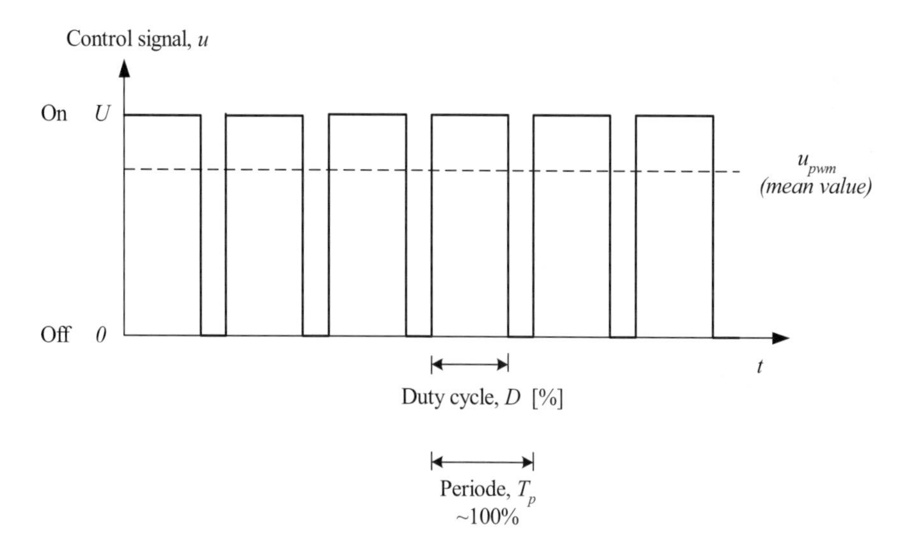

Figure 3.2: Pulse Width Modulation

in the on-state for a certain time-interval of a fixed cycle period. This time-interval (for the on-state) is called the duty cycle, and it is expressed as a number in unit percent. For example, 60% duty cycle means that the PWM-signal is in on-state in 60% of the period (and in the off-state the rest of the period). The duty cycle is equal to the specified analog control signal which is calculated by the PID controller function. In the mean the PWM-signal will become equal to the specified analog control signal, if the cycle period is small compared to the time constant of the process to be controlled.

- In addition to the analog inputs and outputs the process controller typically have *digital inputs* (on/off-signals) to detect signals from switches, buttons, etc., and *digital outputs* which can be used to control relays, lamps, motors etc.

- The process controller typically have ports for serial (RS-232) *communication* with another controller or a computer.

- The controller can be programmed from a panel on the front or on the side of the unit or from a connected PC. The programming includes combining function modules (see below).

- One of the function modules is the PID controller. You will have to enter values for the PID parameters, typically being K_p, T_i and T_d, but other parameters may be used in stead, as the proportional band, PB, which is equal to $100\%/K_p$, cf. Chapter 2.6.5.

- Other function modules include logical functions as AND, OR, SR-flipflop, etc., and arithmetic functions, as addition, multiplication, square root calculation, etc.

- In addition to the common single-loop PID control, the process controller may implement more advanced control methods, as *feedforward control* (cf. Chapter 9.1) *gain scheduling PID control* (4.9.2), *cascade control* (9.2) and *ratio control* (9.3).

- The operator can adjust the setpoint *internally* (locally) on the controller, but it is usually possible to use an *external* setpoint, too, which may come from another controller (as in cascade control) or from a sensor (as in ratio control) or from a computer which calculates the setpoint as a result of a plantwide control strategy.

- The operator can take over the control by switching the process controller from *automatic mode* to *manual mode.* This may be necessary if the control program is to be modified, and at first-time start-up. In manual mode the operator adjusts the controller output (which is used as the manipulating variable of the process) manually or directly, while in automatic mode the control output is calculated according to the control function (typically PID control function).

- Many controllers have the possibility of *auto-tuning*, which means that the process controller – initialized by the operator – calculates proper values of the controller parameters based on some automatically conducted experiment on the process to be controlled. An even more advanced option which may be implemented is an *adaptive PID controller* where the PID parameters are calculated continuously from an estimated process model.

- The operator can define alarm limits of the measurement signal. The alarm can be indicated on the front panel of the controller unit, and the alarm limits can be used by the control program to set the value of a digital output of the process controller to turn on e.g. an alarm lamp.

Figure 3.3 shows a section of the data sheet of the controller ECA600 shown in Figure 3.1.

Controller	
Control functions	P, PD, PI, PID, pPI
Gain	0.01–99.99
Integral time	0.1–9999.9 seconds
Derivative time	0.0–9999.9 seconds
Control action	Direct, reversed
Set point	Internal, external, ramp
Control output	Analogue, pulse
Alarms	Process value, deviation.
Sample time	30–500 ms
Analogue Inputs	
Input ranges	0–20 mA, 4–20 mA, 0–5 V, 1–5 V, 0–10 V, 2–10 V.
Input types	Differential or single ended (jumper selectable).
Input impedance	Current 250 Ω Voltage 200 kΩ
Alarm function for out-of-range signal	Yes, for 4–20 mA, 1–5 V and 2–10 V, when the signal drops below the lower limit.
Functions	First-order software filter, linear / square root.
Resolution	12 bits
Inaccuracy	Max. ± 0.2% of FS within 5–55°C.
Temperature stability	0.01% FS per °C within 5–55°C.
Analogue Outputs	
Output ranges	0–20 mA, 4–20 mA.
Type	Current source
Max. output current	22 mA
Load resistance on current output	Max. 650 Ω
Short circuit protection	Yes
Resolution	12 bits
Output signal break detection	Yes
Inaccuracy	Max. ± 0.2% of FS within 0–50°C.
Digital Inputs	
Type	24 V DC, common digital input ground, current sink, opto-isolated.
Voltage	Max. 35 V, min. -0.5 V.
Logic levels	0 < 3 V (IEC 1131-2, type 1) 1 > 15 V (IEC 1131-2, type 1).
Digital Outputs	
Type	24 VDC, current source.
Load current	Max. 250 mA per output, max. 500 mA total.
Short-circuit current	Max. 500 mA transient current during 1 μs.
Power supply	
AC	115/230 V AC ± 10%, 50–60 Hz, 20 VA or 19 V AC ± 10%, 50–60 Hz, 1 A.
DC	24 V DC ± 10%
Protection	Secondary side of transformer and direct supply fused via thermo type fuse.
Transmitter	Max. 24 V DC/150 mA.
Environmental specifications	
Operating temperature	+5 to +55°C (IEC 68-2-1/2).
Non-operating temperature	-25 to +70°C (IEC 68-2-1/2).
Non-operating damp heat steady state	93% relative humidity at +40°C (IEC 68-2-3).
Protection class	IP20 generally. IP65 for front. IP65 for front against IP65 compliant panel with panel mounting kit.
Electrical environment	Fulfils ElectroMagnetic Compatibility, EMC, directive 89/336/EEC
Order codes	ECA 06–0000 ECA 60–0000 ECA 600–0000 EMA 60–0000

Figure 3.3: A section of the datasheet of the controller ECA600 shown in Figure 3.1.

3.2 PLCs and similar equipment

*PLC*s are common in industrial automation. Figure 3.4 shows a PLC system (Mitsubishi FX2N). PLC is short for Programmable Logical

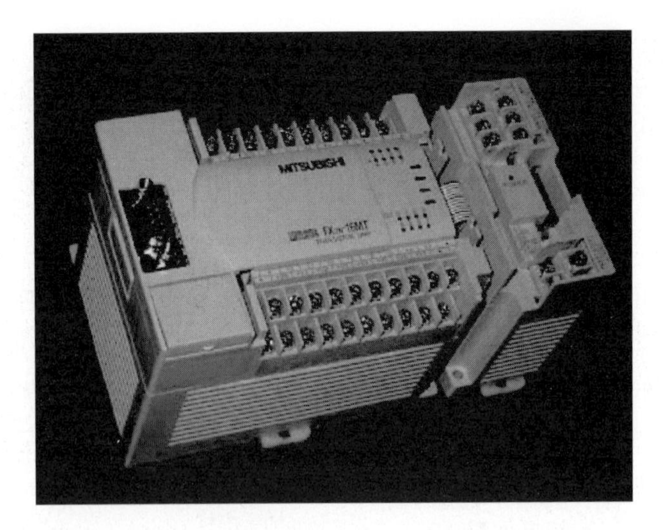

Figure 3.4: A PLC (Programmable Locical Controller). (Mitsubishi FX2N)

Controller. PLC-systems are modular systems for logical (binary) and sequential control of valves, motors, lamps etc. Modern PLCs includes function modules for PID control. The programming is usually executed on a PC, and the PLC-program is then downloaded (transfered) to the CPU in the PLC-system which then can be disconnected from the PC. The program languages are standardized (the IEC 1131-3 standard), but the actual languages which are implemented on commercial PLCs may differ more or less from the standard.

There exists alternatives to PLCs. Figure 3.5 shows National Instruments' Compact FieldPoint, which is denoted a *PAC* - Programmable Automation Controller. This is a modular system similar to a PLC in several aspects. Both logical and sequential control and PID control can be realized in the PAC. The control program is developed in LabVIEW on a PC, and then it is downloaded to the PAC, where it runs independently of the PC.

3.3 SCADA systems and DSC systems

3.3.1 SCADA systems

SCADA systems (Supervisory Control and Data Acquisition) are automation systems where typically PCs are used for supervision and control, but the execution of the control program takes place in a PLC or some other control equipment. In this way the SCADA system implements

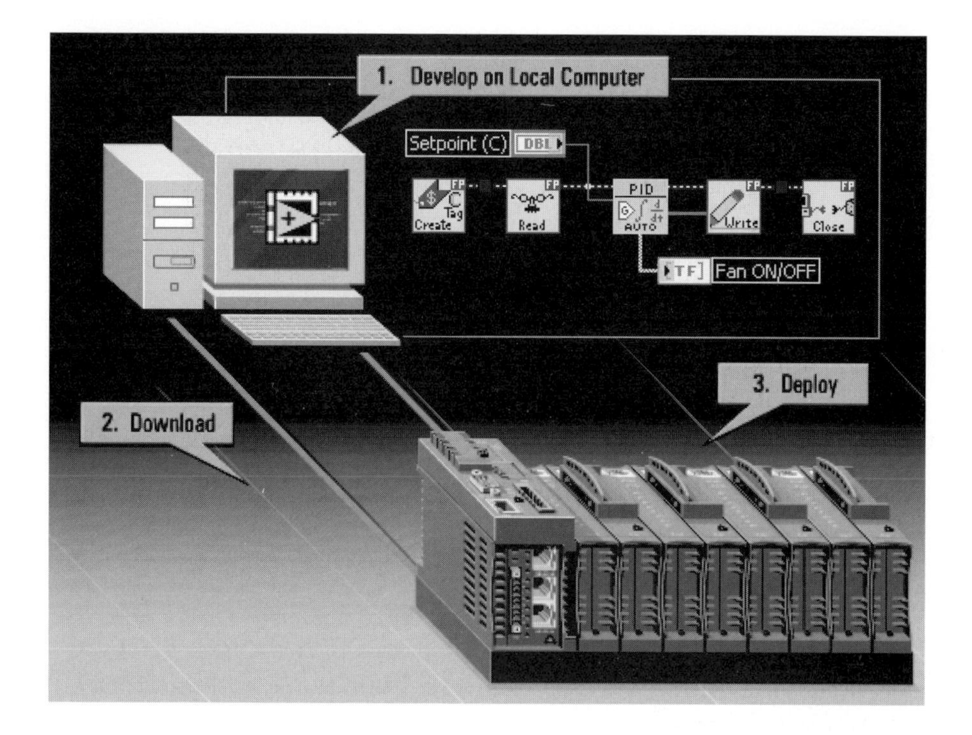

Figure 3.5: Modular control equipment: Compact Fieldpoint, denoted *PAC* - Programmable Automation Controller. (National Instruments)

a distributed control system architecture. Figure 3.6 shows an example of a SCADA-system.

Here are some characteristics of PC-based control systems:

- A set of function modules are available in the SCADA program on the PC, as arithmetic and logical functions, and signal processing functions. The setpoint to be used by the connected PLC or other control equipment may be calculated by the SCADA program according to some optimal control strategy.
- The user can build the screen contents containing process images of tanks, vessels, valves, etc., and bars, diagrams, numeric displays, alarm indicators etc.
- The PCs can communicate with other PCs or other kinds of computers via standard communication means, as RS-232 cables or Ethernet etc.
- The SCADA system has driver programs for a number of PLC-systems (from different vendors) and other I/O-systems

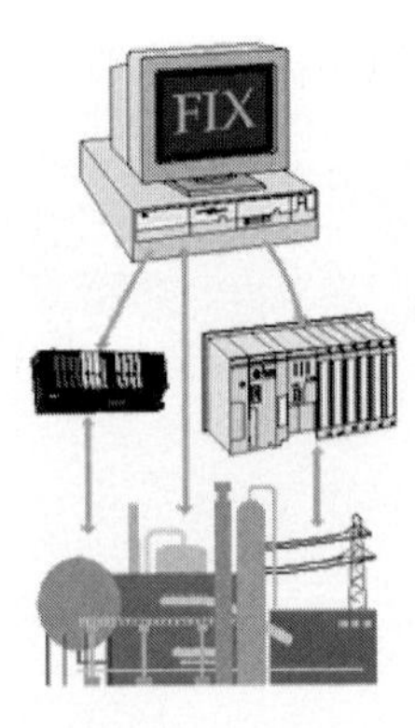

Figure 3.6: SCADA system (FIX, Novotek)

(systems for analog and digital Input/Output). The number of drivers may exceed 100.

- Data can be exchanged between programs in real-time. The OPC standard (OLE for Process Control)[2] has become an important standard for this.

3.3.2 DCS

DCS (Distributed Control Systems) are similar to SCADA systems in that the control equipment is distributed (not centralized) throughout the plant. Special process stations – not standard PLCs or process controllers – executes the control. DCSs can however communicate with PLCs etc. The process stations are mounted in special rooms close to process. The whole plant can be supervised and controlled from control rooms, where the operators communicate with the distributed control equipment, see Figure 3.7.

3.4 Embedded controllers in motors etc.

Producers of electrical and hydraulic servo motors also offers controllers for the motors. These controllers are usually embedded in the motor drive system, as separate physical unit connected to the motor via cables. The controllers implement speed control and/or positional control. Figure 3.8 shows an example of a servo amplifier for DC-motor. The servo amplifier

[2] OLE = Object Linking and Embedding, which is a tecnology developed by Microsoft for distributing objects and data bewteen Windows applications.

Figure 3.7: Control room of a distributed control system (DCS)

have an embedded PI controller which the user can tune via screws on the servo amplifier card.

Figure 3.8: DC-motor with servo amplifier (shown behind the motor) implementing a PI-controller

Chapter 4

Experimental tuning of PID controllers

4.1 Introduction

This chapter describes several methods for experimental tuning of controller parameters in P-, PI- and PID controllers, that is, methods for finding proper values of K_p, T_i and T_d. The methods can be used experimentally on physical systems, but also on simulated systems.

The methods described can be applied only to processes having a time delay or having dynamics of order higher than 3. Here are a few examples of processes (transfer function models) for which the method can *not* be used:

$$H(s) = \frac{K}{s} \qquad \text{(integrator)} \tag{4.1}$$

$$H(s) = \frac{K}{Ts+1} \qquad \text{(first order system)} \tag{4.2}$$

$$H(s) = \frac{K}{(\frac{s}{\omega_0})^2 + 2\zeta\frac{s}{\omega_0} + 1} \qquad \text{(second order system)} \tag{4.3}$$

Controller tuning for processes as above can be executed with a transfer function based method, cf. Chapter 7.

The methods described in this chapter can be regarded as general methods since their procedure is the same, regardless the dynamic properties of the process to be controlled. There are processes for which the methods does not fit well, for example a first order process with a time delay much larger than the time constant. Chapter 7 describes tuning methods which are based on the given dynamic properties of the process as expressed in a transfer function model, and the PID parameters are then tailored for this process. You can expect that such model-based tuning methods will give the control system better performance (as faster control) than if the controller was tuned with a general tuning method. Despite this, the general tuning methods are important because they have proven to work well and because they are simple to use (they do not require an explicit process model).

4.2 A criterion for controller tuning

A reasonable criterion for tuning the controller parameters is that the control system has *fast control with satisfactory stability.* These two requirements – fast control and satisfactory stability – are in general contradictory: Very good stability corresponds to sluggish control (not desirable), and poor stability (not desirable) corresponds to fast control. A tuning method must find a compromise between these two contradictory requirements.

What is meant by satisfactory stability? Simply stated, it means that the response in the process output variable converges to a constant value with satisfactory damping after a time-limited change of the setpoint or the disturbance. Satisfactory damping can be quantified in several ways. Ziegler and Nichols [20] who published famous tuning rules in the 1940s claimed that satisfactory damping corresponds to an amplitude ratio of approximately 1/4 between subsequent peaks in the same direction (due to a step disturbance in the control loop), see Figure 4.1:

$$\frac{A_2}{A_1} = \frac{1}{4} \tag{4.4}$$

Ziegler and Nichols used this as a stability criterion when they derived their PID tuning rules. However, there is no guaranty that the actual amplitude ratio of a given control system becomes 1/4 after tuning with one of the Ziegler and Nichols' methods, but it should not be very different from 1/4.

If you think that the stability of the control loop becomes too bad or too good, you can try to adjust the controller parameters. The first aid, which may be the only adjustment needed, is to adjust the controller gain K_p as follows:

- Too bad stability: Decrease K_p somewhat, for example a 25%

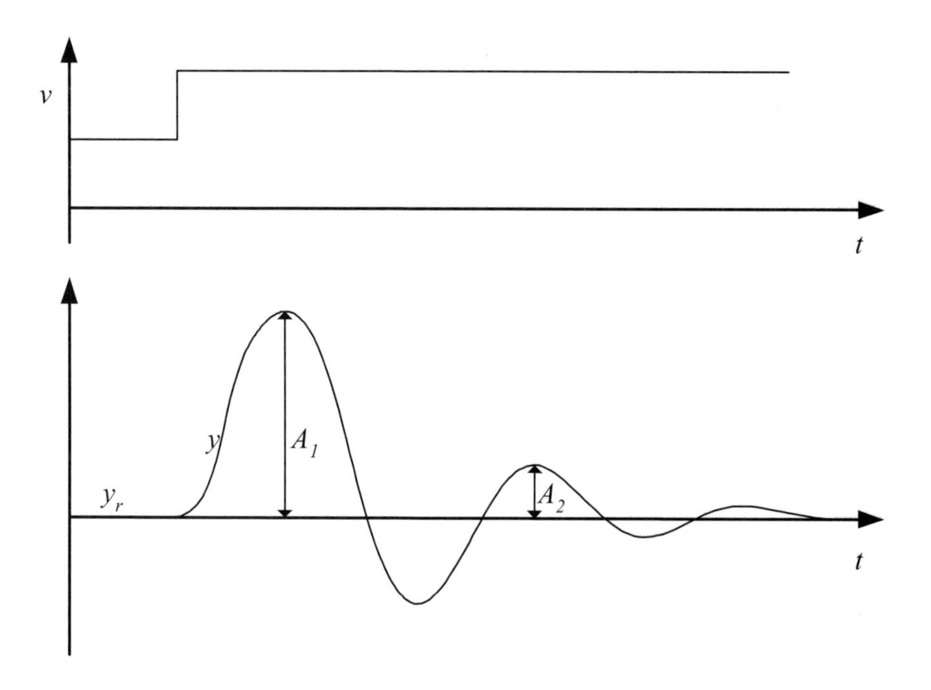

Figure 4.1: Good stability (according to Ziegler and Nichols)

decrease.

- Too good stability (which corresponds to sluggish control): Increase K_p somewhat, for example a 25% increase.

4.3 The P-I-D method

The P-I-D method is a simple and intuitive method (which does not require the control system to have sustained oscillations, as in the Ziegler-Nichols' closed loop method, cf. Section 4.4). The method is based on experiments on the established control system (or on a simulator of the control system), see Figure 4.2. The method is as follows:

1. Bring the process to or close to the normal or specified operation point by adjusting the nominal control signal u_0 (with the controller in manual mode).

2. Controller tuning:

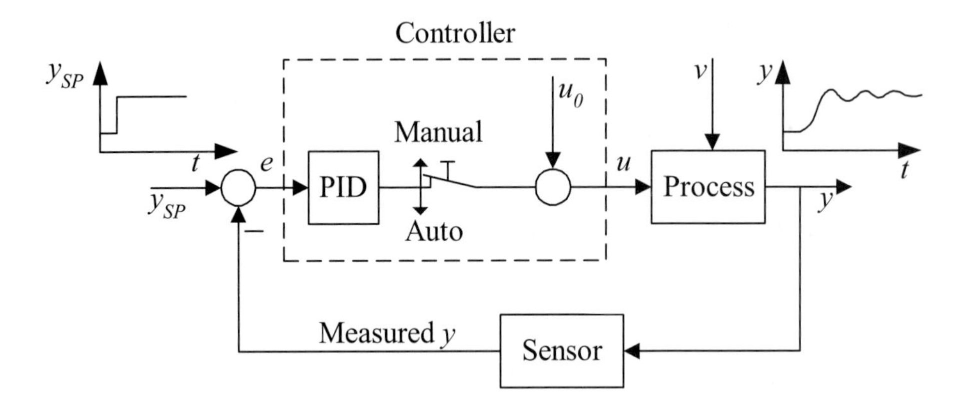

Figure 4.2: The P-I-D method is applied to the established control system.

- **P controller**: Ensure that the controller is a P controller with $K_p = 0$ (set $T_i = \infty$ and $T_d = 0$). Increase K_p until the control loop gets satisfactory stability as seen in the response in the measurement signal after e.g. a step in the setpoint or in the disturbance (exciting with a step in the disturbance may be impossible on a real system, but it possible in a simulator).
  If you do not want to start with $K_p = 0$, you can try $K_p = 1$ (which is a good initial guess in many cases) and then increase or decrease the K_p value until you are content with the stability of the control loop.
- **PI controller**:
  (a) Start by executing the procedure for a P controller (see above).
  (b) Activate the integral term by reducing T_i until the loop gets a little too poor stability. Alternatively, you can jump to the following T_i-value: $T_i = T_p/1.5$, where T_p is the time period of the damped oscillations when using the P controller. Because of the introduction of the I-term, the loop will have a somewhat reduced stability than with the P controller only.
  (c) Adjust K_p (you can try decreasing K_p by 20%) until the stability of the loop is satisfactory.
- **PID controller**:
  (a) Start by executing the procedure for a P controller (see above).
  (b) Then activate both the integral term by reducing T_i – an initial guess is $T_i = T_p/2$ where T_p is the time period of the

damped oscillations for the P controller, and the derivative term by increasing T_d – an initial guess is $T_i/4$.

(c) Adjust K_p (you can try increasing it by 20%) until the stability of the loop is satisfactory.

Example 4.1 ***Controller tuning of a wood-chip level control system with the P-I-D method***

I have used the P-I-D method on the simulator shown in Figure 2.15. The PID parameter values became

$$K_p = 2.1;\ T_i = 10\text{min} = 600s;\ T_d = 2.5\text{min} = 150\text{s} \tag{4.5}$$

Figure 4.3 shows the resulting responses. The control system seems to have satisfactory stability.

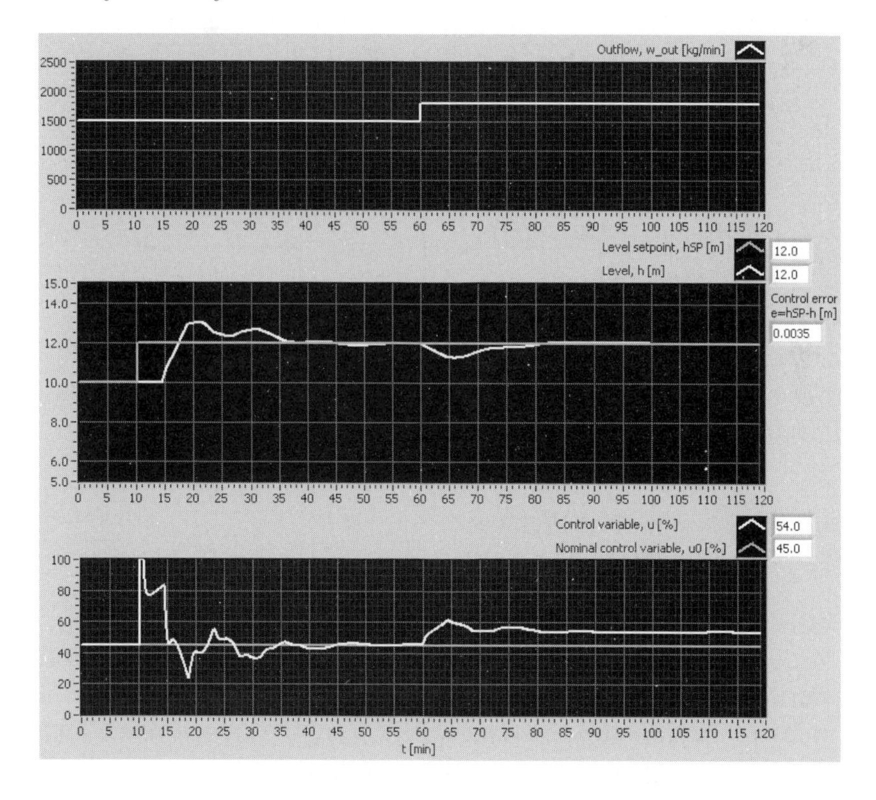

Figure 4.3: Example 4.1: Level control of the wood-chip tank with a P-controller. (The front panel of the simulator is as shown in Figure 2.15.)

[End of Example 4.1]

4.4 Ziegler-Nichols' closed loop method

Ziegler and Nichols published in 1942 a paper [20] where they described two methods for tuning the parameters of P-, PI- and PID controllers. These two methods are the *Ziegler-Nichols' closed loop method* (which is described in this section) and the *Ziegler-Nichols' open loop method* (described in Section 4.6). These methods are still useful despite many years of research on PID tuning, and they form the basis of some auto-tuning methods (auto-tuning is described in Section 4.8).

The method is based on experiments executed on an established control loop (a real system or a simulated system), see Figure 4.4.

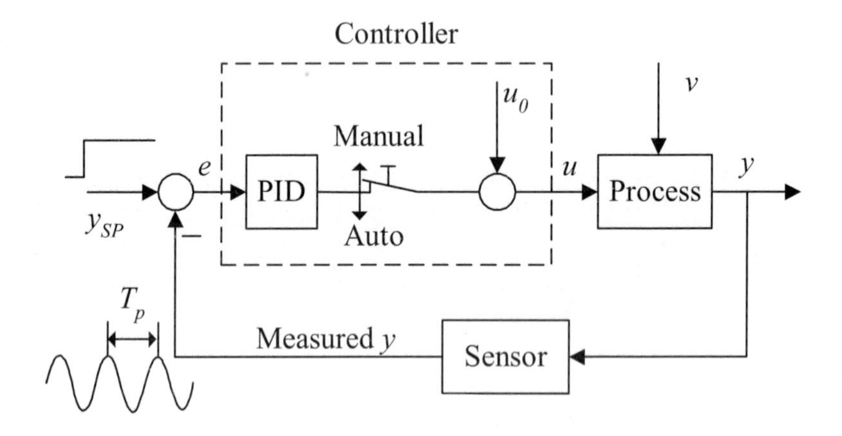

Figure 4.4: The Ziegler-Nichols' closed loop method is executed on an established control system.

The tuning procedure is as follows:

1. Bring the process to (or as close to as possible) the specified *operating point* of the control system to ensure that the controller during the tuning is "feeling" representative process dynamic[1] and to minimize the chance that variables during the tuning reach limits. You can bring the process to the operating point by manually adjusting the control variable, with the controller in manual mode, until the process variable is approximately equal to the setpoint.
2. Turn the PID controller into a *P controller* with gain $K_p = 0$ (set $T_i = \infty$ and $T_d = 0$). Close the control loop by setting the controller

[1] This may be important for nonlinear processes.

in automatic mode.

3. Increase K_p until there are *sustained oscillations* in the signals in the control system, e.g. in the process measurement, after an excitation of the system. (The sustained oscillations corresponds to the system being on the stability limit.) This K_p value is denoted the *ultimate (or critical) gain*, K_{p_u}.

   The excitation can be a step in the setpoint. This step must be small, for example 5% of the maximum setpoint range, so that the process is not driven too far away from the operating point where the dynamic properties of the process may be different. On the other hand, the step must not be too small, or it may be difficult to observe the oscillations due to the inevitable measurement noise.

   It is important that K_{p_u} is found without the actuator being driven into any saturation limit (maximum or minimum value) during the oscillations. If such limits are reached, you will find that there will be sustained oscillations for any (large) value of K_p, e.g. 1000000, and the resulting K_p-value (as calculated from the Ziegler-Nichols' formulas, cf. Table 4.1) is useless (the control system will probably be unstable). One way to say this is that K_{p_u} must be the smallest K_p value that drives the control loop into sustained oscillations.

4. Measure the *ultimate (or critical) period* T_u of the sustained oscillations.

5. *Calculate the controller parameter values* according to Table 4.1, and use these parameter values in the controller.

   The lowpass filter time constant T_f (cf. Section 2.6.7) can be set to

   $$T_f = 0.1T_d \tag{4.6}$$

   (if no other specification exists).

   If the stability of the control loop is poor, try to improve the stability by decreasing K_p.

	K_p	T_i	T_d
P controller	$0.5K_{p_u}$	∞	0
PI controller	$0.45K_{p_u}$	$\frac{T_u}{1.2}$	0
PID controller	$0.6K_{p_u}$	$\frac{T_u}{2}$	$\frac{T_u}{8} = \frac{T_i}{4}$

Table 4.1: Formulas for the controller parameters in the Ziegler-Nichols' closed loop method.

Example 4.2 ***The Ziegler-Nichols' closed loop method***

Figure 4.5 shows the signals in the simulated wood-chip level control system shown in Figure 2.15 (page 32). The system was excited by a step in the setpoint from 10m to 10.5m. The ultimate gain was $K_{p_u} = 3.1$, and

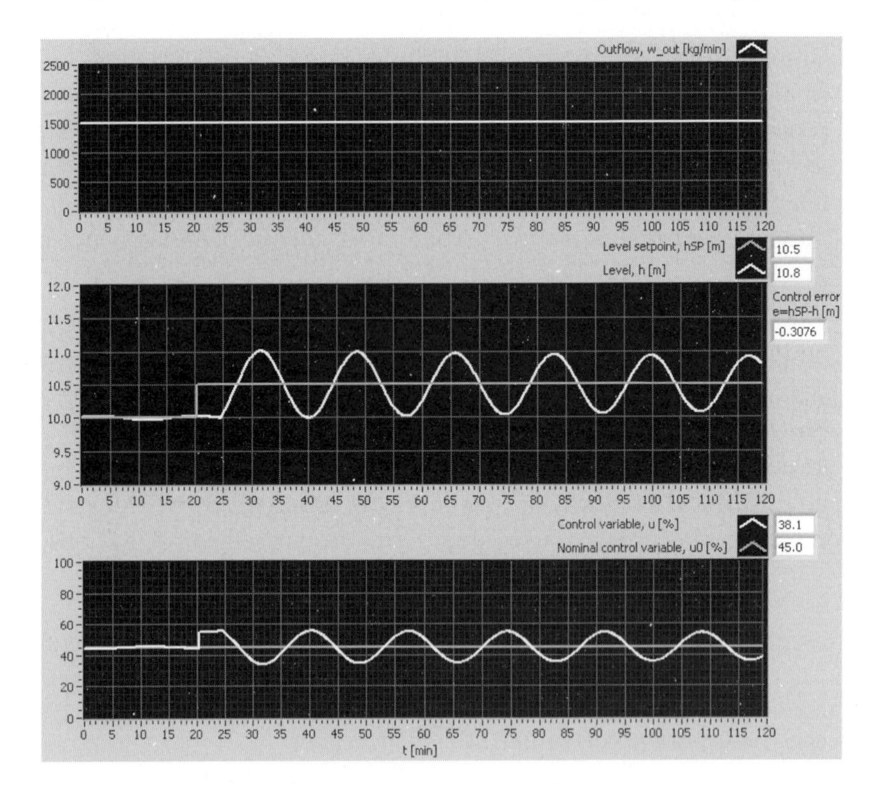

Figure 4.5: Example 4.2: The tuning phase of the Ziegler-Nichols' closed-loop method. (The front panel of the simulator is as shown in Figure 2.15.)

the ultimate period is approximately $T_u = 18\text{min}$. From Table 4.1 we get the following PID parameters:

$$K_p = 1.86;\ T_i = 9\text{min} = 540\text{s};\ T_d = 2.25\text{min} = 135\text{s} \tag{4.7}$$

Figure 4.6 shows signals of the control system with the above PID parameter values. The control system has satisfactory stability. The amplitude ratio in the damped oscillations is less than 1/4, that is, which means that the stability is a little better than prescribed by Ziegler and Nichols'.

[End of Example 4.2]

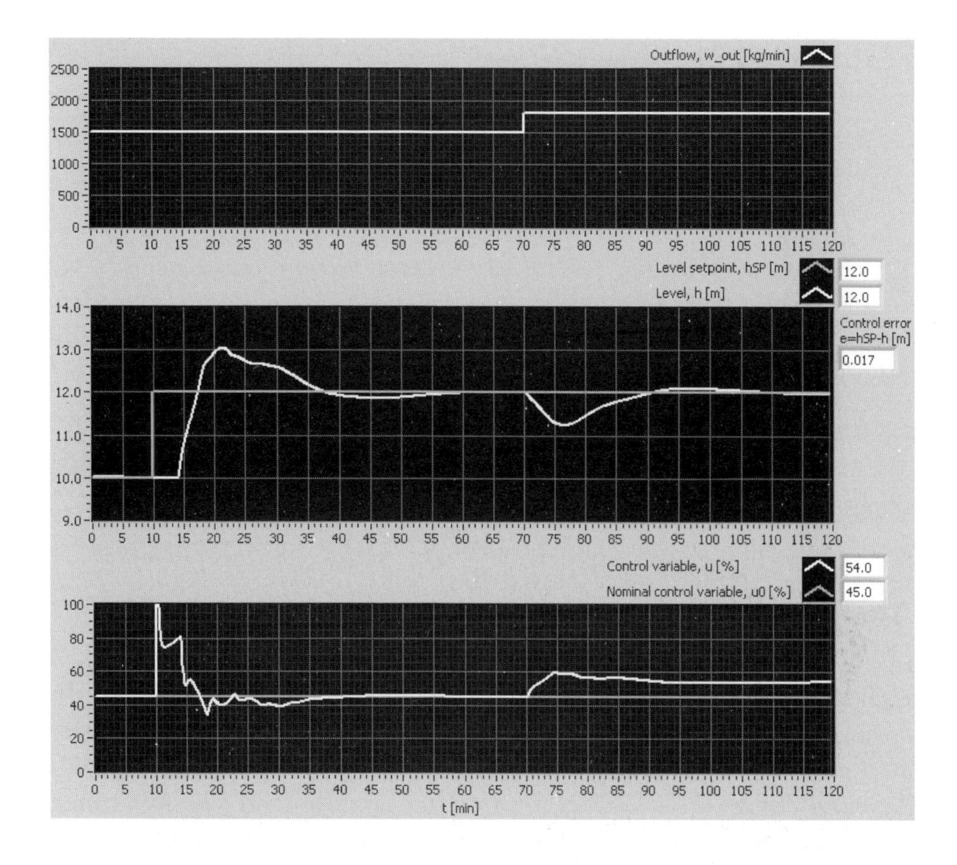

Figure 4.6: Example 4.2: Time responses with PID parameters tuned using the Ziegler-Nichols' closed loop method

Some comments to the Ziegler-Nichols' closed loop method

1. *You do not know in advance the amplitude of the sustained oscillations.* The amplitude depends partly of the initial value of the process measurement. By using the Åstrøm-Hägglund's tuning method described in Section 4.5 in stead of the Ziegler-Nichols' closed loop method, you have full control over the amplitude, which is beneficial, of course.

2. *For sluggish processes it may be time consuming* to find the ultimate gain in physical experiments. The Åstrøm-Hägglund's method reduces this problem since the oscillations come automatically.

3. If the operating point varies and if the process dynamic properties depends on the operating point, you should consider using some kind of *adaptive control or gain scheduling*, where the PID parameter are adjusted as functions of the operating point.

If the controller parameters shall have fixed value, they should be tuned in the worst case as stability is regarded. This ensures proper stability if the operation point varies. The worst operating point is the operation point where the process gain has its greatest value and/or the time delay has its greatest value.

4. *The responses in the control system may become unsatisfactory* with the Ziegler-Nichols' method. 1/4 decay ratio may be too much, that is, the damping in the loop is too small. A simple re-tuning in this case is to reduce the K_p somewhat, for example by 20%.

   A possibly better way to re-tune the controller for better stability is described by Ziegler and Nichols in [20]. They suggested to decrease K_p, $1/Ti$ and T_d with the same factor, for example 10%.[2]

In the beginning

The Ziegler and Nichols' methods have definitely proven to be useful, but they actually met some resistance in the beginning. In [2] Ziegler reports from a meeting in the American Society of Mechanical Engineers (ASME): "*The questions at the end were pretty bitter because they (the 'old-timers') could not stomach this ultimate sensitivity*[3]. *The questions got worse and worse and I was answering them. Finally a little guy in the back of the room got up. He was from Goodyear. Since he was on the committee he had received an advance copy of the paper. He stuttered some, and stammered out for all to hear: 'We had one process in our plant, a very bad one, and so I tried this method and it just worked perfectly.' That broke up the meeting.*"

4.5 Åstrøm-Hägglund's On/off method

Åstrøm-Hägglund's On/off method can be regarded as a practical implementation of the Ziegler-Nichols' closed loop method described in Chapter 4.4. There are a few practical problems with the Ziegler-Nichols' method:

- It may be time-consuming to find the least controller gain K_p which gives sustained oscillations.

[2] Note: Decreasing $1/Ti$ is the same as increasing T_i.

[3] which implies that the control system is on the stability limit and *oscillates*

- You do not have full control over the amplitude of the oscillations.

Both these problems are eliminated is in the Åstrøm-Hägglund's' method [21]. The method is based on using an On/off controller in place of the PID controller to be tuned, see Figure 4.7. The On/off controller is the

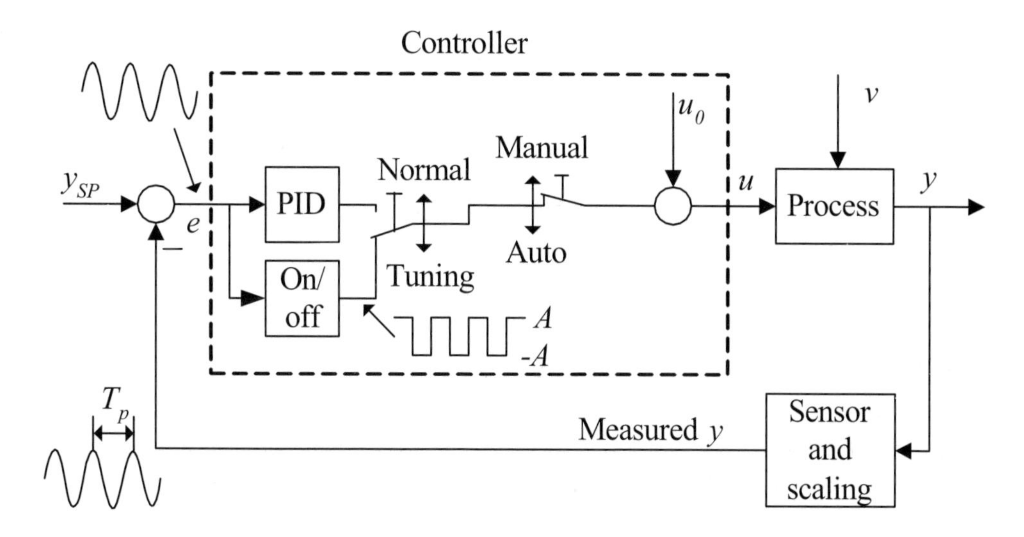

Figure 4.7: Configuration of the control loop in Åstrøm-Hägglund's On/off method for tuning a PID controller

same as described in Section 2.6.3. Due to the On/off controller the sustained oscillations in control loop will come *automatically*. These oscillations will have approximately the same period as if the Ziegler-Nichols' closed loop method were used, and the ultimate gain K_{p_u} can be easily calculated. The method is as follows.

1. Bring the process to (or as close to as possible) the specified *operating point* of the control system to ensure that the controller during the tuning is "feeling" representative process dynamic[4] and to ensure that the signals during the tuning can vary without meeting limits due to being in a non-representative operating point. You bring the process to the operating point by manually adjusting the control variable, with the controller in manual mode, until the process variable is approximately equal to the setpoint.

2. Set the amplitude A of the On/off controller, cf. Chapter 2.6.3, to a reasonable value – not too small and not too large – for example 10% of the range of the control signal.

[4]This may be important for nonlinear processes.

3. Switch the On/off controller into the loop. This causes sustained oscillations to appear automatically in the control loop. It is not necessary to excite the control loop externally for the oscillations to come (thus, the setpoint can be constant).

4. Read off the amplitude E of the oscillations of the input signal to the On/off controller, which is the control error, and calculate the *equivalent gain* as follows:

$$K_e = \frac{A_u}{A_e} \tag{4.8}$$

where A_u is

$$A_u = \frac{4A}{\pi} \tag{4.9}$$

and A_e is to be selected among these following alternatives, depending on the type of input signal e (control error) to the On/off controller:

- If the oscillations are *sinusoidal* (this is the most common signal form) with amplitude E[5], set

$$A_e = E \tag{4.10}$$

- If the oscillations are *triangular* (this is more seldom, but exists in integrator systems as the wood-chip tank) with amplitude E, set

$$A_e = \frac{8E}{\pi^2} \tag{4.11}$$

A few words about the background of the formulas above:

- (4.8) calculates the equivalent gain of the On/off controller as the ratio between equivalent amplitudes in the output signal and the input signal of the On/off controller.
- A_u is the amplitude of the first harmonic of a Fourier series expansion of the square pulse train at the output of the On/off controller.
- A_e in (4.10) is the amplitude in the sinusoidal control error.
- A_e in (4.11) is the first harmonic of a Fourier series expansion of the triangular pulse train at the output of the On/off controller.

5. Read off the *ultimate period* T_p as the period of the sustained oscillations (T_p can be read off fro any signal in the the control loop).

[5]The amplitude may be measured as half the distance between the maximum value and the minimum value.

6. Calculate the controller parameters of a P, PI or PID controller according to the Ziegler-Nichols' closed loop method, cf. Table 4.1, using $K_{p_u} = K_e$ and T_p.

7. Once the PID parameters have been entered into the controller, activate the PID controller. (The On/off controller has finished its job.)

Example 4.3 ***Controller tuning using Åstrøm-Hägglund's method***

Figure 4.8 shows the signals in the controller tuning phase for the wood-chip level control system shown in Figure 2.15 (page 32). The

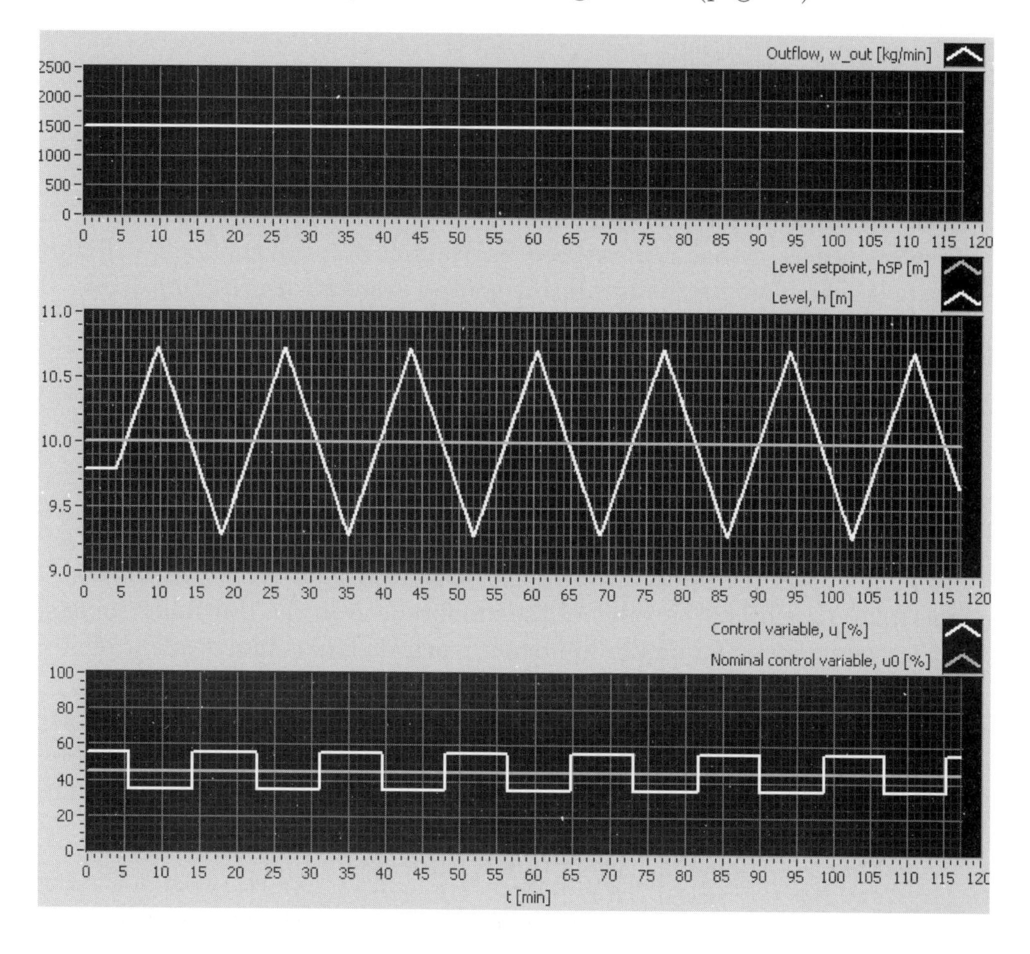

Figure 4.8: Example 4.3: Oscillations in the control system in the tuning phase of the Åstrøm-Hägglund's method

amplitude A in the On/off controller is set to 10%. Then oscillations in the control error e are triangular with amplitude $E = 0.75$m. This value must be transformed to a corresponding value in % using the measurement function. (It is necessary transform to percent because percent is the unit of the signal into the On/off controller.) Since $0 - 15$m corresponds to $0 - 100\%$, 0.75m corresponds to $5.0\% = E$, which we insert into (4.11) since the oscillations are triangular. (4.8) gives

$$K_e = \frac{A_u}{A_e} = \frac{\frac{4A}{\pi}}{\frac{8E}{\pi^2}} = \frac{\frac{4 \cdot 10\%}{\pi}}{\frac{8 \cdot 5.0\%}{\pi^2}} = 3.14 \tag{4.12}$$

(which is close to $K_{p_u} = 3.1$ found using the Ziegler-Nichols' closed loop method in Example 4.2).

The ultimate period T_p is read off to 18min = 1080s (same as as in Example 4.2).

Finally, inserting the values of K_{p_u} and T_p into the formulas in Table 4.1 gives the following PID controller parameters:

$$K_p = 1.88;\ T_i = 9\text{min} = 540\text{s};\ T_d = 2.25\text{min} = 135\text{s} \tag{4.13}$$

Since these PID values are very similar to those found using the Ziegler-Nichols' closed loop method in Example 4.2, you can look at Figure 4.6 to see simulated responses in the control system with the tuned PID controller.

[End of Example 4.3]

How to control the amplitude of the oscillations

You will obtain the same value for the equivalent gain K_e of the On/off controller no matter the value of the amplitude A in the controller (unless some maximum or minimum limits are reached). K_e is independent of A because K_e gets a value which is (quite) equal to the value of K_{p_u} we would have found using the Ziegler-Nichols' closed loop method, and the K_{p_u} value is of course independent of A. This fact can be used to control the amplitude of the oscillations. Let us here assume that the signal e into the On/off controller is sinusoidal, but the conclusion still holds if the oscillations are triangular. From (4.8)-(4.10),

$$K_e = \frac{4A}{\pi E} \tag{4.14}$$

Two different A values A_1 and A_2 will, since K_e has a fixed value, results in two different amplitude values E_1 and E_2, but so that

$$K_e = \frac{4A_1}{\pi E_1} = \frac{4A_2}{\pi E_2} \tag{4.15}$$

From (4.15) we get a formula for a new A value, A_{new}, from a specified new E value, E_{new}:

$$A_{\text{new}} = A_{\text{old}} \frac{E_{\text{new}}}{E_{\text{old}}} \tag{4.16}$$

For example, if we want to halve the amplitude E, we set $E_{\text{new}} = E_{\text{old}}/2$, giving $A_{\text{new}} = A_{\text{old}}/2$. The ultimate period T_u of the sustained oscillations is independent of the value of A.

4.6 Ziegler-Nichols' open loop method

4.6.1 Ziegler-Nichols' open loop method used experimentally

The Ziegler-Nichols' open loop method is based on the *process step response.* The PID parameters are calculated from the response in the process measurement y_m after a step with height U in the control variable u, see Figure 4.9. The term "process" here means all blocks in the control except the controller. The step response experiment is executed on the uncontrolled process, so the control loop is open (no feedback).

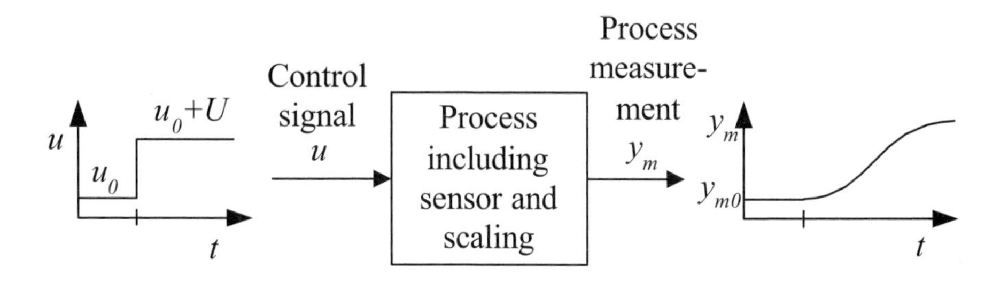

Figure 4.9: The Ziegler-Nichols' open loop method is based on the step response of the uncontrolled process

The method is as follows.

1. If the control loop is closed (i.e. feedback), the loop must be opened. This can be done by setting the controller in manual mode.
2. Bring the process to the operation point by adjusting the control variable manually. This is done by adjusting u_0 in Figure 4.9.
3. Excite the process via a step of amplitude U in the control variable u. The step should be "small" so that the process is not brought too

far from the operation point, but of course the step must large enough to give an observable response in process measurement y_m. A step amplitude of $U = 10\%$ can be a reasonable value, but the amplitude must be chosen individually in each case.

4. Read off the following characteristic parameters from the step response in y_m:

   - *Equivalent dead-time or lag L*
   - *Rate or slope R*

   Figure 4.10 which shows the relevant part of a typical step response. In the figure the time axis starts at the step time. The annotation "0.0" along the y-axis corresponds to y_{m0} in Figure 4.9. L is the time from the step time to the point of intersection between the "0.0"-line and the steepest tangent. R is the slope of the steepest tangent.

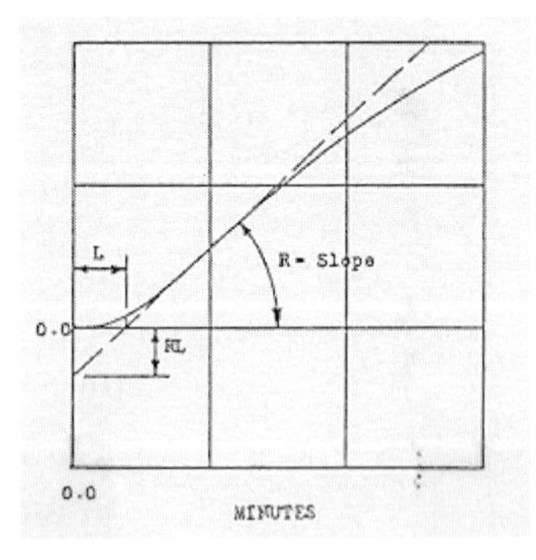

Figure 4.10: Ziegler-Nichols' open loop method: The equivalent dead-time L and rate R read off from the process step response. (The figure is a reprint from [20] with permission.)

5. Calculate the controller parameters according to Table 4.2.

6. After the controller parameters have been calculated and entered into the PID controller, the control loop is closed (by setting the controller in automatic mode).

Example 4.4 ***Controller tuning using Ziegler-Nichols' open loop method***

In this example the Ziegler-Nichols' open loop method is applied to the wood-chip tank shown in Figure 2.15. The step in the control variable is

	K_p	T_i	T_d
P controller	$\frac{1}{LR/U}$	∞	0
PI controller	$\frac{0.9}{LR/U}$	$3.3L$	0
PID controller	$\frac{1.2}{LR/U}$	$2L$	$0.5L = \frac{T_i}{4}$

Table 4.2: Ziegler-Nichols' open loop method: Formulas for the controller parameters.

$U = 10\%$ from $u_0 = 45\%$ to 55%. Figure 4.11 shows the response in the level measurement. We read off

$$L = 4.2\text{min} \tag{4.17}$$

(which is in good accordance with the time delay on the conveyor belt of 4.17min used in the simulator).

R can be calculated from the slope of the level response. From Figure 4.11 we will find that the level increases approximately 1.7m during 10min, giving a slope of $R_1 = 1.7/10 = 0.17\text{m/min}$. However, the slope R must be expressed in unit %/min. The level sensor transforms $0 - 15\text{m}$ to $0 - 100\%$, corresponding to a measurement gain of $K_m = 6.67\%/\text{m}$. Therefore,

$$R = R_1 K_m = 0.17\text{m/min} \cdot 6.67\%/\text{m} = 1.13\%/\text{min} \tag{4.18}$$

Using these values for L and R in Table 4.2 gives the following PID parameters:

$$K_p = 2.52;\ T_i = 8.4\text{min} = 504\text{s};\ T_d = 2.1\text{min} = 126\text{s} \tag{4.19}$$

Figure 4.12 shows simulated responses with the PID controller with these above parameters. There is a step in the setpoint and a step in the disturbance (the outflow). The control system has satisfying stability, but the stability is a little reduced as compared to the case of Ziegler-Nichols' closed loop method, cf. Figure 4.6.

[End of Example 4.4]

4.6.2 Ziegler-Nichols' open loop method with transfer function models

If the process to be controlled is a first order system with time delay or an integrator with time delay, it is simple to find formulas for controller tuning based on the Ziegler-Nichols' open loop method. The tuning method can then be applied without experiments or simulations.

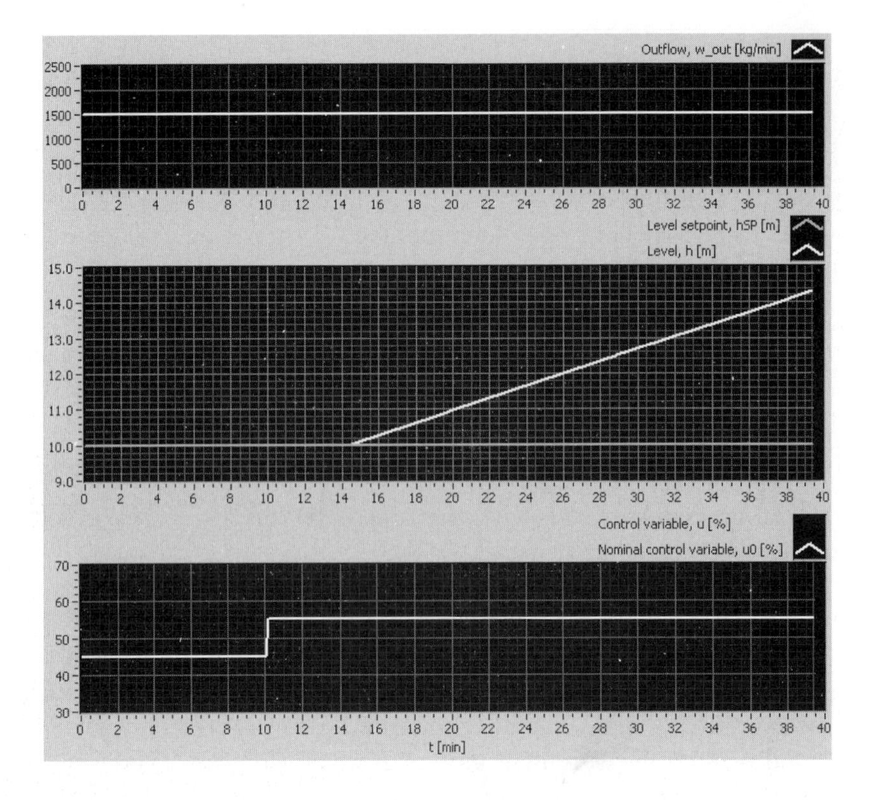

Figure 4.11: Example 4.4: Ziegler-Nichols' open-loop method applied to the wood-chip tank

- **First order system with time delay**: Assume that the transfer function from control variable u to process measurement y_m is

$$\frac{y_m(s)}{u(s)} = H(s) = \frac{K}{Ts+1}e^{-\tau s} \tag{4.20}$$

where K is the gain, T is the time constant and τ is the time delay. According to Ziegler-Nichols' open loop method u is assumed to be step of amplitude U. It can be shown that the steepest slope of the step response in y_m is

$$R = \frac{KU}{T} \tag{4.21}$$

The time delay is of course

$$L = \tau \tag{4.22}$$

R and L can now be used in Table 4.2 to calculate controller parameters.

For example, the PID parameters are

$$K_p = \frac{1.2T}{\tau K};\ T_i = 2\tau;\ T_d = 0.5\tau \tag{4.23}$$

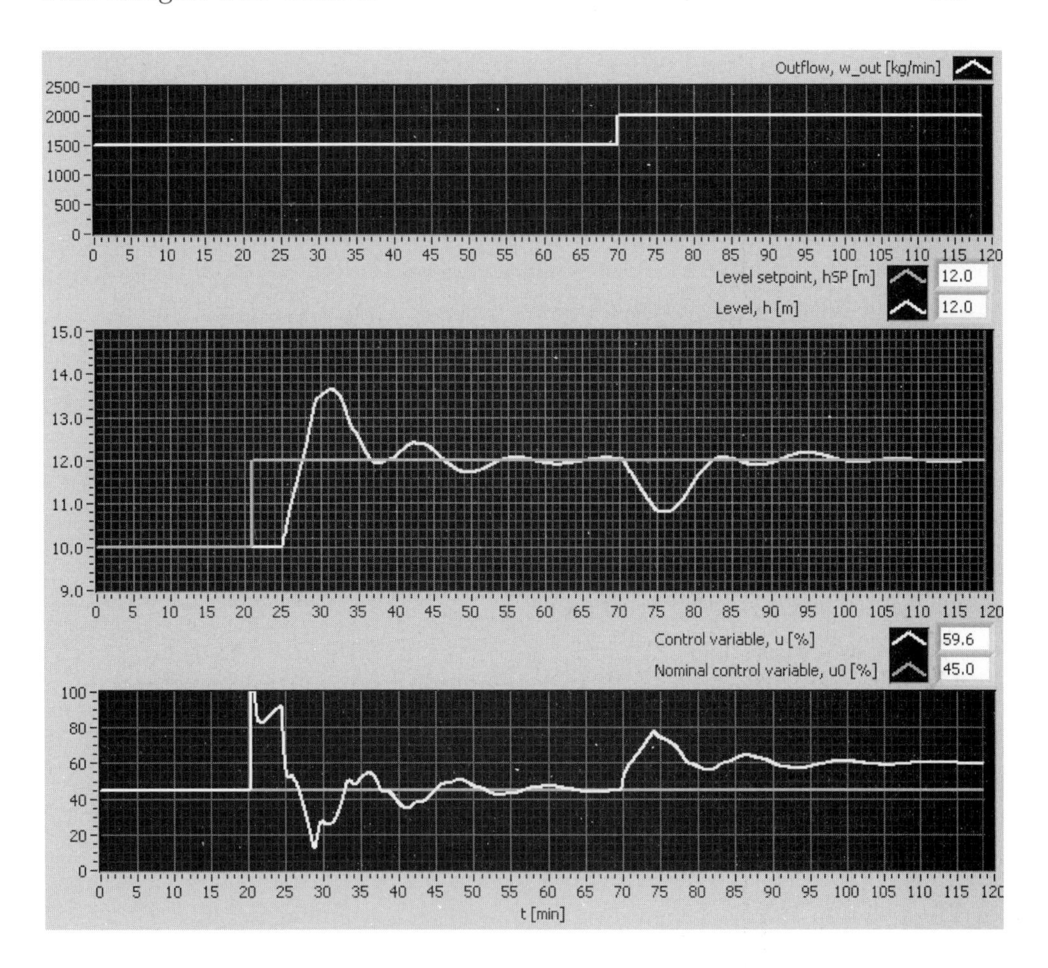

Figure 4.12: Example 4.4: Time responses with PID parameters tuned using the Ziegler-Nichols' open loop method

- **Integrator with time delay**: The transfer function from control variable u to process measurement y_m is

$$\frac{y_m(s)}{u(s)} = H(s) = \frac{K}{s}e^{-\tau s} \tag{4.24}$$

where K is the gain and τ is the time delay. u is assumed to be step of amplitude U. It can be shown that the slope of the step response in y_m (this response is a ramp) is

$$R = KU \tag{4.25}$$

The time delay is

$$L = \tau \tag{4.26}$$

R and L can now be used in Table 4.2.

Example 4.5 ***Ziegler-Nichols' open loop method with transfer function model***

A level control system for a wood-chip tank including conveyor belt is described in Example 2.3. Mass balance gives the following mathematical model of the tank, cf. Example 2.7:

$$\dot{h}(t) = \frac{1}{\rho A}\left[u(t-\tau) - w_{out}(t)\right] \tag{4.27}$$

Taking the Laplace transform of this differential equation gives

$$sh(s) - h_0 = \frac{1}{\rho A}\left[K_s e^{-\tau s} u(s) - w_{out}(s)\right] \tag{4.28}$$

which leads to the following transfer function from control variable u (control signal acting on the inlet screw) to the level h:

$$\frac{h(s)}{u(s)} = H_p(s) = \frac{K_s}{\rho A s} e^{-\tau s} \tag{4.29}$$

The transfer function from level h to level measurement h_m is

$$\frac{h_m(s)}{h(s)} = H_m(s) = K_m \tag{4.30}$$

Combining (4.29) and (4.30) gives the following transfer function from u to h_m:

$$\frac{h_m(s)}{u(s)} = H_p(s)H_m(s) = \frac{K_m K_s}{\rho A s} e^{-\tau s} = \frac{K}{s} e^{-\tau s} \tag{4.31}$$

which is on the form (4.24). We get

$$R = KU = \frac{K_m K_s}{\rho A} U \tag{4.32}$$

$$L = \tau \tag{4.33}$$

R and L can now be used in Table 4.2. For example, a PID controller will have the following parameter values (the process parameters can be seen at the front panel of the simulator shown in Figure 2.15 (page 32):

$$K_p = \frac{1.2}{LR/U} \tag{4.34}$$

$$= \frac{1.2}{\tau \frac{K_m K_s}{\rho A} U/U} \tag{4.35}$$

$$= \frac{1.2\rho A}{\tau K_m K_s} \tag{4.36}$$

$$= \frac{1.2 \cdot 145\frac{\text{kg}}{\text{m}^3} \cdot 13.4\text{m}^2}{4.17\text{min} \cdot 6.67\frac{\%}{\text{m}} \cdot 33.36\frac{\text{kg/ min}}{\%}} \tag{4.37}$$

$$= 2.51 \tag{4.38}$$

$$T_i = 2L = 2\tau = 2 \cdot 4.17 = 8.3\text{min} \tag{4.39}$$

$$T_d = 0.5L = 0.5\tau = 0.5 \cdot 4.17 = 2.1\text{min} \tag{4.40}$$

These parameter values are in very good accordance with the values found "experimentally" in Example 4.4.

[End of Example 4.5]

Adjustment of controller parameters using Ziegler-Nichols' open loop method

Ziegler-Nichols' open loop method is a good starting point for deciding how the controller parameters can be adjusted if there is a change of parameters in the control loop. Let us assume that the process gain K may be changed and that the time delay τ in the process may be changed. We assume that the slope R of the process step response is proportional to K and (of course) also proportional to the step amplitude U of the input signal acting on the process:

$$R = aKU \tag{4.41}$$

(It is not important for the discussion below what is the actual value of a.) Furthermore, we assume that the time delay or lag L is independent of K, while it is approximately equal to the process dead time τ:

$$L = \tau \tag{4.42}$$

Assuming as an example that the controller is a PID controller, we get from Table 4.2

$$K_p = \frac{1.2}{LR/U} = \frac{1.2}{a\tau K} \tag{4.43}$$

$$T_i = 2\tau \tag{4.44}$$

$$T_d = 0.5\tau \tag{4.45}$$

One example: Assume that the process gain K is doubled. (4.43) – (4.45) says that the controller gain K_p should be *halved*, while T_i and T_d remains unchanged.

Another example: Assume that the process dead time τ doubled. Then, according to (4.43) – (4.45) K_p should be *halved* and both T_i and T_d should be *doubled*.

4.7 Consequences of adjusting controller parameters

4.7.1 Introduction

After you have tuned the controller parameters using e.g. one of the Ziegler-Nichols' methods, it may still be a need for adjusting the controller parameters. The reason for this may be that the original tuning did not give satisfactory results, as poor stability of the control system, or that there are later changes of the dynamic properties of the controlled process.

We will now take a brief look at typical consequences of adjusting the three parameters of a PID controller. The level control system for the wood-chip tank will be used as an example. The front panel of the simulator front panel is shown in 2.15. In the outset the PID parameters are

$$K_p = 1.86;\ T_i = 9\text{min} = 540\text{s};\ T_d = 2.25\text{min} = 135\text{s} \tag{4.46}$$

which we found using the Ziegler-Nichols' closed loop method in Example 4.13. Figure 4.6 shows the response with the above PID parameters after step in the disturbance (the outflow w_{out}).

In the following examples we consider responses due to a step in the disturbance w_{out} from 1500kg/min to 1800kg/min for *adjusted controller* parameter values. Only changes in one direction for each PID parameter will be shown.

4.7.2 Increasing K_p

Figure 4.14 shows responses due to a step in w_{out} (disturbance) from 1500kg/min to 1800kg/min for K_p increased from 1.86 to 2.7. *The stability is reduced due to the increased K_p.*

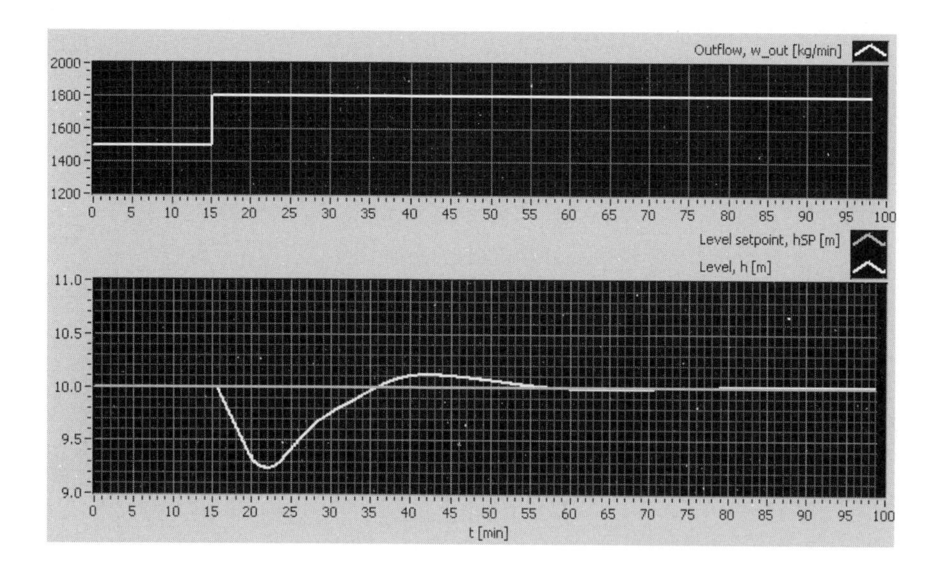

Figure 4.13: Level control of wood-chip tank: Response in level with PID parameters after step in disturbance (outflow w_{out}) for PID parameters found from Ziegler-Nichols' closed loop method

4.7.3 Reducing T_i

If T_i is *reduced*, the integration of the control error runs faster, but unfortunately the stability of the control system is reduced. Figure 4.15 shows responses due to a step in w_{out} (disturbance) from 1500kg/min to 1800kg/min for T_i reduced from 9.0min to 4.0min. *The stability is reduced due to the reduced* T_i.

4.7.4 Increasing T_d

In general the derivative term can increase both the quickness and the stability of a control system. But it is important that T_d has a proper value. Increasing T_d from a proper value may actually reduce the stability because the control variable is more (or too) sensitive to changes in the control error causing "overcompensation". In addition the control variable becomes more sensitive to high frequent measurement noise, cf. Chapter 2.6.7.

Figure 4.16 shows responses due to a step in w_{out} (disturbance) from 1500kg/min to 1800kg/min for T_d increased from 2.25min to 4.0min. *The stability is reduced due to the increased* T_d.

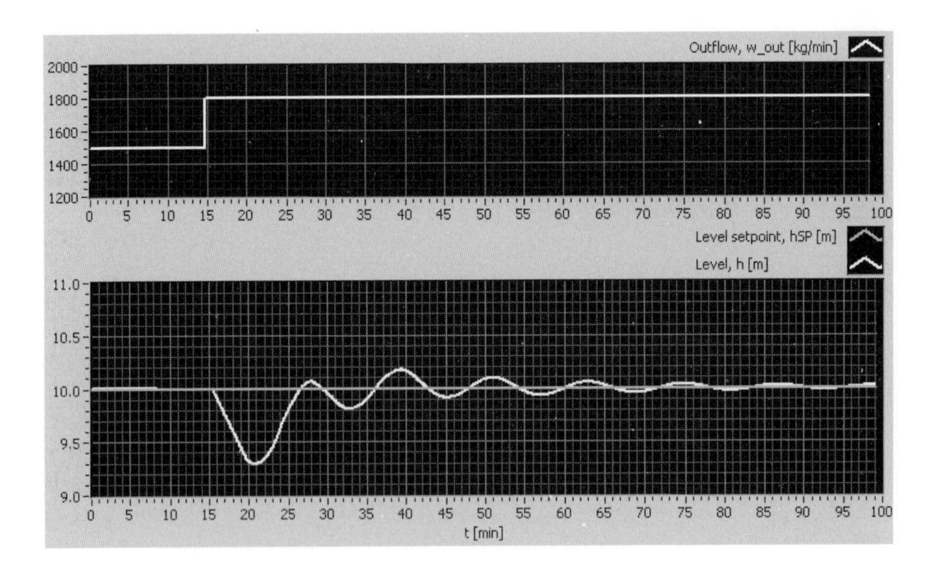

Figure 4.14: Level control of wood-chip tank: Response in level with K_p increased from 1.86 to 2.7

4.8 Auto-tuning

Auto-tuning is automatic tuning of controller parameters in one experiment. It is common that commercial controllers offers auto-tuning. The operator starts the auto-tuning via some button or menu choice on the controller. The controller then executes automatically a pre-planned experiment on the uncontrolled process or on the control system depending on the auto-tuning method implemented. Below are described a couple of auto-tuning methods.

Auto-tuning based On/off control

The Åstrøm-Hägglund's On/off method for tuning PID controllers, cf. Section 4.5, is used as the basis of auto-tuning in some commercial controllers.[6] The principle of this method is as follows:

- When the auto-tuning phase is started, an On/off controller is used as the controller in the control loop, see Figure 4.7. Due to this On/off controller there are sustained oscillations in control loop, cf. Chapter 4.5, and these oscillations come automatically.

[6] E.g. the ECA600 PID controller by ABB.

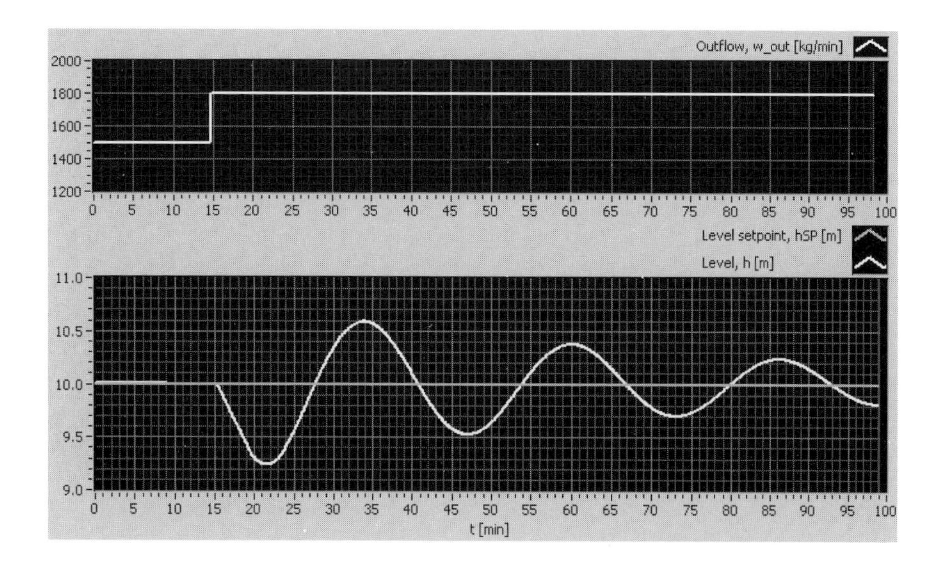

Figure 4.15: Level control of wood-chip tank: Response in level for T_i reduced from 9.0min to 4.0min.

- Once the controller has measured information about amplitude and period of the oscillations in the process measurement – and a few periods may give sufficient information – the PID parameters are calculated automatically. Immediately thereafter the PID controller with the tuned parameters is switched into the control loop.

Auto-tuning based on estimated process model

Commercial software tools [7] exist for auto-tuning based on en estimated process model developed from a sequence of logged data – or time-series – of the control variable u and process measurement y_m. The process model is a "black-box" input-output model in the form of a transfer function. The controller parameters are calculated automatically on the basis of the estimated process model. The time-series of u and y_m may be logged from the system being in closed loop or in open loop:

- **Closed loop**, with the control system being excited via the setpoint, y_{SP}, see Figure 4.17. The closed loop experiment may be used when the controller should be re-tuned, that is, the parameters should be optimized.

[7] som MultiTune (norsk) og ExpterTune

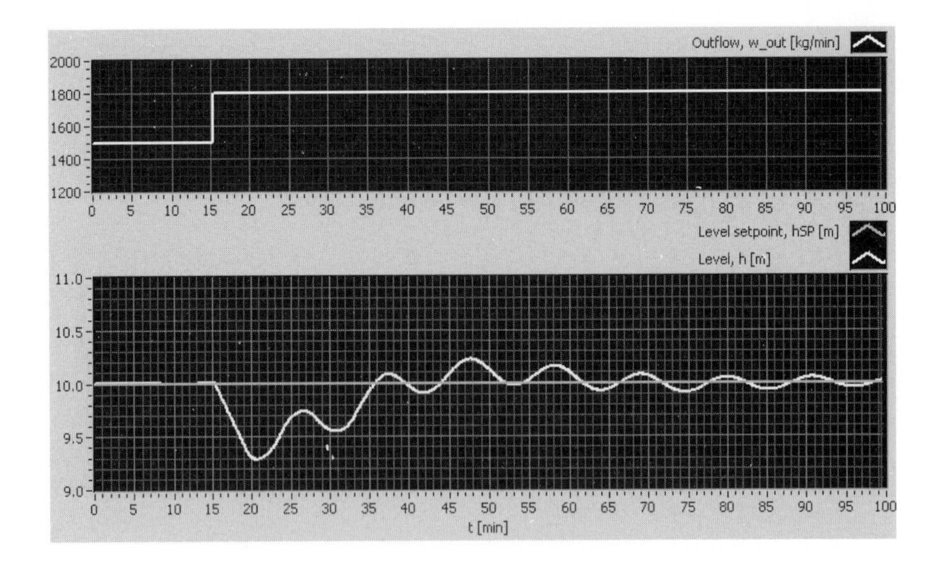

Figure 4.16: Level control of wood-chip tank: Response in level for T_d increased from 2.25min to 4.0min.

- **Open loop**, with the process, which in this case is *not* under control, being excited via the control variable, u, see Figure 4.18. This option must be made if there are no initial values of the controller parameters.

Manual model based controller tuning

If you have the right software tools at your disposal you can perform modeling and model based controller tuning yourself. Here is a short description of what is needed:

- **Tools for estimating input-output models from logged data (time-series) of control signal and process measurement**: Examples of tools are MATLAB's System Identification Toolbox and LabVIEW's System Identification Toolkit. These tools contains powerful functions for estimating input-output models. Particularly useful and user-friendly are the estimation functions based on subspace methods: n4sid in MATLAB and State Space Estimation in LabVIEW. The subspace based DSR-toolbox [3] for MATLAB is also powerful for such modeling. From the estimated modeling, a transfer function model in the form of a discrete-time z-transfer function, say $H_p(z)$, can be derived.

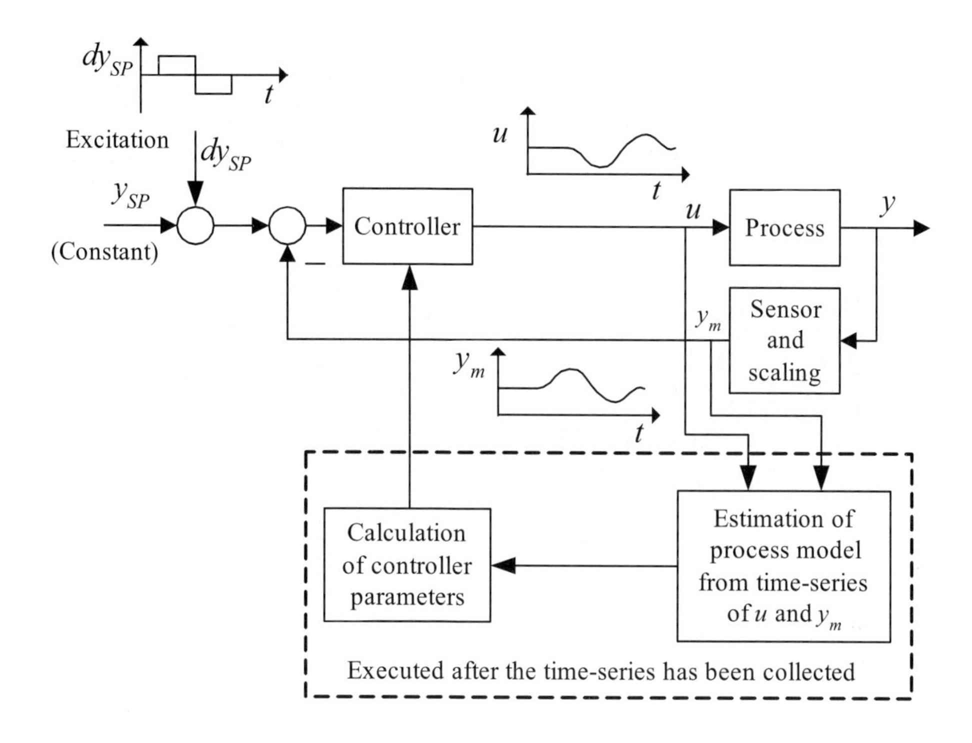

Figure 4.17: Auto-tuning based on closed loop excitation via the setpoint

- **Tools for simulation of discrete-time dynamic systems**: Examples of tools are MATLAB's Control System Toolbox and SIMULINK, and LabVIEW's Simulation Module and Control Design Toolkit. There are even simulation tools in MATLAB's System Identification Toolbox and LabVIEW's System Identification Toolkit. With such tools you can simulate control loops consisting of a z-transfer function $H_c(z)$ of the controller and a process transfer function $H_p(z)$, and tune controller parameters "experimentally" on the simulator, using e.g. the Ziegler-Nichols' closed loop method.

  With tools for frequency response analysis, as in MATLAB's Control System Toolbox and LabVIEW's Control Design Toolkit, you can combine simulations with frequency response analysis and design of control system.

Example 4.6 ***Controller tuning from estimated model***

The n4sid function in MATLAB's System Identification Toolbox is used to estimate a process model (a z-transfer function) from time-series of control signal u and process measurement y_m from a simulated process. The

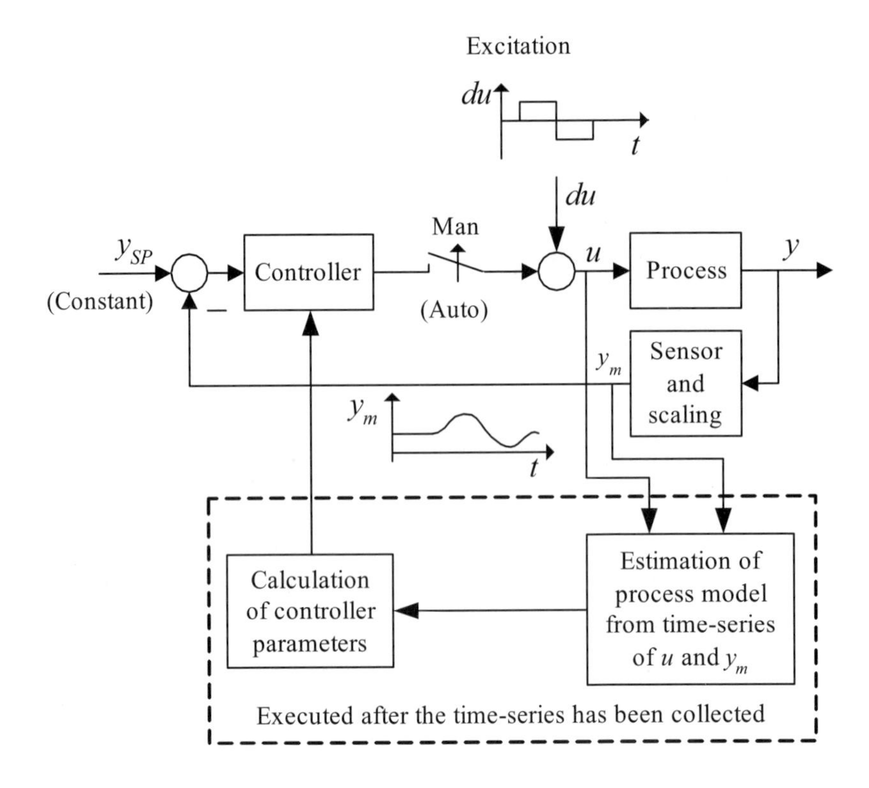

Figure 4.18: Auto-tuning based on open loop excitation via the control variable

simulated process is a first order system with gain 2 and time constant 1s in series with a time delay of 0.4s. A closed loop system experiment is performed. The controller is a PI controller with the following original parameter values:

$$K_p = 0.50;\ T_i = 3.0\text{s} \tag{4.47}$$

(it was observed that responses in the control system are somewhat sluggish, so it is hoped that improvements can be made after re-tuning).

Figure 4.19 shows the control signal u and process measurement y_m from the simulation (excitation). The time-series of u and y_m are used in the n4sid function to estimate a process model. The Ziegler-Nichols' closed loop method were then used in a simulator based on the step simulation function of the MATLAB's Control System Toolbox to calculate PI parameters. The result is

$$K_p = 0.89;\ T_i = 1.37\text{s} \tag{4.48}$$

Figure 4.20 shows simulated responses in process measurement after a step in the setpoint. Two cases are shown: Simulated response for PI controller with original parameter values (4.47), and for PI controller with re-tuned

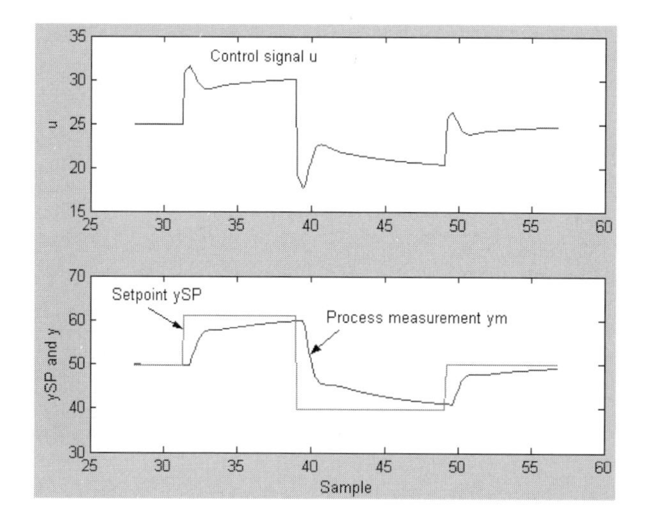

Figure 4.19: Example 4.6: Control signal u and process measurement y_m used for estimation of process model

parameter values (4.48). The re-tuning has clearly given an improvement of the quickness of the control loop, and the stability of the control loop is satisfactory.

[End of Example 4.6]

4.9 PID tuning when process dynamics varies

4.9.1 Introduction

A well tuned PID controller has parameters which are adapted to the dynamic properties to the process, so that the control system becomes fast and stable. If the process dynamic properties varies without re-tuning the controller, the control system

- gets *reduced stability* or
- becomes *more sluggish*.

Problems with variable process dynamics can be solved as follows:

- **The controller is tuned in the most critical operation point**, so that when the process operates in a different operation point, the

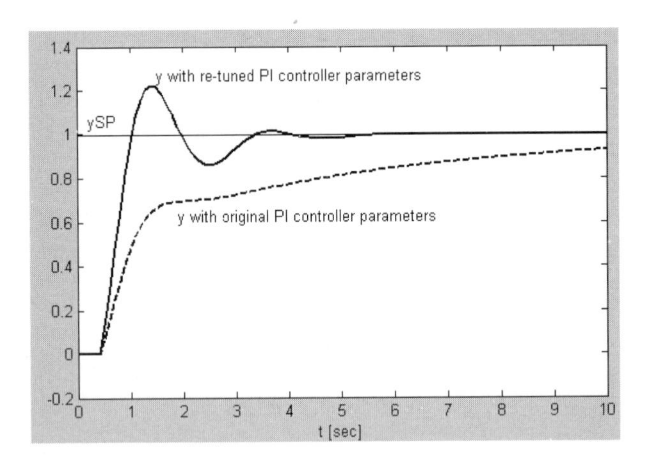

Figure 4.20: Example 4.6: Response in process output for control system with original and re-tuned PI controller parameters

stability of the control system is just better — at least the stability is not reduced. However, if the stability is too good the tracking quickness is reduced, giving more sluggish control.

- **The controller parameters are varied in the "opposite" direction of the variations of the process dynamics**, so that the performance of the control system is maintained, independent of the operation point. Two ways to vary the controller parameters are:
  - *PID controller with gain scheduling.* This is described in detail in Section 4.9.2.
  - *Model-based adaptive controller.* This is described briefly in Section 4.9.4.

Commercial control equipment is available with options for gain scheduling and/or adaptive control.

4.9.2 Gain scheduling PID controller

Figure 4.21 shows the structure of a control system for a process which may have varying dynamic properties, for example a varying gain. The *Gain scheduling variable GS* is some measured process variable which at every instant of time expresses or represents the dynamic properties of the process. As you will see in Example 4.7, GS may be the mass flow through a liquid tank.

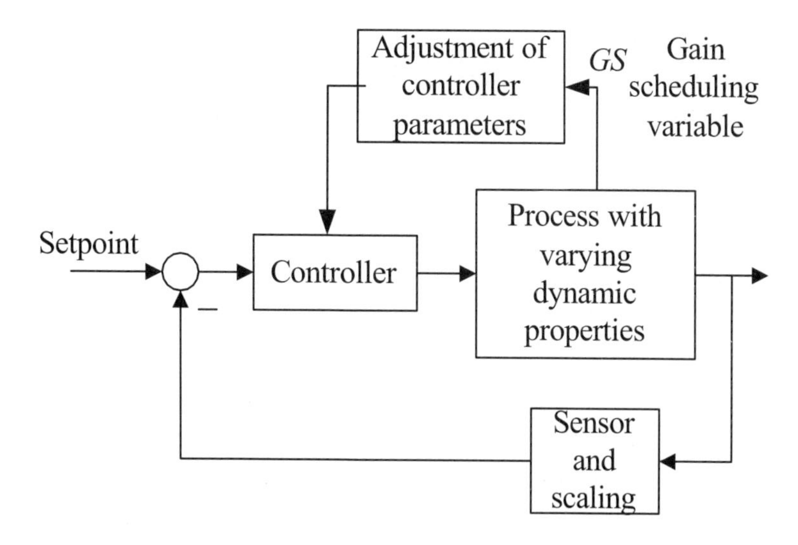

Figure 4.21: Control system for a process having varying dynamic properties. The GS variable expresses or represents the dynamic properties of the process.

Assume that proper values of the PID parameters K_p, T_i and T_d are found using for example Ziegler-Nichols' closed loop method for a set of values of the GS variable. These PID parameter values can be stored in a parameter table – the gain schedule – as shown in Table 4.3. From this table proper PID parameters are given as functions of the gain scheduling variable, GS.

GS	K_p	T_i	T_d
P_1	K_{p_1}	T_{i_1}	T_{d_1}
P_2	K_{p2}	T_{i_2}	T_{d_2}
P_3	K_{p_3}	T_{i_3}	T_{d_3}

Table 4.3: Gain schedule or parameter table of PID controller parameters.

There are several ways to express the PID parameters as functions of the GS variable:

- **Piecewise constant controller parameters**: An interval is defined around each GS value in the parameter table. The controller parameters are kept constant as long as the GS value is within the interval. This is a simple solution, but is seems nonetheless to be the most common solution in commercial controllers.

  When the GS variable changes from one interval to another, the controller parameters are changed abruptly, see Figure 4.22 which illustrates this for K_p, but the situation is the same for T_i and T_d. In

Figure 4.22 it is assumed that GS values toward the left are critical with respect to the stability of the control system. In other words: It is assumed that it is safe to keep K_p constant and equal to the K_p value in the left part of the the interval.

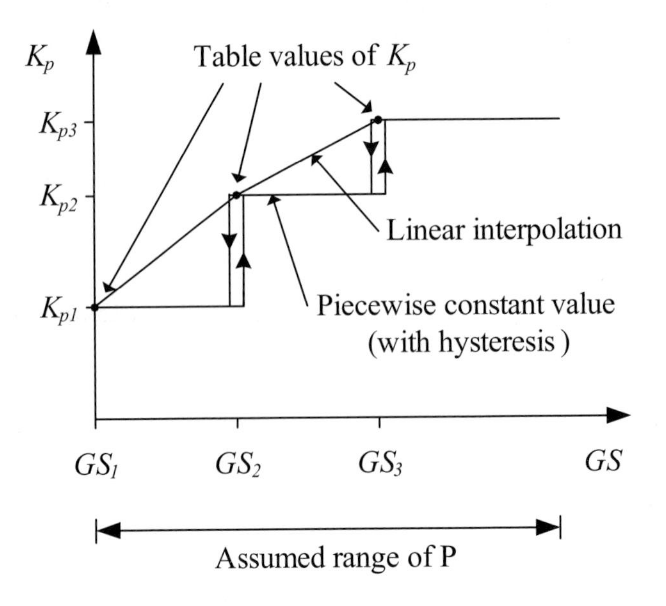

Figure 4.22: Two different ways to interpolate in a PID parameter table: Using piecewise constant values and linear interpolation

Using this solution there will be a disturbance in the form of a step in the control variable when the GS variable shifts from one interval to a another, but this disturbance is probably of negligible practical importance for the process output variable. Noise in the GS variable may cause frequent changes of the PID parameters. This can be prevented by using a hysteresis, as shown in Figure 4.22.

- **Piecewise interpolation**, which means that a linear function is found relating the controller parameter (output variable) and the GS variable (input variable) between to adjacent sets of data in the table. The linear function is on the form

$$K_p = a \cdot GS + b \tag{4.49}$$

where a and b are found from the two corresponding data sets:

$$K_{p_1} = a \cdot GS_1 + b \tag{4.50}$$

$$K_{p_2} = a \cdot GS_2 + b \tag{4.51}$$

(Similar equations applies to the T_i parameter and the T_d parameter.) (4.50) and (4.51) constitute a set of two equations with two unknown variables, a and b.[8]

- Other interpolations may be used, too, for example a polynomial function fitted exactly to the data or fitted using the least squares method.

Example 4.7 ***Gain schedule based PID temperature control at variable mass flow***

Figure 4.25 shows the front panel of a simulator for a temperature control system for a liquid tank with variable mass flow, w, through the tank. The control variable u controls the power to heating element. The temperature T is measured by a sensor which is placed some distance away from the heating element. There is a time delay from the control variable to measurement due to imperfect blending in the tank.

The process dynamics We will initially, both in simulations and from analytical expressions, that the dynamic properties of the process *varies with the mass flow* w. The response in the temperature T is simulated for the following two open loop cases (i.e., not feedback control):

- A step in u of amplitude 10% from 31.5% to 41.5% at mass flow $w = 12\text{kg/min}$, which in this context is a relatively *small* value, see Figure 4.23.

- A step in u of amplitude 10%, from 63.0 % to 73.0 % at $w = 24\text{kg/min}$, which in this context is a relatively *large* value, see Figure 4.24.

The simulations show that the following happens when the mass flow w is reduced (from 24 to 12kg/min): The gain process K is larger, the time constant T_t is larger, and the time delay τ is larger. (These terms assumes that system is a first order system with time delay. The simulator is based on such a model. The model is described below.)

Let us see if the way the process dynamics seems to depend on the mass flow w as seen from the simulations, can be confirmed from a

[8] The solution is left to you.

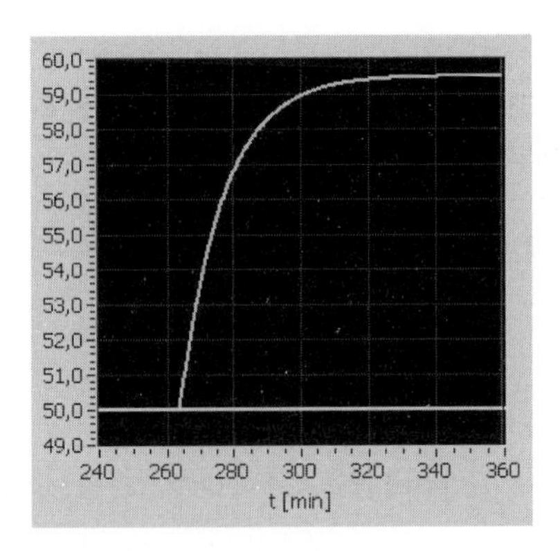

Figure 4.23: Response in temperature T after a step in u of amplitude 10% from 31.5% to 41.5% at the small mass flow $w = 12$kg/min

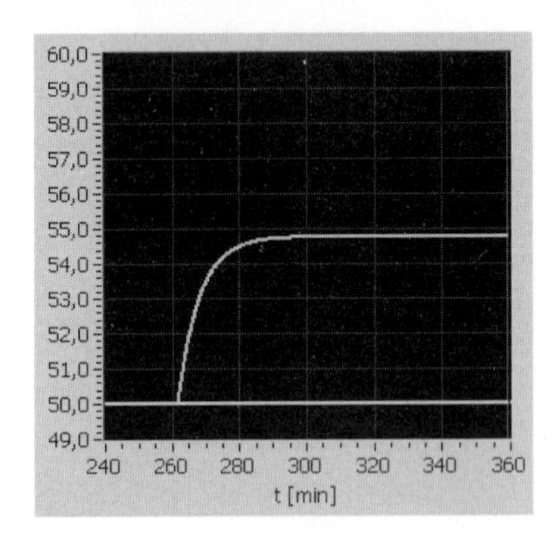

Figure 4.24: Response in temperature T after a step in u of amplitude 10% from 63.0% to 73.0% at the large mass flow $w = 24$kg/min

mathematical process model.[9] Assuming perfect stirring in the tank to have homogeneous conditions in the tank, we can set up the following energy balance for the liquid in the tank:

$$c\rho V\dot{T}_1(t) = K_P u(t) + cw\left[T_{in}(t) - T_t(t)\right] \tag{4.52}$$

where T_1 [K] is the liquid temperature in the tank, T_{in} [K] is the inlet temperature, c [J/(kg K)] is the specific heat capacity, V [m^3] is the liquid volume, ρ [kg/m^3] is the density, w [kg/s] is the mass flow (same out as in), K_P [W/%] is the gain of the power amplifier, u [%] is the control variable, $c\rho VT_1$ is (the temperature dependent) energy in the tank. It is assumed that the tank is isolated, that is, there is no heat transfer through

[9] Well, it would be strange if not. After all, we will be analyzing the same model as used in the simulator.

the walls to the environment. To make the model a little more realistic, we will include a time delay τ [s] to represent inhomogeneous conditions in the tank. Let us for simplicity assume that the time delay is inversely proportional to the mass flow. Thus, the temperature T at the sensor is

$$T(t) = T_1(t - \underbrace{\frac{K_\tau}{w}}_{\tau}) \tag{4.53}$$

where τ is the time delay and K_τ is a constant. Let us study the transfer function from u to T. Taking the Laplace transform of (4.52) gives

$$c\rho V\left[sT_1(s) - T_{1_0}\right] = K_P u(s) + cw\left[T_{in}(s) - T_t(s)\right] \tag{4.54}$$

where T_{1_0} is the initial value of T. Rearranging (4.54) yields the following model

$$T_1(s) = \frac{\frac{\rho V}{w}}{\frac{\rho V}{w}s + 1}T_{1_0} + \frac{\frac{K_P}{cw}}{\frac{\rho V}{w}s + 1}u(s) + \frac{1}{\frac{\rho V}{w}s + 1}T_{in}(s) \tag{4.55}$$

Taking the Laplace transform of (4.53) gives

$$T(s) = e^{-\frac{K_\tau}{w}s}T_1(s) \tag{4.56}$$

Substituting $T_1(s)$ in (4.56) by $T_1(s)$ from (4.55) yields the following transfer function $H_u(s)$ from u to T:

$$T(s) = \frac{\overbrace{\frac{K_P}{cw}}^{K}}{\underbrace{\frac{\rho V}{w}}_{T_t}s + 1}e^{-\overbrace{\frac{K_\tau}{w}}^{\tau}s}u(s) \tag{4.57}$$

$$= \underbrace{\frac{K}{T_t s + 1}e^{-\tau s}}_{H_u(s)}u(s) \tag{4.58}$$

Thus,

$$K = \frac{K_P}{cw} \tag{4.59}$$

$$T_t = \frac{\rho V}{w} \tag{4.60}$$

$$\tau = \frac{K_\tau}{w} \tag{4.61}$$

This confirms the observations in the simulations: Reduced mass flow w implies larger process gain, larger time constant, and larger time delay.

Heat exchangers and blending tanks in a process line where the production rate or mass flow varies, have similar dynamic properties as the tank in this example.

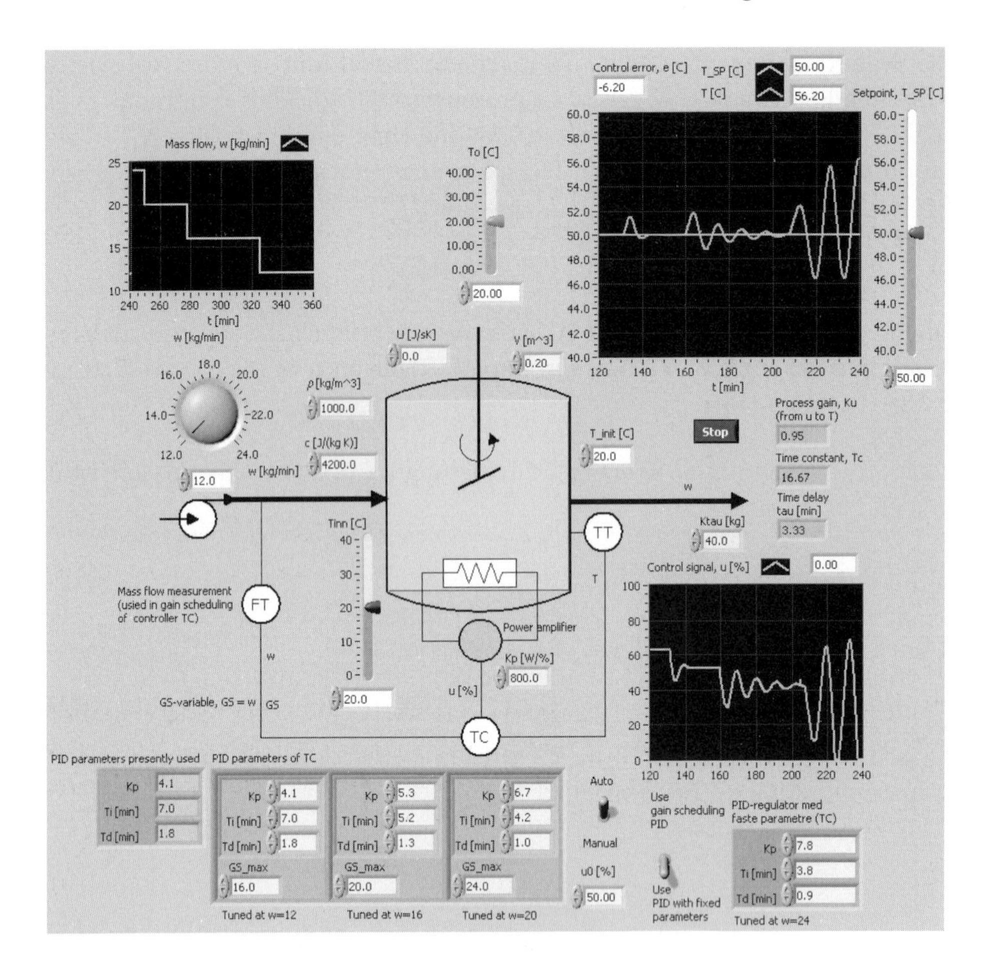

Figure 4.25: Example 4.7: Simulation of temperature control system with PID controller with fixed parameters tuned at maximum mass flow, which is $w = 24\text{kg/min}$

Control without gain scheduling (with fixed parameters) Let us look at temperature control of the tank. The mass flow w varies. In which operating point should the controller be tuned if we want to be sure that the stability of the control system is not reduced when w varies? In general the stability of a control loop is reduced if the gain increases and/or if the time delay of the loop increases. (4.59) and (4.61) show how the gain and time delay depends on the mass flow w. According to (4.59) and (4.61) the PID controller should be tuned at minimal w. If we do the opposite, that is, tune the controller at the maximum w, the control system may actually become unstable if w decreases.

Let us see if a simulation confirms the above analysis. Figure 4.25 shows a temperature control system. The PID controller is in the example tuned with the Ziegler-Nichols' closed loop method for a the maximum w value, which here is assumed 24kg/min. The PID parameters are

$$K_p = 7.8;\ T_i = 3.8\text{min};\ T_d = 0.9\text{min} \tag{4.62}$$

Figure 4.25 shows what happens at a stepwise reduction of w: The stability becomes worse, and the control system becomes *unstable* at the minimal w value, which is 12kg/min.

Instead of using the PID parameters tuned at maximum w value, we can tune the PID controller at minimum w value, which is 12kg/min. The parameters are then

$$K_p = 4.1;\ T_i = 7.0\text{min};\ T_d = 1.8\text{min} \tag{4.63}$$

The control system will now be stable for all w values, but the system behaves sluggish at large w values. (Responses for this case is however not shown here.)

Control with gain scheduling Let us see if gain scheduling maintains the stability for varying mass flow w. The PID parameters will be adjusted as a function of a measurement of w since the process dynamics varies with w. Thus, w is the gain scheduling variable, GS:

$$GS = w \tag{4.64}$$

A gain schedule consisting of three PID parameter value sets will be used. Each set is tuned using the Ziegler-Nichols' closed loop method at the following GS or w values: 12, 16 and 20kg/min. These three PID parameter sets are shown down to the left in Figure 4.25. The PID parameters are held piecewise constant in the GS intervals. In each interval, the PID parameters are held fixed for an increasing $GS = w$ value, cf. Figure 4.22.[10] Figure 4.26 shows the response in the temperature for decreasing values of w. The simulation shows that the *stability of the control system is maintained even if w decreases.*

[End of Example 4.7]

4.9.3 Adjusting PID parameters from process model

In Section 4.9.2 the adjustment of the PID parameters was based on interpolating between PID parameter values in a parameter table. However, a table with interpolation is not the only way the adjustment can

[10] The simulator uses the inbuilt gain schedule in LabVIEW's PID Control Toolkit.

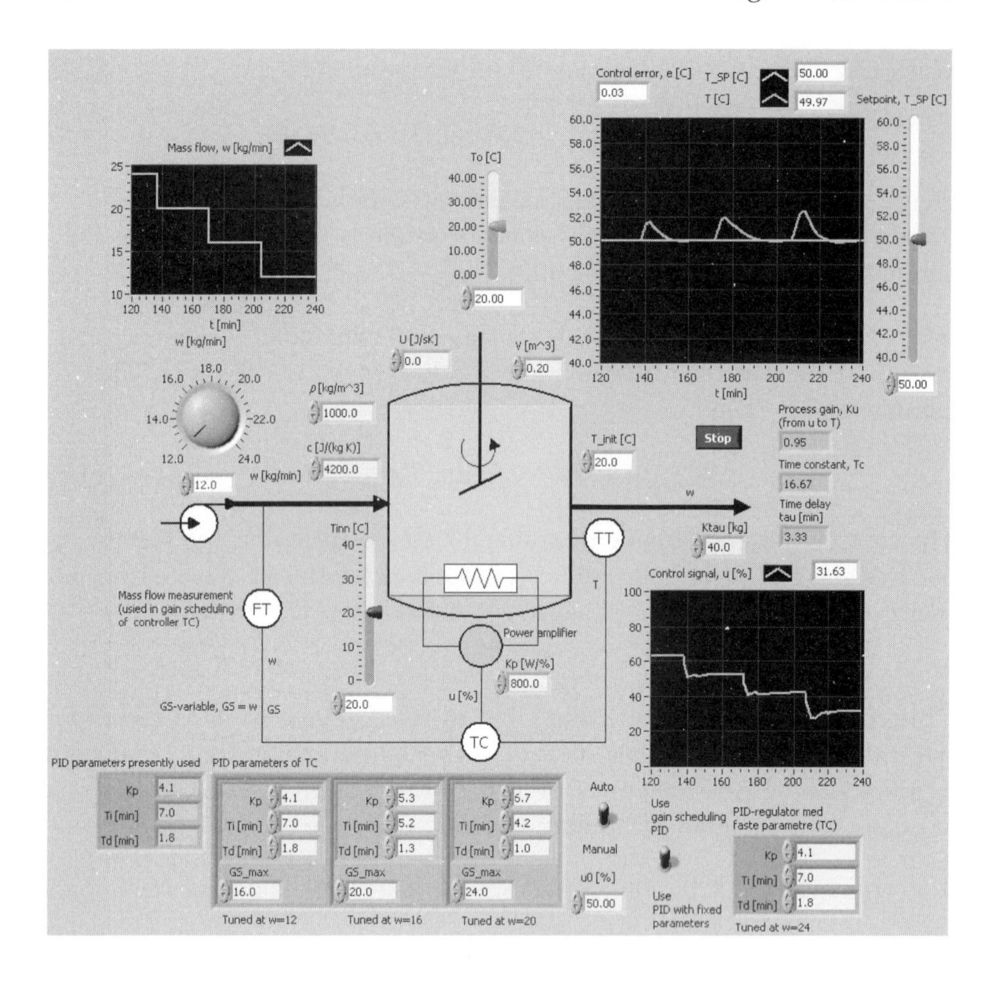

Figure 4.26: Example 4.7: Simulation of temperature control system with a gain schedule based PID controller

be implemented. By studying the process model we may find a function for parameter adjustment without having to make tuning in a number of operating points. Assume as an example that the process gain K is a function of a process variable P:

$$K = f_K(P) \tag{4.65}$$

In many control loops the stability of the loop is maintained if the loop gain K_L, which is the product of the gain of each subsystems in the loop, is constant, say K_{L_0}. In other words, the stability is maintained if

$$K_L = K_p K K_s = K_p f_K(P) K_s = K_{L_0} \tag{4.66}$$

where K_p is the controller gain (of a P or PI or PID controller) and K_m is the measurement gain (including a scaling function). For a given P value, say P_1,

$$K_{p_1} f_K(P_1) K_s = K_{L_0} \tag{4.67}$$

where K_{p_1} is assumes to be a proper K_p value (found using some tuning method) when $P = P_1$. By dividing (4.66) by (4.67) we get

$$\frac{K_p f_K(P) K_s}{K_{p_1} f_K(P_1) K_s} = \frac{K_{L_0}}{K_{L_0}} = 1 \tag{4.68}$$

from which we get the following formula for adjusting the controller gain K_p:

$$K_p = K_{p_1} \frac{f_K(P_1)}{f_K(P)} \tag{4.69}$$

Adjusting K_p according to (4.69) ensures that the stability of the control loop is maintained for any P value.

Example 4.8 ***Model based adjustment of level controller***

Figure 4.27 shows a level control system for a cylindrical tank. You will

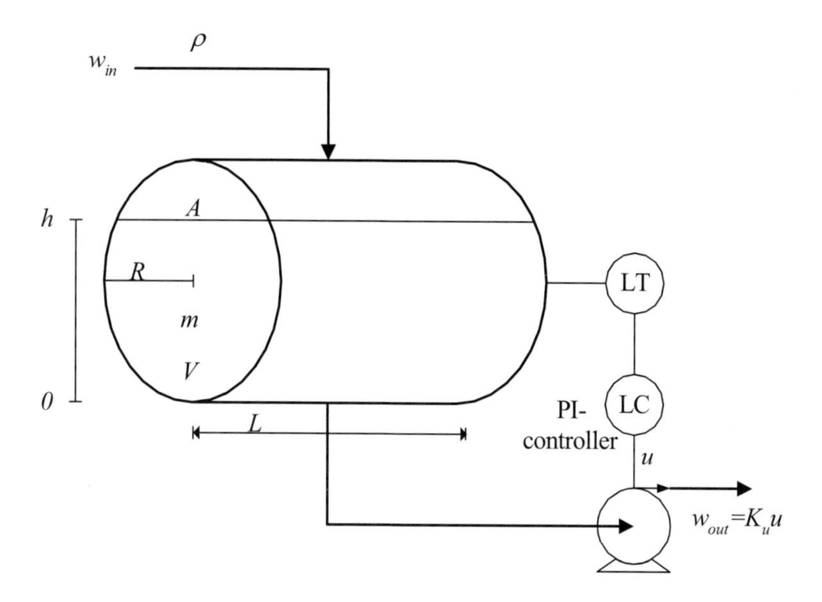

Figure 4.27: Example 4.8: Level control of a cylindric tank. The cross sectional area is a function of the level.

now see that the process gain K varies with the level. This implies that the controller gain K_p should vary. Mass balance for the liquid of the tank (we assume homogeneous conditions) is

$$\frac{dm}{dt} = \rho \frac{dV}{dt} = \rho A \frac{dh}{dt} = w_{in} - w_{out} = w_{in} - K_u u \tag{4.70}$$

It can be shown that the cross sectional area A is a function of the level h as follows:

$$A(h) = 2L\sqrt{R^2 - (R-h)^2} \tag{4.71}$$

From (4.70) we find that the transfer function from a deviation Δu in the control signal to the corresponding deviation Δh in the level is[11]

$$\frac{\Delta h(s)}{\Delta u(s)} = H(s) = -\frac{K_u}{\rho A(h)s} \tag{4.72}$$

giving the following process gain

$$K = -\frac{K_u}{\rho A(h)} = -\frac{K_u}{2\rho L\sqrt{R^2 - (R-h)^2}} = f_K(h) \tag{4.73}$$

The controller gain should be adjusted according to (4.69), which in this case gives

$$K_p = K_{p_1}\frac{f_K(h_1)}{f_K(h)} = K_{p_1}\frac{\left[-\frac{K_u}{\rho A(h_1)}\right]}{\left[-\frac{K_u}{\rho A(h)}\right]} = K_{p_1}\frac{A(h)}{A(h_1)} \tag{4.74}$$

where f_K is given by (4.73) and $A(h)$ is given by (4.71). K_{p_1} is a K_p value of a P or PI controller (the PID controller is not a good choice for this level control system since the process has pure integrator dynamics) tuned at some level h_1. (For example, h_1 may correspond to half of the maximum level.) K_p can be found by trial and error, or better: from transfer function based controller tuning, cf. Chapter 7. For example, (4.74) says that if the cross sectional area is halved (which gives doubled process gain), K_p should be halved. The integral time T_i in a PI controller can be unchanged in this case.

[End of Example 4.8]

4.9.4 Adaptive controller

In an adaptive control system, see Figure 4.28, a mathematical model of the process to be controlled is continuously estimated from samples of the control signal (u) and the process measurement (y_m). The model is typically a transfer function model. Typically, the structure of the model is fixed. The model parameters are estimated continuously using e.g. the least squares method. From the estimated process model the parameters of a PID controller (or of some other control function) are continuously

[11] The transfer function is here actually relating the deviation variables about an operating point since the process model is nonlinear.

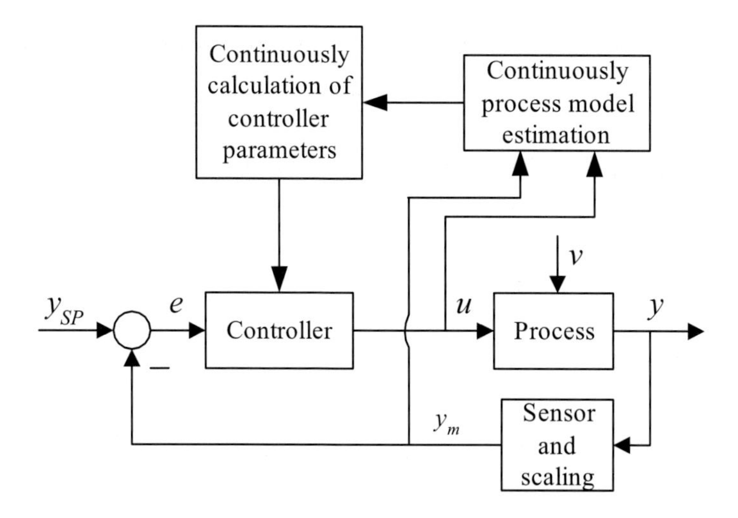

Figure 4.28: Adaptive control system

calculated so that the control system achieves specified performance in form of for example stability margins, poles, bandwidth, or minimum variance of the process output variable[22]. Adaptive controllers are commercially available, for example the ECA60 controller (ABB).

Chapter 5

Discrete-time PID controller

5.1 Introduction

In former days, as in the 1930s, PID controllers were implemented with pneumatic components. Later electronic components were used. Today, dedicated control computers are used. A computer operates in discrete time. A computer which executes PID control calculates a new value of the control variable at each new time step. This calculation is implemented in a computer program which may be written in principle in any programming language, as C, Delphi, Visual Basic, MATLAB or LabVIEW. This chapter describes briefly the main components of a discrete-time control loop, and you will learn how to develop a discrete-time PID controller function in the form of an algorithm which can be programmed.

In commercial control equipment the discrete-time PID control function is already implemented, and the tuning parameters available for the user is the well-known P-, I,- and D-parameters defined in Chapter 1. It may still be useful to know the details of discrete-time PID controllers if you want to understand the background of the discrete-time PID control function which may be shown in the documentation of the control equipment.

5.2 Computer based control loop

Figure 5.1 shows a control loop where controller is implemented in a computer.[1] The computer registers the process measurement signal via an AD converter (from analog to digital). The AD converter produces a numerical value which represents the measurement. As indicated in the block diagram this value may also be scaled, for example from volts to percent. The resulting digital signal, $y(t_k)$, is used in the control function, which is in the form of a computer algorithm or program calculating the value of the control signal, $u(t_k)$.

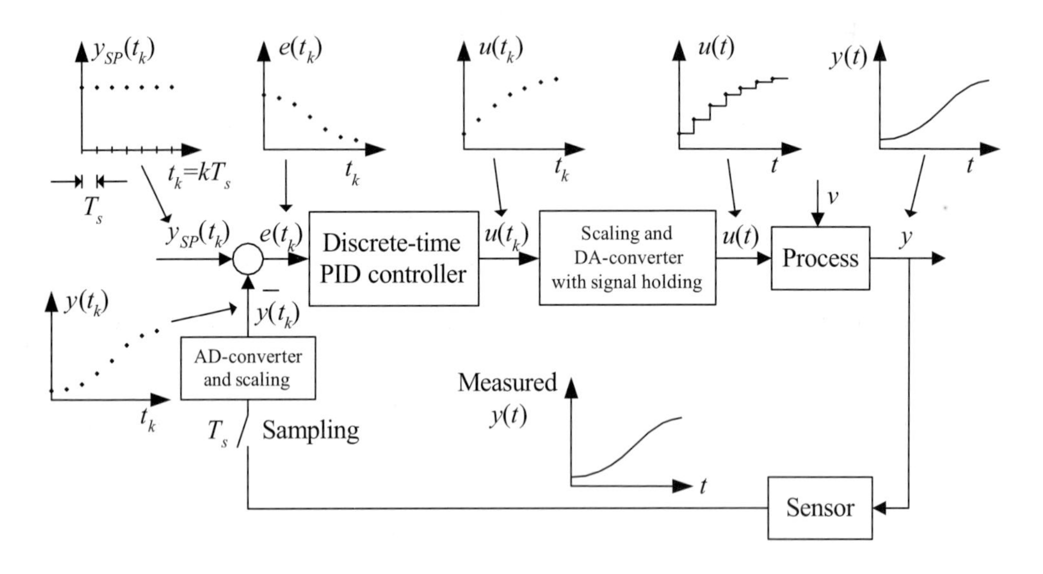

Figure 5.1: Control loop where the controller function is implemented in a computer

The control signal is scaled, for example from percent to milliamperes, and sent to the DA converter (from digital to analog) where it is held constant during the present time step. Consequently the control signal becomes a staircase signal. The time step or the sampling interval, T_s [s], is usually small compared to the time constant of the actuator (e.g. a valve) so the actuator does not feel the staircase form of the control signal.

A typical value of T_s in commercial controllers is $T_s = 0.1$s. With $T_s = 0.1$ the sampling and calculation of the control value is executed 10 times per second.

[1] In this Chapter a somewhat less detailed nomenclature is used compared to other Chapters of this book. For example y is used to represent the scaled measurement signal, while in other chapters the symbol y_m is used.

A few words about the symbols used in Figure 5.1. For example in $u(t_k)$, the integer k is the time step index. The time interval between subsequent time steps is T_s. Time step k represents the present time step, while $k-1$ represents the previous time step. Alternative ways of writing $u(t_k)$ are $u(k)$ and $u(kT_s)$.

5.3 Development of discrete-time PID control function

5.3.1 PID control function on absolute form

There are two main forms of a discrete-time PID control function or algorithm:

- The *absolute* algorithm, also denoted the *positional* algorithm
- The *incremental* algorithm, also denoted the *velocity* algorithm

We will no derive the absolute algorithm. The incremental algorithm is described in Section 5.3.2.

The starting point is the continuous-time PID control function presented in Section 2.6.7. It is repeated here:

$$u = u_0 + \underbrace{K_p e_p}_{u_p} + \underbrace{\frac{K_p}{T_i}\int_0^t e\,d\tau}_{u_i} + \underbrace{K_p T_d \frac{de_{d_f}}{dt}}_{u_d} \tag{5.1}$$

$$= u_0 + u_p + u_i + u_d \tag{5.2}$$

where the derivative error term e_{d_f} is given by

$$e_{d_f}(s) = \frac{1}{T_f s + 1} e_d(s) \tag{5.3}$$

and

$$e_p = w_p y_{SP} - y \tag{5.4}$$

$$e_d = w_d y_{SP} - y \tag{5.5}$$

w_p and w_d are setpoint weights in the P-term and in the D-term, respectively.

The discrete-time PID control function will be derived by discretizing the above continuous-time PID control function. The aim is to derive a formula for the control variable $u(t_k)$. There are several ways to make the discretization. We will use a procedure where the additive terms of (5.1) are discretized individually. Finally, $u(t_k)$ is calculated using the summation (5.2).

Discretizing the nominal control variable u_0 yields

$$u_0(t_k) = u_0 = \text{ constant} \tag{5.6}$$

Discretizing the P-term yields

$$u_p(t_k) = K_p e_p(t_k) \tag{5.7}$$

The I-term can be written

$$u_i(t_k) = \frac{K_p}{T_i} \int_0^{t_k} e(t)\, dt \tag{5.8}$$

The integral term in (5.8) must be calculated using some numerical method. It seems to be a common practice to use the *Euler backward method* , and we will use this method here. In the Euler backward method the calculation of the integral is based on holding the integrand at a constant value throughout the duration of the time step. Generally the Euler backward is as follows:

$$\text{Euler backward:} \int_{t_{k-1}}^{t_k} x(t)\, dt \approx T_s x(t_k) \tag{5.9}$$

That is, the time integral of the function $x(t)$ between time $t_{k-1} = (k-1)\,T_s$ and time $t_k = kT_s$ is approximated by $T_s x(t_k)$, which is a rectangular approximation of the area under the curve, which is the exact integral. This is illustrated in Figure 5.2.

Let us now return to the discretization of the I-term (5.8): We divide (5.8) into two parts, as follows:

$$u_i(t_k) = \underbrace{\frac{K_p}{T_i} \int_0^{t_{k-1}} e(t)\, dt}_{u_i(t_{k-1})} + \frac{K_p}{T_i} \int_{t_{k-1}}^{t_k} e(t)\, dt \tag{5.10}$$

$$= u_i(t_{k-1}) + \frac{K_p}{T_i} \int_{t_{k-1}}^{t_k} e(t)\, dt \tag{5.11}$$

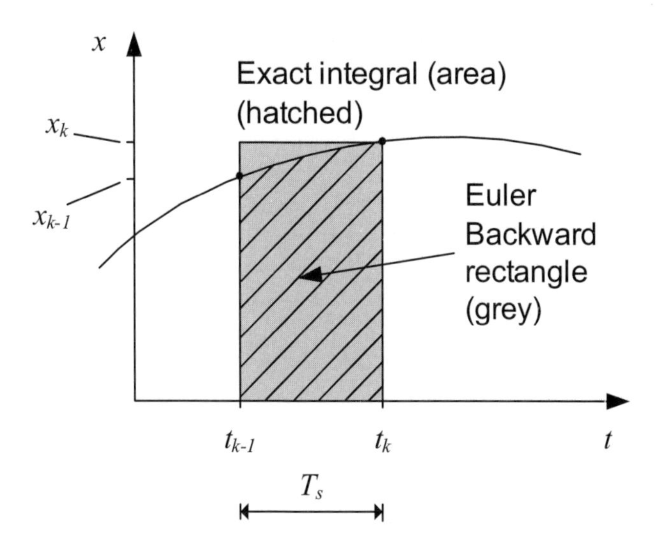

Figure 5.2: Euler backward method of numerical calculation of an integral is a rectangular approximation of the integral (the area under the curve)

Applying the Euler backward method (5.9) on the integral in (5.11) yields

$$u_i(t_k) = u_i(t_{k-1}) + \frac{T_s K_p}{T_i} e(t_k) \tag{5.12}$$

which is the discrete version of the I-term.

The D-term is

$$u_d(t_k) = K_p T_d \frac{de_{d_f}(t_k)}{dt} \tag{5.13}$$

Again we may use the Euler backward method, this time as a numerical method of calculating the time-derivative term in (5.13). In general the Euler backward method applied to the time-derivative is

$$\text{Euler backward method: } \frac{dx(t_k)}{dt} \approx \frac{x(t_k) - x(t_{k-1})}{T_s} \tag{5.14}$$

That is, the time-derivative is calculated as the difference between the value of the present time-step and the previous time-step, divided by the length of one time-step. Applying the Euler backward method on (5.13) yields

$$u_d(t_k) = K_p T_d \frac{e_{d_f}(t_k) - e_{d_f}(t_{k-1})}{T_s} \tag{5.15}$$

We have to discretize the filtering function (5.3), too. From (5.3) we get, by cross-multiplying,

$$T_f s e_{d_f}(s) + e_{d_f}(s) = e_d(s) \tag{5.16}$$

Taking the inverse Laplace transformation gives the following differential equation (where time t is represented by t_k):

$$T_f \frac{de_{d_f}(t_k)}{dt} + e_{d_f}(t_k) = e_d(t_k) \tag{5.17}$$

Applying the Euler backward method (5.14) on the time-derivative yields the following filtering algorithm:

$$e_{d_f}(t_k) = \frac{1}{(T_s/T_f)+1} e_{d_f}(t_{k-1}) + \frac{T_s/T_f}{(T_s/T_f)+1} e_d(t_k) \tag{5.18}$$

Finally, we now have all terms of (5.2).

To sum up, here is the algorithm constituting the discrete-time PID control function, ready for programming:

Discrete-time PID control function:

Read the measurement $y(t_k)$ from the AD converter (5.19)

P-term:

$$e_p(t_k) = w_p y_{SP}(t_k) - y(t_k) \tag{5.20}$$

$$u_p(t_k) = K_p e_p(t_k) \tag{5.21}$$

I-term:

$$e(t_k) = y_{SP}(t_k) - y(t_k) \tag{5.22}$$

$$u_i(t_k) = u_i(t_{k-1}) + \frac{T_s K_p}{T_i} e(t_k) \tag{5.23}$$

D-term including lowpass filter:

$$e_d(t_k) = w_d y_{SP}(t_k) - y(t_k) \tag{5.24}$$

$$e_{d_f}(t_k) = \frac{1}{(T_s/T_f)+1} e_{d_f}(t_{k-1}) + \frac{T_s/T_f}{(T_s/T_f)+1} e_d(t_k) \tag{5.25}$$

$$u_d(t_k) = K_p T_d \frac{e_{d_f}(t_k) - e_{d_f}(t_{k-1})}{T_s} \tag{5.26}$$

Total control value:

$$u(t_k) = u_0 + u_p(t_k) + u_i(t_k) + u_d(t_k) \tag{5.27}$$

Write the control value $u(t_k)$ to the DA converter (5.28)

Assign for use in next time-step:

$$u_i(t_{k-1}) := u_i(t_k) \tag{5.29}$$

$$e_{d_f}(t_{k-1}) := e_{d_f}(t_k) \tag{5.30}$$

$$\text{Start with (5.19) in next time-step} \tag{5.31}$$

When starting this algorithm for the first time, you can set $k = 1$. The value of $u_i(t_0)$ can initially be set to 0 if it assumed that u_0 in (5.27) has a proper value. If you choose not to include the u_0-term in (5.27) you can set $u_i(t_0)$ equal to u_0. The initial value of $e_{d_f}(t_0)$ can be set to 0.

Integral anti wind-up

In Section 2.7.2 integral anti wind-up was described. Integral anti wind-up can be realized in the absolute algorithm (5.20)–(5.27) as follows:

1. Calculate an intermediate value of the control variable $u(t_k)$ according to (5.20)–(5.27), but do not send this value to the DA converter.

2. Check if the intermediate $u(t_k)$ is greater than the maximum value $u_{\max}$ (typically 100%) or less than the minimum value $u_{\min}$ (typically 0%). If $u(t_k)$ is exceeding one of these limits, calculate the I-term $u_i(t_k)$ once more, but now with

$$u_i(t_k) = u_i(t_{k-1}) \tag{5.32}$$

   (which implies that the I-term is fixed), and calculate $u(t_k)$ once more according to (5.27) using $u_i(t_k)$ given by (5.32).

3. Write $u(t_k)$ to the DA converter.

Bumpless transfer

Figure 5.3 shows a block diagram of a control loop. Suppose the controller is switched from automatic to manual mode, or from manual to automatic mode (this will happen during maintenance, for example). It is important that the control signal does not jump much. In other words, the transfer between modes must be bumpless, ideally. Bumpless transfer can be realized as follows:

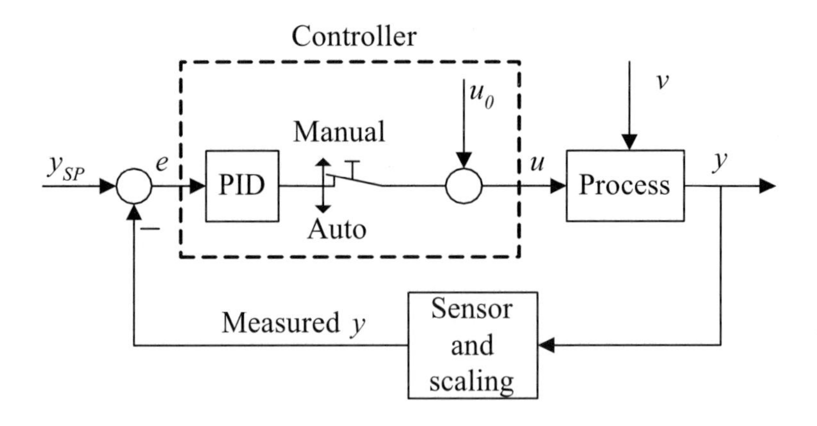

Figure 5.3: Block diagram of a control loop

- **Bumpless transfer from automatic to manual mode:** In manual mode it is the nominal control signal u_0 which controls the process. We assume that the control variable u has a proper value, say u_{good}, so that the control error is small immediately before the switch to manual mode. At the switching moment set u_0 equal to u_{good}.

  While the controller is in manual mode, each of the P-, I-, and the D-term is given zero value.

- **Bumpless transfer from manual to automatic mode:** At the switching from manual to automatic mode each of the P- , I- , and D-terms is set to zero value.

  If the controller is implemented so that the control variable is calculated by

  $$u(t_k) = u_p(t_k) + u_i(t_k) + u_d(t_k) \tag{5.33}$$

  that is, if the nominal (manual control variable) u_0 is not included in the PID control function, the I-term should be given a proper value at the switching. This can be obtained by setting $u_i(t_k)$ equal to the manually tuned control value, u_0, at the switching.

5.3.2 Incremental PID control function

The PID control function (5.20)–(5.27) is on *absolute* form. The control function can be written in an alternative form denoted the *incremental* or velocity form. The incremental form is based on splitting the calculation of the control value into two steps:

1. First the incremental control value Δu_k is calculated.
2. Then the total or absolute control value is calculated with $u_k = u_{k-1} + \Delta u_k$.

One way to find the expression of the increment Δu_k is by differentiating the expression (5.1) and then approximating the derivatives using the Euler backward method given by (5.14). We will not do these calculations in detail here, but the result is as follows:

Discrete-time PID control function on incremental form:

$$\begin{aligned} \Delta u(t_k) &= K_p \left[e_p(t_k) - e_p(t_{k-1}) \right] + \frac{K_p T_s}{T_i} e(t_k) \\ &\quad + \frac{K_p T_d}{T_s} \left[e_{d_f}(t_k) - 2e_{d_f}(t_{k-1}) + e_{d_f}(t_{k-2}) \right] \qquad (5.34) \\ u_k &= u_{k-1} + \Delta u_k \qquad (5.35) \end{aligned}$$

where $e_p(t_k)$ is given by (5.20) and $e_{df}(t_k)$ is given by (5.25), where $e_d(t_k)$ is given by (5.24). To obtain bumpless transfer from manual to automatic mode, the nominal (manually adjusted) control value u_0 is used as the initial value in (5.35).

The incremental PID control function is particularly useful if the actuator is controlled by an incremental signal. A step-motor is such an actuator (the motor itself implements the numerical integral (5.35)). A benefit of the absolute PID algorithm (5.20)–(5.27) as compared to the incremental algorithm is that it is easier to implement modifications which involves only the P- or the I- or the D-term. Another benefit is that it is somewhat easier to develop and to understand the absolute algorithm.

5.4 How the sampling interval influences loop stability and tuning

If the time-step or sampling interval T_s is sufficiently small, tuning a discrete-time PID controller is in most cases nothing different from tuning a continuous-time PID controller. For example, you can probably use the Ziegler-Nichols' closed loop method in the ordinary way. However, if the sampling time T_s is relatively large, the discrete-time nature of the controller becomes more important, and it may be necessary to cope with it, as explained in the following.

Let us inspect to some detail what actually happens in the DA converter, cf. Figure 5.1. The calculated control value is held fixed during the present time-step or sampling interval. This holding implies that the control signal is *time delayed* approximately $T_s/2$, see Figure 5.4. This delay influences

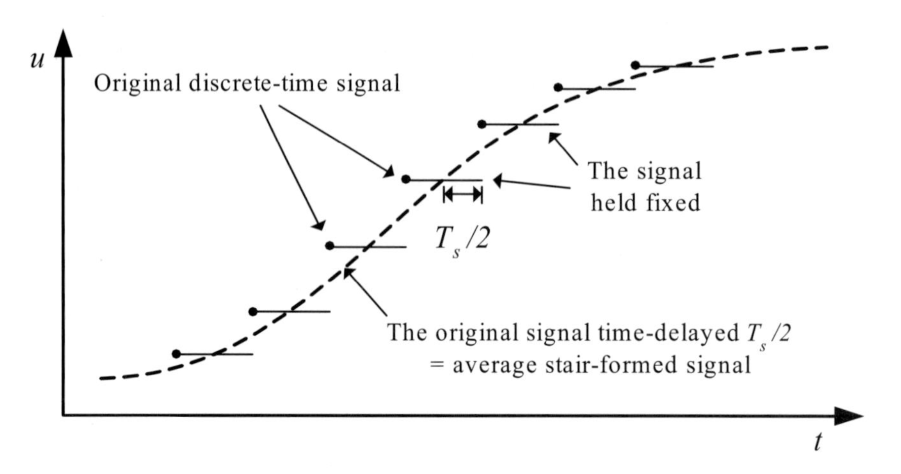

Figure 5.4: The DA-converter holds the calculated control signal throyghout the sampling interval, thereby introducing an approximate time-delay of $T_s/2$.

the stability of the control loop. Suppose we have tuned a continuous-time PID controller, and apply these PID parameters on a discrete-time PID controller. Then the control loop will get worse stability because of the approximate delay of $T_s/2$. How much is the stability reduced? As a rule of thumb, the stability reduction is small and tolerable if the time delay is less than *one tenth of the response-time* of the control system as it would have been with a continuous-time controller or a controller having very small sampling time:

$$\frac{T_s}{2} \leq \frac{T_r}{10} \tag{5.36}$$

which gives

$$T_s \leq \frac{T_r}{5} \tag{5.37}$$

(The response time is here the 63% rise time which can be read off from the setpoint step response. For a system the having dominating time constant T, the response-time is approximately equal to this time constant.) If the bandwidth of the control system is ω_b [rad/s] (assuming that the PID parameters have been found using a continuous-time PID controller), the response-time of the control system can be estimated by

$$T_r \approx \frac{1}{\omega_b} \tag{5.38}$$

(5.37) can be used as an upper limit of the sampling interval T_s if T_s is to be chosen by the user.[2]

If you want to include the effect of the discrete-time nature of the PID controller when tuning the controller on a simulator or when analyzing the discrete-time control loop, you can select one of the following options:

- *Exact method*: The PID controller is represented by a *discrete-time model.* This model may be the PID algorithm itself or the z-transfer function of the controller. z-transfer functions belong to the theory of discrete-time systems. This theory is not covered by the present book.[3]
- *Approximate method*: The PID controller is represented by a *continuous-time model* which may be in the form of an s-transfer function of the controller, cf. Section 2.6.7. To include a model of the sample-and-hold effect of the DA converter, the transfer function of a *time delay of* $T_s/2$ can be used. This transfer function is

$$H_{\text{sh}}(s) = e^{-\frac{T_s}{2}s} \tag{5.39}$$

  $H_{\text{sh}}(s)$ may be included as a factor in the controller transfer function (alternatively as a factor in the process transfer function).

Example 5.1 ***The importance of the sampling time***

Let us look at responses in a simulated control system. In the simulator the PID controller is represented by its discrete-time algorithm (thus, it is exactly represented in the simulator). The process to be controlled has the following transfer function from control variable u to process output variable y:

$$\frac{y(s)}{u(s)} = \frac{K}{(T_1 s + 1)(T_2 s + 1)} e^{-\tau s} = H_u(s) \tag{5.40}$$

[2] It can be shown from frequency based stability analysis that the reduction of the phase margin is approximately 5° if the limit (5.37) is used, and the PID parameters are found assuming a continuous-time PID controller.

[3] Analysis of discrete-time control systems is described in a document available from the home page of this book at http://techteach.no.

where

$$K = 1;\ T_1 = 2\text{s};\ T_2 = 1\text{s};\ \tau = 1\text{s} \quad (5.41)$$

The PID parameters have been found as

$$K_p = 2.2;\ T_i = 2.8\text{s};\ T_d = 0.7\text{s} \quad (5.42)$$

(using the Ziegler-Nichols' closed loop with a very small T_s). The response time T_r is found by simulation to be $T_r = 2.0$s. The upper limit of T_s according to (5.37) then becomes

$$T_{s_{\max}} = \frac{2.0}{5} = 0.4\text{s} \quad (5.43)$$

Let us initially set T_s equal to the upper limit 0.4s. Figure 5.5 shows for example the response in y due to a step in the setpoint with discrete-time PID controller with parameters equal to (5.42). We observe that the

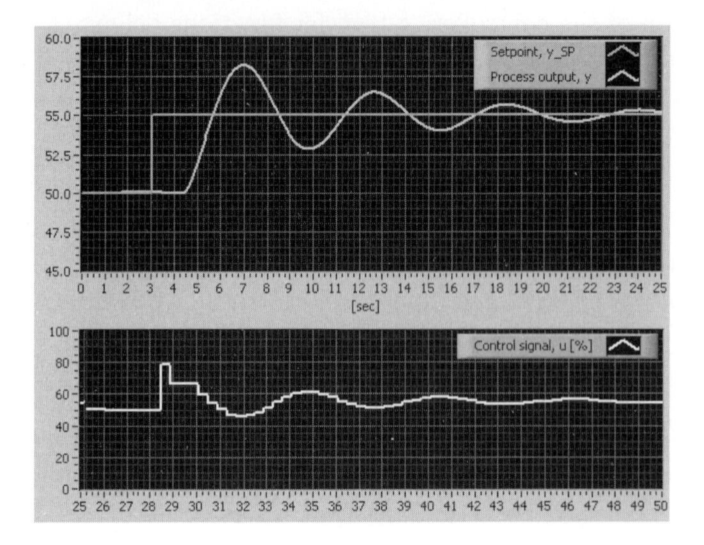

Figure 5.5: Step response in y with discrete-time PID controller with parameters equal to (5.42). $T_s = 0.4$s which is on the limit given by (5.37).

control system has *acceptable stability*.

Now let us increase the sampling time T_s from 0.4s to 1.0s without changing the PID parameters. Figure 5.6 shows for example the response in y due to a step in the setpoint. The control system is now unstable, due to the large T_s.

Finally, let us re-tune the PID controller when $T_s = 1.0$s. The PID parameters becomes (using the Ziegler-Nichols' closed loop method)

$$K_p = 1.7;\ T_i = 3.5\text{s};\ T_d = 0.83\text{s} \quad (5.44)$$

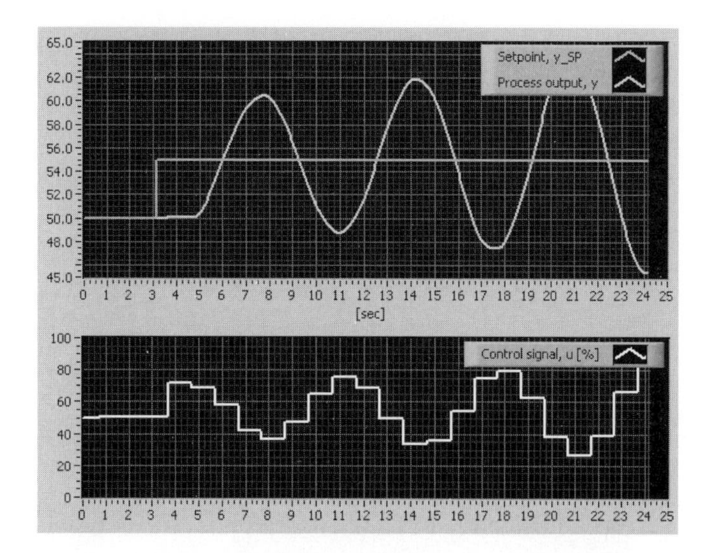

Figure 5.6: Step response in y. The time-step T_s has a relatively large value of 1.0s, but with unchanged PID parameters. The control system is unstable!

Figure 5.7 shows that the control system now is stable, but the stability is not good. The stability could be improved by reducing the controller gain K_p somewhat.

[End of Example 5.1]

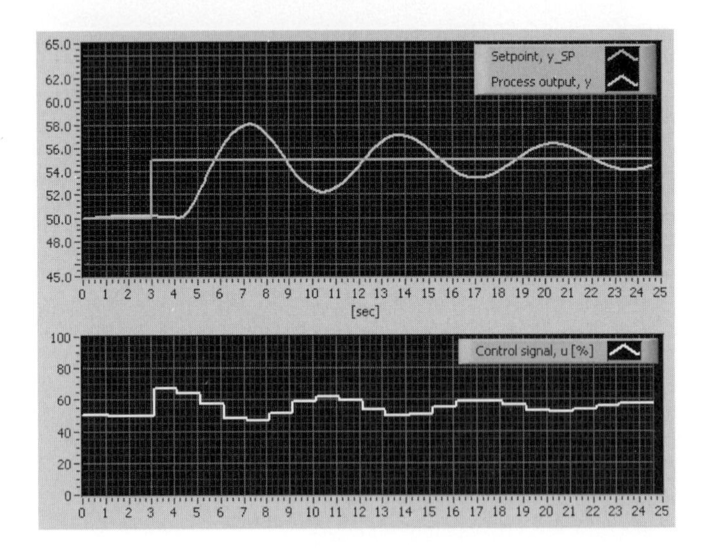

Figure 5.7: Step response in y. The time-step T_s is 1.0s, but the PID controller has been retuned. The control system is now stable.

Chapter 6

Analysis of feedback control systems

6.1 Introduction

In this chapter various methods of analysis of control systems are described:

- simulation,
- analytical calculation of time responses,
- frequency response analysis,
- stability analysis.

A theoretical analysis of a control system assumes that a mathematical model of the control system exists, that is, a model of the controller, the process and the sensor constituting the control system (control loop). Most analysis methods assume *linear* models. However, practical control systems are nonlinear due to phenomena as saturation, hysteresis, stiction, nonlinear signal scaling etc. Such nonlinearities can influence largely the dynamic behaviour of the control system.

To perform "linear" analysis of a non-linear model, this model must be linearized [7] about some operating point. Thus, the results of the analysis will be valid at or close to the operation point where the linearization was made. This fact limits the usefulness of a theoretical analysis of a given nonlinear control system using linear systems methods, but the results may still be useful, particularly if the system most of the time operates close to the chosen or specified operating point.

Although a "linear" analysis of a given nonlinear control system may have limited value, you will get much general knowledge about the behaviour of control systems through analysis of examples of linear control systems.

6.2 About using simulators

If you have a mathematical model of a given control system, you should definitely *run simulations* as a part of the analysis. This applies for both linear and nonlinear control systems. Actually, you may get all the answers you need by just running simulations. The types of answers may concern response-time, static control error, responses due to process disturbances and measurement noise, and effects of parameter variations.

Figure 6.1 shows a detailed block diagram of a control system (the units shown are just examples of units). Such a block diagram can be used directly in block diagram based (graphical) simulation tools, as SIMULINK and LabVIEW. The individual blocks may represent transfer functions, nonlinear elements, etc.

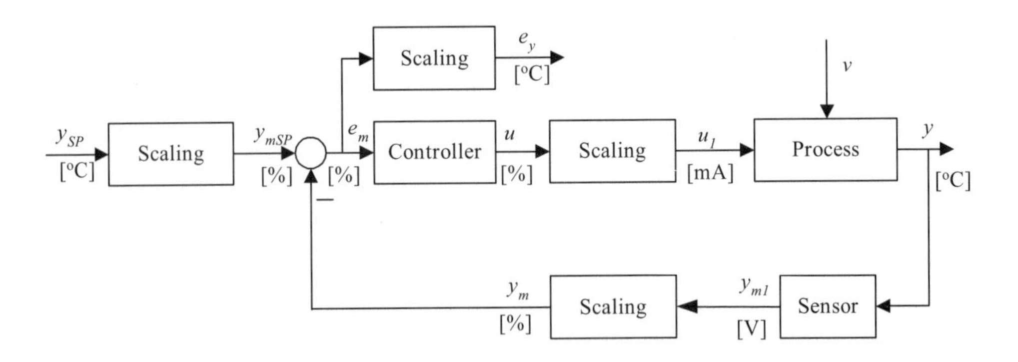

Figure 6.1: Detailed block diagram of a control system. Such block diagrams can be used directly in block diagram based simulation tools as SIMULINK and LabVIEW.

6.3 Setpoint tracking and disturbance compensation

This section concerns theoretical analysis of control systems.

6.3.1 Introduction

Figure 6.2 shows a principal block diagram of a control system. There are

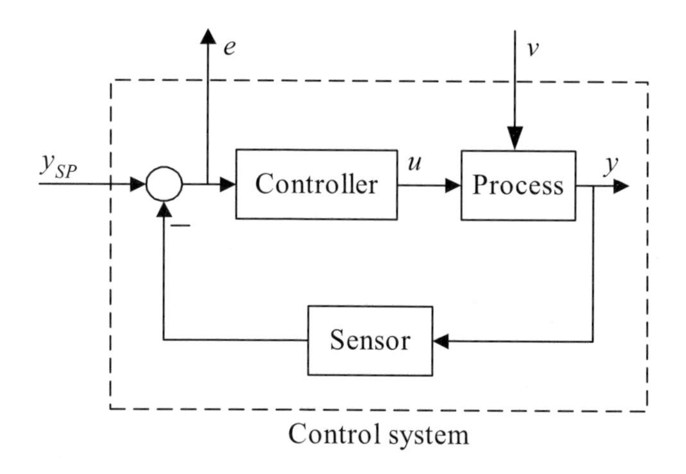

Figure 6.2: Principal block diagram of a control system

two input signals to the control system, namely the setpoint y_{SP} and the disturbance v. The value of the control error e is our primary concern (it should be small, preferably zero). Therefore we can say that e is the (main) output variable of the control system. The value of e expresses the *performance* of the control system: The less e, the higher performance. e is influenced by y_{SP} and v. Let us therefore define the following two conceptions:

- The **setpoint tracking property** of the control system concerns the relation between y_{SP} and e.
- The **disturbance compensation property** of the control system concerns the relation between v and e.

Totally, the setpoint tracking and disturbance compensation properties determine the *performance* of the control system.

6.3.2 Analysis based on differential equation models

A (theoretical) analysis of control system performance (read: calculation of the control error) can be made on the basis of a time domain model of the control system, that is, on basis of the differential- and/or integral equations constituting the model. The model may be nonlinear. Dynamic time responses may be calculated from these equations by solving the equations analytically (which may be quite difficult except for very simple models), or numerically. Numerical calculations can be realized in a computer program using e.g. the Euler forward method or one of the Runge-Kutta methods [7], or using some simulation tool[1]. In this section we confine ourselves to *static analysis* of control systems based on manual calculations using the differential- and/or integral equations constituting the model of the control system.

Control systems often operates under approximate static condition, which means that all signals are approximately constant. Therefore a static analysis may be quite relevant. Static analysis can be executed by hand as explained below.

1. Given models of the elements of the control loop (process, controller, sensor, scalings). Substitute the control variable u in the process model by u from the controller function. The resulting (combined) model constitute the *control system model.*

2. If the controller function contains an integrator (as in the PID controller): Transform the control system model so that the integral term "disappears" by differentiating the entire model with respect to time. (The purpose of this is to simplify the mathematical calculations. It is easier to use a model containing only time derivatives – and no integrals – in a static analysis.)

3. Derive a static model of the control system by assuming that all variables have constant values. This assumption implies that the time-derivatives can be set equal to zero and that time delays can be neglected.

4. Calculate the static control error, e_s, from the static model of the control system.

Before we look at an example, I will emphasize that a static analysis does not express anything about the stability property or dynamic property of

[1] Such tools are described in a document available from the home page of the book on http://techteach.no.

the system. Therefore we should not use static analysis for controller design, only for control system analysis. For example, a controller gain designed from a static control system model so that the control error is smaller than some specified limit, may cause instability of the control system.

Example 6.1 ***Static analysis of wood-chip tank level control system with P controller***

A wood-chip tank with level control system is described in Example 2.3 (page 19). A mathematical model of the tank is developed in Example 2.7 (page 29).

Below we follow the items of the static analysis defined above.

1. The process model is given by (2.20), which is repeated here:

$$\rho A\dot{h}(t) = K_s u(t-\tau) - w_{out}(t) \tag{6.1}$$

Let us use a P controller which has the following controller function:

$$u(t) = u_0 + K_p e_m(t) = u_0 + K_p[h_{m_{SP}}(t) - h_m(t)] \tag{6.2}$$

where e_m is the control error in % and $h_{m_{SP}}$ is level setpoint in % (the measurement unit). We substitute $u(t)$ in the process model by u from the controller function:

$$\rho A\dot{h}(t) = K_s[u_0 + K_p e_m(t-\tau)] - w_{out}(t) \tag{6.3}$$

which constitutes a model of the control system.

2. The controller function contains no integral term. Therefore, this point is not applicable in this example.

3. The static model of the control system becomes (subindex s is for "static")

$$\underbrace{\rho A\dot{h}_s(t)}_{0} = K_s(u_0 + K_p e_{m_s}) - w_{out_s} \tag{6.4}$$

4. From (6.4) we find

$$e_{m_s} = \frac{\frac{w_{out_s}}{K_s} - u_0}{K_p} \text{ [\%]} \tag{6.5}$$

Let us calculate the error in meters: The measurement gain is K_m, so $e_{m_s} = K_m e_s$, which inserted into (6.5) yields

$$e_s = \frac{w_{out_s}}{K_s K_p K_m} - \frac{u_0}{K_p K_m} \text{ [m]} \tag{6.6}$$

Assume that the nominal control value, u_0, has the correct value

$$u_0 = \frac{w_{out_s}}{K_s} \tag{6.7}$$

(so that u_0 compensates for the outflow w_{out_s}). Now, the static control error e_s in (6.6) becomes zero. This is confirmed by the first part of the simulation shown in Figure 2.17 (page 36).

Assume now that u_0 has an incorrect value. e_s given by (6.6) is then different from zero. What is the value of e_s? The numerical values in (6.6) are $K_p = 1.55$, $w_{out_s} = 1800\text{kg/min}$, $K_s = 33.36(\text{kg/min})/\%$, $K_m = 6.67\%/\text{m}$, $u_0 = 45.0\%$, which gives $e_s = 0.87\text{m}$. Is this in accordance with the simulated response in Example 2.9? Although Figure 2.17 indicates static control error of $e_s = 0.83\text{m}$ at $t = 120\text{s}$, the actual value as $t \to \infty$ is 0.87m (the responses has not converged completely at $t = 120\text{s}$). Thus, the calculated value of e_s is in accordance with the simulated value.

A final question: According to (6.6) the e_s becomes less if K_p increases. So why not set K_p to a very large value?[2]

[End of Example 6.1]

The next example shows how the differential equation based analysis can be performed when the controller contains an integral term (as in a PID controller).

Example 6.2 ***Static analysis of wood-chip tank level control system with PI controller***

The controller is now a PI controller:

$$u(t) = u_0 + K_p e_m(t) + \frac{K_p}{T_i}\int_0^t e_m(t^*)\,dt^* \tag{6.8}$$

1. We substitute $u(t)$ in process model (6.1) by $u(t)$ in (6.8):

$$\rho A\dot{h}(t) = K_s\left[u_0 + K_p e_m(t-\tau) + \frac{K_p}{T_i}\int_0^{t-\tau} e(t^*)\,dt^*\right] - w_{out}(t) \tag{6.9}$$

which is the model of the control system.

[2] Because the stability of the control system is reduced as K_p is increased, and the system becomes unstable if K_p is too large.

2. The controller function contains an integral. To "get rid of" the integral, we time-differentiate (6.9):

$$\rho A\ddot{h}(t) = K_s \left[\dot{u_0} + K_p \dot{e}_m(t-\tau) + \frac{K_p}{T_i} e_m(t-\tau) \right] - \dot{w_{out}}(t) \quad (6.10)$$

3. We assume that all variables have constant values, so that the time-derivatives can be set to zero. Furthermore, the time delay can be neglected. (6.10) now becomes

$$0 = \frac{K_u K_p}{T_i} e_{m_s} \quad (6.11)$$

4. From (6.11) $e_{m_s} = 0$, so

$$e_s = 0 \quad (6.12)$$

So we have found that the control error becomes zero (independent of the value of u_0, w_{out} and setpoint h_{SP}). This is in accordance with the simulation shown in Figure 2.19.

[End of Example 6.2]

6.3.3 Transfer function based analysis of setpoint tracking and disturbance compensation

Introduction

If the control system model is linear, whether the model is linear originally or it is linear due to linearization, it is convenient to base the analysis of the control system on a transfer function model of the system.

In transfer function based analysis it is normally assumed that the initial values of the output signals (of the transfer functions) are zero.

Calculation of control error in setpoint tracking and disturbance compensation. Sensitivity function

We assume that the control system has a transfer function-based block diagram as shown in Figure 6.3. This block diagram corresponds to the block diagram shown in Figure 6.1. In the block diagram $U_0(s)$ represents the Laplace transform of the nominal control variable u_0. In practice u_0 is constant, giving

$$U_0(s) = \frac{u_0}{s} \tag{6.13}$$

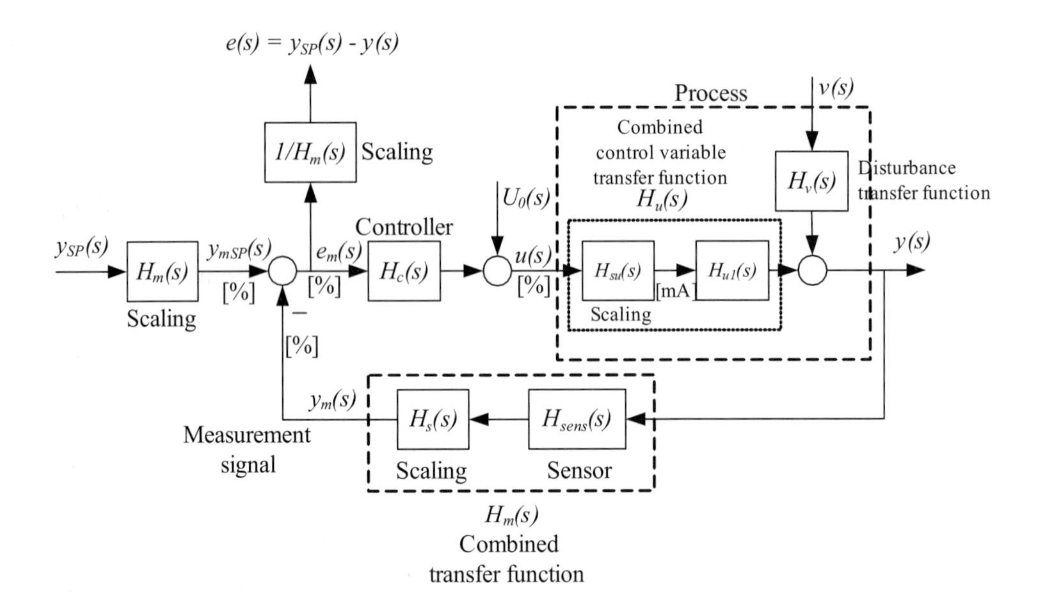

Figure 6.3: Transfer function based block diagram of a control system. (The units, e.g. %, are typical examples of units.)

We regard the setpoint y_{SP} and the disturbance v as input variables and the control error e as the output variable of the system. Thus, we will derive the transfer function from y_{SP} to e and the transfer function from v to e. From the block diagram we the can write the following expressions for $e(s)$:

$$e(s) = \frac{1}{H_m(s)} e_m(s) \tag{6.14}$$

$$= \frac{1}{H_m(s)} \left[y_{m_{SP}}(s) - y_m(s) \right] \tag{6.15}$$

$$= \frac{1}{H_m(s)} \left[H_m(s) y_{SP}(s) - H_m(s) y(s) \right] \tag{6.16}$$

$$= y_{SP}(s) - y(s) \tag{6.17}$$

$$= H_v(s)v(s) + H_u(s)u_0(s) + H_u(s)H_c(s)e_m(s) \tag{6.18}$$

$$= H_v(s)v(s) + H_u(s)u_0(s) + H_u(s)H_c(s)H_s(s)e(s) \tag{6.19}$$

Solving for $e(s)$ gives

$$\begin{aligned} e(s) &= \frac{1}{1+H_c(s)H_u(s)H_m(s)}\left[y_{SP}(s) - H_v(s)v(s) - H_u(s)U_0(s)\right] \\ &= \underbrace{\frac{1}{1+L(s)}}_{S(s)}\left[y_{SP}(s) - H_v(s)v(s) - H_u(s)U_0(s)\right] \quad (6.20) \\ &= S(s)\left[y_{SP}(s) - H_v(s)v(s) - H_u(s)U_0(s)\right] \quad (6.21) \\ &= \underbrace{S(s)y_{SP}(s)}_{e_{SP}(s)} - \underbrace{S(s)H_v(s)v(s)}_{e_v(s)} - \underbrace{S(s)H_u(s)U_0(s)}_{e_{u_0}(s)} \quad (6.22) \\ &= e_{SP}(s) + e_v(s) + e_{u_0}(s) \quad (6.23) \end{aligned}$$

which is a transfer functions based model of the control system. $S(s)$ is the *sensitivity function*:

$$S(s) = \frac{1}{1+L(s)} \quad (6.24)$$

where

$$L(s) \equiv H_c(s)H_u(s)H_m(s) \quad (6.25)$$

is the *loop transfer function* which is the product of the transfer functions in the loop. From (6.22) we can calculate the control error for any setpoint signal, any disturbance signal and any nominal control signal (assuming we know their Laplace transform).

In the following we discuss the various terms in (6.23).

- **The response in the error due to the setpoint**: The response in the control error due to the *setpoint* is

$$e_{SP}(s) = S(s)y_{SP}(s) = \frac{1}{1+L(s)}y_{SP}(s) \quad (6.26)$$

which gives a quantitative expression of the *tracking property* of the control system. The *static* tracking is given by static error when y_{SP} is constant. This error can be calculated as follows:[3]

$$\begin{aligned} e_{SP} &= \lim_{t\to\infty} e_{SP}(t) = \lim_{s\to 0} s\cdot e_{SP}(s) \quad (6.27) \\ &= \lim_{s\to 0} s\cdot S(s)y_{SP}(s) = \lim_{s\to 0} s\cdot S(s)\frac{y_{SP_s}}{s} = S(0)y_{SP_s} \quad (6.28) \end{aligned}$$

Roughly speaking that the tracking property of the control system are good if the sensitivity function N has small (absolute) value – ideally zero.

[3] Here the Final Value Theorem of the Laplace transform is used, cf. Appendix B.1.

- **The response in the error due to the disturbance**: The response in the control error due to the disturbance is

$$e_v(s) = -S(s)H_v(s)v(s) = -\frac{H_v(s)}{1+L(s)}v(s) \tag{6.29}$$

which expresses the *compensation property* of the control system. The *static* compensation property is given by

$$\begin{aligned} e_{v_s} &= \lim_{t\to\infty} e_v(t) = \lim_{s\to 0} s \cdot e_v(s) && (6.30)\\ &= \lim_{s\to 0} s \cdot [-S(s)H_v(s)v(s)] && (6.31)\\ &= \lim_{s\to 0} s \cdot \left[-S(s)H_v(s)\frac{v_s}{s}\right] && (6.32)\\ &= -S(0)H_v(0)v_s && (6.33) \end{aligned}$$

From (6.33) we see that the *compensation property is good if the sensitivity function S has a small (absolute) value (close to zero).*

- **The response in the error due to the nominal control variable**: The response in the control error due to the nominal control variable or signal u_0 is

$$e_{u_0}(s) = -S(s)H_u(s)U_0(s) = -\frac{H_u(s)}{1+L(s)}u_0(s) \tag{6.34}$$

If u_0 is constant (which is the typical case), its Laplace transform is

$$u_0(s) = \frac{U_0}{s} \tag{6.35}$$

which can be used for $u_0(s)$ in (6.34).

The tracking transfer function

The *tracking transfer function* $T(s)$ – or simply the tracking function – is the transfer function from the setpoint y_{SP} to the process output variable y:

$$y(s) = T(s)y_{SP}(s) \tag{6.36}$$

From the block diagram in Figure 6.3, or by setting $e_{y_{SP}}(s) \equiv y_{SP}(s) - y(s)$ for $e_{y_{SP}}(s)$ in (6.26), we can find the tracking function $T(s)$ as the transfer function from y_{SP} to y:

$$\frac{y(s)}{y_{SP}(s)} = T(s) = \frac{H_c(s)H_u(s)H_m(s)}{1+H_c(s)H_u(s)H_m(s)} = \frac{L(s)}{1+L(s)} = 1 - S(s) \tag{6.37}$$

The *static* tracking property is given by the static tracking ratio $T(0)$:

$$y_s = \lim_{t\to\infty} y(t) = \lim_{s\to 0} s \cdot y(s) \tag{6.38}$$

$$= \lim_{s\to 0} s \cdot T(s) y_{SP}(s) = \lim_{s\to 0} s \cdot T(s) \frac{y_{SP_s}}{s} \tag{6.39}$$

$$= T(0) y_{SP_s} \tag{6.40}$$

The *tracking property is good if the tracking function T has (absolute) value equal to or close to 1* (since then y will be equal to or close to y_{SP}).

In some contexts it is useful to be aware that the sum of the tracking function and the sensitivity function is always 1:

$$T(s) + S(s) = \frac{L(s)}{1 + L(s)} + \frac{1}{1 + L(s)} \equiv 1 \tag{6.41}$$

Example 6.3 ***Transfer function based analysis of speed control system***

Figure 6.4 shows a block diagram of a speed control system of a motor. We

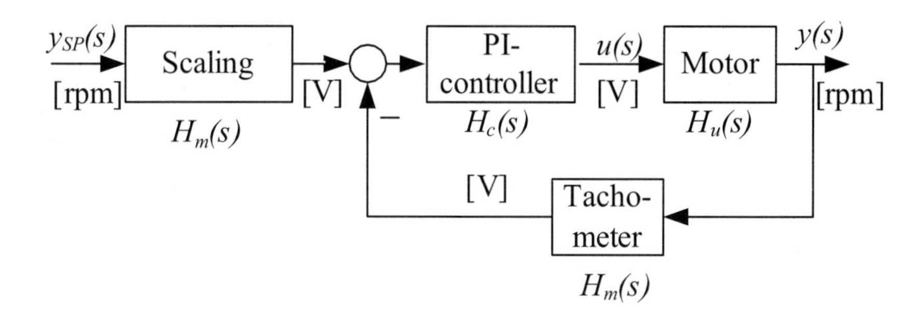

Figure 6.4: Speed control system of a motor

assume there is no disturbance acting on the motor, and that the nominal control signal is 0. The speed setpoint is y_{SP_s} (constant). The controller is a PI controller with proper values of K_p and T_i. The controller transfer function is

$$H_c(s) = K_p \frac{T_i s + 1}{T_i s} \tag{6.42}$$

The motor transfer function is assumed

$$H_u(s) = \frac{K_u}{T_u s + 1} \tag{6.43}$$

The measurement transfer function of the sensor (tachometer) is

$$H_m(s) = K_m \tag{6.44}$$

We shall find the static control error e_s from (6.28). We start by calculating the static sensitivity function $S(0)$. (6.28) includes $S(s)$:

$$\begin{aligned} S(s) &= \frac{1}{1+L(s)} && (6.45)\\ &= \frac{1}{1+K_p\frac{T_is+1}{T_is}\cdot\frac{K_u}{T_us+1}\cdot K_m} && (6.46)\\ &= \frac{T_iT_us^2+T_is}{T_iT_us^2+(K_pK_uK_m+1)\,s+K_pK_uK_m} && (6.47)\end{aligned}$$

The static control error becomes, according to (6.28),

$$\begin{aligned} e_s &= \lim_{t\to\infty} e_{y_{SP}}(t) && (6.48)\\ &= \lim_{t\to\infty} S(s)\frac{y_{SPLs}}{s} && (6.49)\\ &= 0 && (6.50)\end{aligned}$$

Thus, the static tracking property is perfect.

The tracking transfer function becomes

$$T(s) = \frac{L(s)}{1+L(s)} = \frac{K_pK_uK_m}{T_iT_us^2+(K_pK_uK_m+1)\,s+K_pK_uK_m} \tag{6.51}$$

The static tracking function is

$$T(0) = 1 \tag{6.52}$$

Thus,

$$y_s = T(0)y_{SP_s} = 1\cdot y_{SP_s} = y_{SP_s} \tag{6.53}$$

So the static error is zero.

Figure 6.5 shows a simulated response in $y(t)$ due to a step in the setpoint y_{SP}. (The parameter values are $K_p = 2$, $T_i = 0.3$, $K_u = 1$, $T_u = 1$.) The simulation confirms that the static error is zero.

[End of Example 6.3]

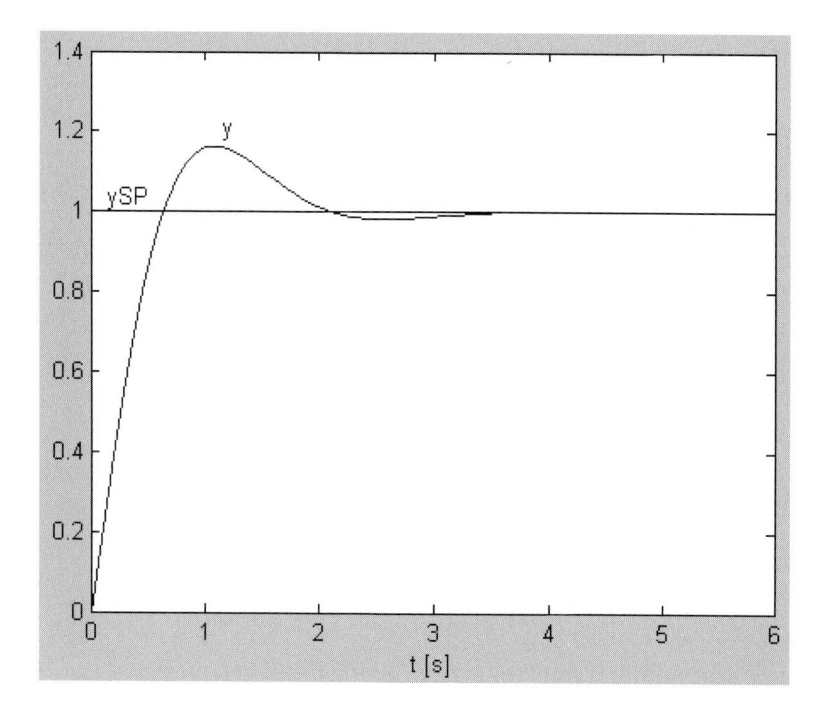

Figure 6.5: Example 6.3: Simulated speed response. The setpoint is a step.

The relation between steady-state control error and number of integrators in the loop

We shall derive a relation between *the steady-state control error and the number of integrators in the control loop.* An integrator has the transfer function of $\frac{1}{s}$. We assume that the transfer functions in the block diagram shown in Figure 6.3 are on the following forms:

$$H_u(s) = \frac{K_u(1 + b_1 s + b_2 s^2 + \cdots)}{s^U(1 + a_1 s + a_2 s^2 + \cdots)} \tag{6.54}$$

(The number of integrators is U.)

$$H_v(s) = \frac{K_v(1 + d_1 s + \cdots)}{s^V(1 + c_1 s + \cdots)} \tag{6.55}$$

(The number of integrators is V.)

$$H_c(s) = \frac{K_c(1 + f_1 s + \cdots)}{s^C(1 + e_1 s + \cdots)} \tag{6.56}$$

(The number of integrators is C.)

$$H_m(s) = K_m \tag{6.57}$$

The control error is given by (6.22). By applying the Final Value Theorem to (6.22), where (6.54), (6.55) and (6.56) is used, we find that the steady-state (static) control error is

$$\begin{aligned} e_s &= \lim_{t\to\infty} e(t) = \lim_{s\to 0} se(s) && (6.58)\\ &= \lim_{s\to 0} s\left[S(s)y_{SP}(s) - S(s)H_v(s)v(s) - S(s)H_u(s)U_0(s)\right] && (6.59)\\ &= \underbrace{\lim_{s\to 0} s\frac{s^{(C+U)}}{s^{(C+U)}+K_L}y_{SP}(s)}_{e_{y_{SP}}(s)} + \underbrace{\lim_{s\to 0} s\frac{-K_v s^{(C+U-V)}}{s^{(C+U)}+K_L}v(s)}_{e_v(s)} && (6.60)\\ &\quad \underbrace{+\lim_{s\to 0} s\frac{-K_u s^{C}}{s^{C}+K_L}U_0(s)}_{e_{u_0}(s)} && (6.61) \end{aligned}$$

where

$$K_L = K_c K_u K_m \text{ (the loop gain)} \tag{6.62}$$

In practice the nominal control signal is constant, let us say u_0, so that

$$U_0(s) = \frac{u_0}{s} \tag{6.63}$$

in (6.61). Using (6.61) we can assume various types of signals in y_{SP} and v, e.g. step or ramp, and calculate the steady-state control error in each case.[4]

Let us concentrate on the most important case, namely that both y_{SP} and v and u_0 have *constant* values (different from zero). Thus,

$$y_{SP}(s) = \frac{y_{SP_s}}{s} \text{ and } v(s) = \frac{v_s}{s} \text{ and } U_0(s) = \frac{u_0}{s} \tag{6.64}$$

Inserting these values into (6.60) yields

$$e_s = \underbrace{\lim_{s\to 0}\frac{s^{(C+U)}}{s^{(C+U)}+K_L}y_{SP_s}}_{e_{y_{SP}}(s)} + \underbrace{\lim_{s\to 0}\frac{-K_v s^{(C+U-V)}}{s^{(C+U)}+K_L}v_s}_{e_v(s)} + \underbrace{\lim_{s\to 0}\frac{-K_u s^{C}}{s^{C}+K_L}u_0}_{e_{u_0}(s)} \tag{6.65}$$

From (6.65) we can conclude as follows:

1. A sufficient condition for the static control error to be zero ($e_s = 0$) with constant setpoint, disturbance and nominal control variable, is

$$C + U - V \geq 1 \tag{6.66}$$

[4] In some cases we will get expressions as $\lim_{s\to 0} s^0$, which has value of 1, not 0.

It is common that there are as many integrations in $H_v(s)$ as in $H_u(s)$, that is, $V = U$. In this case, *a sufficient condition for the static control error to be zero ($e_s = 0$) with constant setpoint, disturbance and nominal control variable, is that the number of integrations C in $H_c(s)$ (the controller) is at least one.* This is achieved for a PI controller and a PID controller.

One observation from (6.60) is that $S(s)$ has $C + U$ numbers of factors of the type s in its numerator polynomial. In other words: It has $C + U$ zeros in origin.

2. (6.65) shows that e_s will be zero even if there are no integrations ($C = 0$) in $H_c(s)$, which corresponds to using a P controller, under the following conditions: (1) There is at least one integration ($U = 1$) in the control variable transfer function $H_u(s)$, *and* (2) The nominal control variable is correctly tuned, that is $u_0 = -v_s K_v / K_u$. However, in practice condition 2 is not satisfied completely. Consequently, *there will in practice always be a static control error different from zero when the controller lacks integral action.*

   In the situations where the static error e_s is different from zero, *this error will decrease if the loop gain K_L is increased*, e.g. if the controller gain K_p is increased. (The drawback of increasing the loop gain is that the control loop gets reduced stability.)

Example 6.4 ***Control error in a speed control system***

See Example 6.3. The speed setpoint is a step at $t = 0$. We calculated the static error e_s to be zero. Can we confirm this result by using (6.65)? In Example 6.3, $v_s = 0$ and $u_0 = 0$. The controller has one integrator, thus $R = 1$. The transfer function $H_u(s)$ contains no integrator, thus $U = 0$. The loop gain is

$$K_L = \frac{K_p K_u K_m}{T_i} = 6.67 \tag{6.67}$$

(6.65) gives

$$e_s = \lim_{s \to 0} \frac{s^{(C+U)}}{s^{(C+U)} + K_L} y_{SP_s} = \lim_{s \to 0} \frac{s^{(1+0)}}{s^{(1+0)} + K_L} y_{SP_s} = 0 \tag{6.68}$$

Thus, the result in Example 6.3 is confirmed!

Now, let us assume that the setpoint is a *ramp signal*:

$$y_{SP}(t) = K_1 t \tag{6.69}$$

where $K_1 = 1$ is the slope. The Laplace transformed setpoint is

$$y_{SP}(s) = \frac{K_1}{s^2} \tag{6.70}$$

What is the steady-state control error? Since the setpoint is a ramp, we must now use (6.60), not (6.65):

$$e_s = \lim_{s\to 0} s \frac{s^{(C+U)}}{s^{(C+U)} + K_L} y_{SP}(s) = \lim_{s\to 0} s \frac{s^{(1+0)}}{s^{(1+0)} + K_L} \frac{K_1}{s^2} = \frac{K_1}{K_L} = 0.15 \tag{6.71}$$

So the steady-state (static) error is different from zero. Figure 6.6 shows a simulation which which confirms this.[End of Example 6.4]

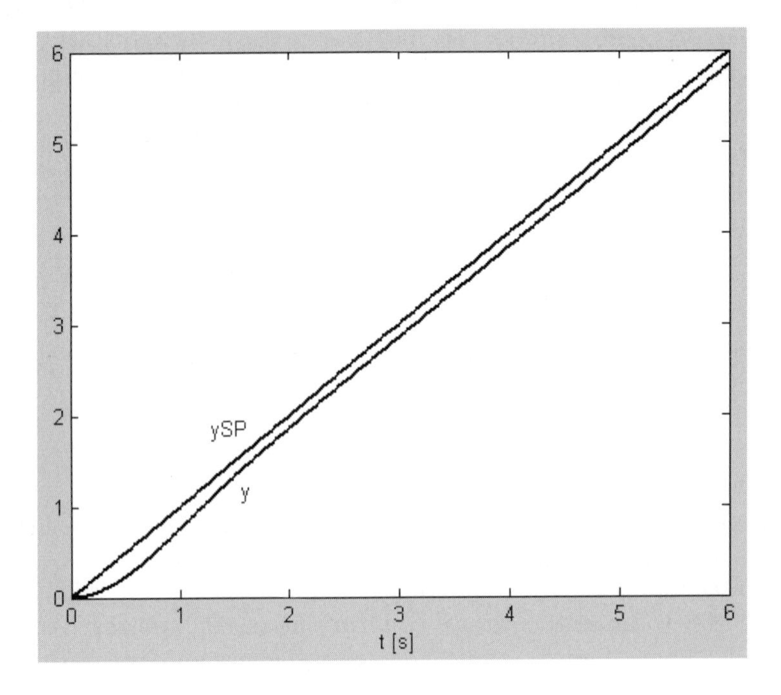

Figure 6.6: Example 6.4: The speed setpoint y_{SP} and the speed y using a PI-controller

Example 6.4 demonstrated that the steady-state control error may become different from zero if the setpoint is more "dynamic" than just a constant (in the example the setpoint was a ramp). Using (6.71) it can be seen that, with a ramped setpoint, the error can be zero if the controller has two integrators ($C = 2$) in stead of one. But in practice a controller does not have more than one integrators. It can be shown that using several integrators requires several differenciators to ensure the stability of the control loop, and using more than one differenciator should be avoided since there would be a large amplification of high frequent (measurement) noise. One way to obtain zero or small control error when the setpoint is "dynamic" is to use feedforward from the setpoint. Feedforward control is described in Chapter 9.1.

6.3.4 Frequency response analysis of setpoint tracking and disturbance compensation

Introduction

Frequency response analysis of control systems expresses the tracking and compensation property under the assumption that the setpoint and the disturbance are *sinusoidal signals* or *frequency components* of a compound signal. The structure of the control system is assumed to be as shown in Figure 6.3. The Laplace transformed control error is given by (6.22), which is repeated here:

$$e(s) = \underbrace{S(s)y_{SP}(s)}_{e_{SP}(s)} \underbrace{-S(s)H_v(s)v(s)}_{e_v(s)} \underbrace{-S(s)H_u(s)U_0(s)}_{e_{u_0}(s)} \tag{6.72}$$

where $S(s)$ is the sensitivity function which is given by

$$S(s) = \frac{1}{1 + L(s)} \tag{6.73}$$

where $L(s)$ is the loop transfer function. In the following we will study both $S(s)$ and the tracking ratio $T(s)$ which is given by

$$T(s) = \frac{L(s)}{1 + L(s)} = \frac{y(s)}{y_{SP}(s)} \tag{6.74}$$

Frequency response analysis of setpoint tracking

From (6.72) we see we that the response in the control error due to the setpoint is

$$e_{SP}(s) = S(s)y_{SP}(s) \tag{6.75}$$

By plotting the *frequency response* $S(j\omega)$ we can easily calculate how large the error is for a given frequency component in the setpoint: Assume that the setpoint is a sinusoid of amplitude Y_{SP} and frequency ω. Then the steady-state response in the error is

$$e_{SP}(t) = Y_{SP} \left| S(j\omega) \right| \sin \left[\omega t + \arg S(j\omega) \right] \tag{6.76}$$

Thus, the error is small and consequently the tracking property is good if $|S(j\omega)| \ll 1$, while the error is large and the tracking property poor if $|S(j\omega)| \approx 1$.

The tracking property can be indicated by the tracking function $T(s)$, too. The response in the process output due to the setpoint is

$$y(s) = T(s)y_{SP}(s) \tag{6.77}$$

Assume that the setpoint is a sinusoid of amplitude Y_{SP} and frequency ω. Then the steady-state response in the process output due to the setpoint is

$$y(t) = Y_{SP}\,|T(j\omega)| \sin\left[\omega t + \arg T(j\omega)\right] \tag{6.78}$$

Thus, $|T(j\omega)| \approx 1$ indicates that the control system has good tracking property, while $|T(j\omega)| \ll 1$ indicates poor tracking property.

Since both $S(s)$ and $T(s)$ are functions of the loop transfer function $L(s)$, cf. (6.73) and (6.74), there is a relation between $L(s)$ and the tracking property of the control system. Using (6.73) and (6.73) we can conclude as follows:

$$\text{Good setpoint tracking: } |S(j\omega)| \ll 1,\ |T(j\omega)| \approx 1,\ |L(j\omega)| \gg 1 \tag{6.79}$$

$$\text{Poor setpoint tracking: } |S(j\omega)| \approx 1,\ |T(j\omega)| \ll 1,\ |L(j\omega)| \ll 1 \tag{6.80}$$

Figure 6.7 shows typical Bode plots of $|S(j\omega)|$, $|T(j\omega)|$ and $|L(j\omega)|$. Usually we are interested in the amplitude gains, not the phase lags. Therefore plots of $\arg S(j\omega)$, $\arg T(j\omega)$ and $\arg L(j\omega)$ are not shown nor discussed here. The bandwidths indicated in the figure are defined below.

The *bandwidth* of a control system is the frequency which divides the frequency range of good tracking and poor tracking. From (6.79) and (6.80) and Figure 6.7 we can list the following three candidates for a definition of the bandwidth:

- ω_t, which is the frequency where the amplitude gain of the tracking function has value $1/\sqrt{2} \approx 0.71 = -3$ dB. This definition is in accordance with the usual bandwidth definition of lowpass filters. The ω_t bandwidth is also called the -3 dB bandwidth $\omega_{-3\text{dB}}$.
- ω_c, which is the frequency where the amplitude gain of the loop transfer function has value $1 = -0$ dB. ω_c is called the *crossover frequency* of L.

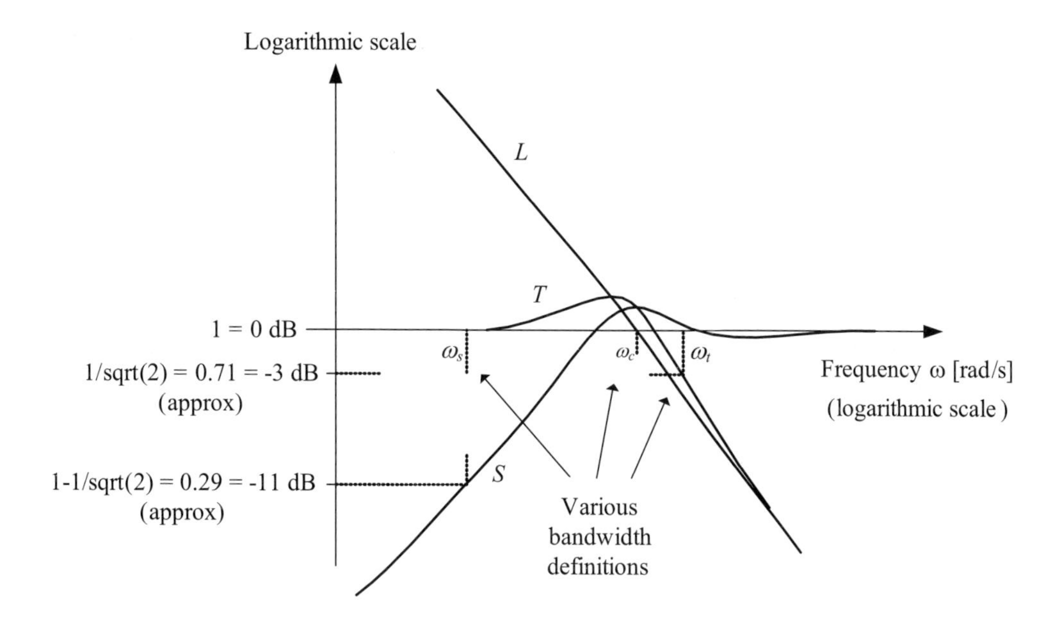

Figure 6.7: Typical Bode plots of $|S(j\omega)|$, $|T(j\omega)|$ and $|L(j\omega)|$

- ω_s, which is the frequency where the amplitude gain of the sensitivity function has value $1 - 1/\sqrt{2} \approx 1 - 0.71 \approx 0.29 \approx -11$ dB. This definition is derived from the -3 dB bandwidth of the tracking function: Good tracking corresponds to tracking gain between $1/\sqrt{2}$ and 1. Now recall that the sensitivity function is the transfer function from setpoint to control error, cf. (6.75). Expressed in terms of the control error, we can say that good tracking corresponds to sensitivity gain $|S|$ less than $1 - 1/\sqrt{2} \approx -11$ dB $\approx 0,29$. The frequency where $|S|$ is -11 dB is denoted the *sensitivity bandwidth*, ω_s.

Of the three bandwidth candidates defined above the sensitivity bandwidth ω_s is most closely related to the control error. Therefore ω_s may be claimed to be the most convenient bandwidth definition as far as the tracking property of a control system concerns. In addition ω_s is a convenient bandwidth related to the compensation property of a control system (this will be discussed in more detail soon). However, the crossover frequency ω_c and the -3 dB bandwidth are the commonly used bandwidth definitions.

As indicated in Figure 6.7 the numerical values of the various bandwidth definitions are different (this is demonstrated in Example 6.5).

If you need a (possibly rough) estimate of the *response time* T_r of a control system, which is time it takes for a step response to reach 63% of its steady-state value, you can use

$$T_r \approx \frac{k}{\omega_s} \text{ [s]} \tag{6.81}$$

where ω_s is the -3 dB bandwidth in rad/s.[5] k can be set to some value between 1.5 and 2.0, say 2.0 if you want to be conservative.

Example 6.5 ***Frequency response analysis of setpoint tracking***

See the block diagram in Figure 6.3. Assume the following transfer functions:

PID controller:

$$H_c(s) = K_p \left(1 + \frac{1}{T_i s} + \frac{T_d s}{T_f s + 1}\right) \tag{6.82}$$

Process transfer functions (second order with time delay):

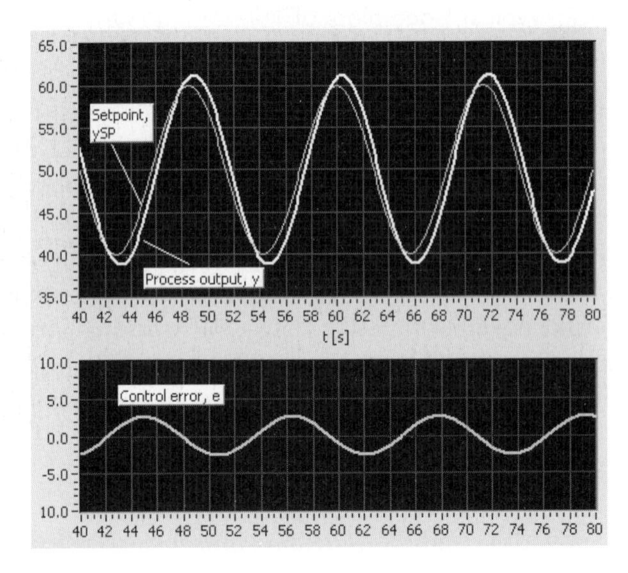

Figure 6.8: Example 6.5: Simulated responses of the control system. The setpoint y_{SP} is sinuoid of frequency $\omega_1 = 0.55$ rad/s.

$$H_u(s) = \frac{K_u}{(T_1 s + 1)(T_2 s + 1)} e^{-\tau s} \tag{6.83}$$

$$H_v(s) = \frac{K_v}{(T_1 s + 1)(T_2 s + 1)} e^{-\tau s} \tag{6.84}$$

[5] How can you find the excact value of the response time? Simulate!

Sensor with scaling:

$$H_s(s) = K_s \tag{6.85}$$

The parameter values are $K_p = 4.3$, $T_i = 1.40$, $T_d = 0.35$, $T_f = 0.1T_d = 0.035$, $K_u = 1$, $K_d = 1$, $T_1 = 2$, $T_2 = 0.5$, $\tau = 0.4$, $K_s = 1$. (The PID parameter values are calculated using the Ziegler-Nichols' closed loop method, cf. Section 4.4.) The operation point is at setpoint value 50%, with disturbance $v = 10\%$ (constant), and nominal control signal = 40%.

Figure 6.8 shows simulated responses in the process output y and in the the control error $e = y_{SP} - y$ when the setpoint y_{SP} is a sinusoid of amplitude 10% (about a bias of 50%) and frequency $\omega_1 = 0.55$rad/s. The frequency of the sinusoidal is chosen equal to the sensitivity bandwidth ω_s. The amplitude of the control error should be $0.29 \cdot 10\% = 2.9\%$, and this is actually in accordance with the simulation, see Figure 6.8.

Figure 6.9 shows Bode plots of $|S(j\omega)|$, $|T(j\omega)|$ and $|L(j\omega)|$.

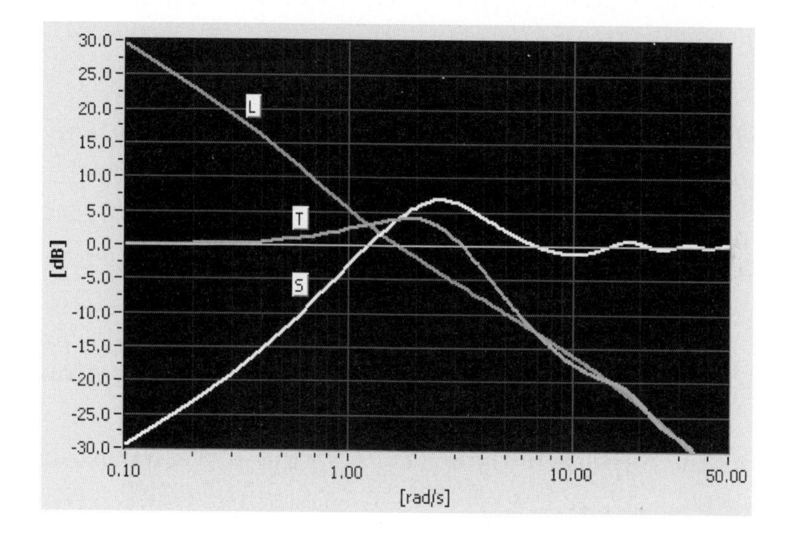

Figure 6.9: Example 6.5: Bode plots of $|L(j\omega)|$, $|T(j\omega)|$ and $|S(j\omega)|$

Let us compare the various bandwidth definitions. From Figure 6.9 we find

- -3 dB bandwidth: $\omega_t = 3.8$ rad/s
- Crossover frequency: $\omega_c = 1.7$ rad/s
- Sensitivity bandwidth: $\omega_s = 0.55$ rad/s

These values are actually quite different. (As commented in the text above this example, it can be argued that the ω_s bandwidth gives the most expressive measure of the control system dynamics.)

Finally, let us read off the response time T_r. Figure 6.10 shows the response in y due to a step in y_{SP}. From the simulation we read off $T_r \approx 1.1$s. The estimate (6.81) with $k = 2$ gives $T_r \approx 2/\omega_t = 2/3.8 = 0.53$, which is about half the value of the real (simulated) value.

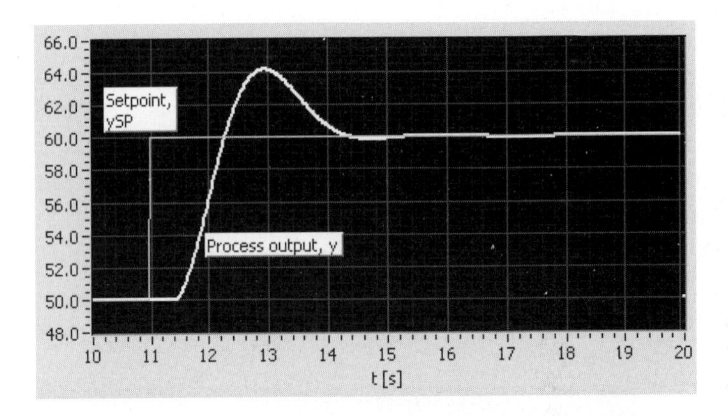

Figure 6.10: Example 6.5: Step response in process output y after a step in setpoint y_{SP}

[End of Example 6.5]

Frequency response analysis of disturbance compensation

(6.72) gives the response in the control error due to the disturbance. It is repeated here:

$$e_v(s) = -S(s)H_v(s)v(s) \tag{6.86}$$

Thus, the sensitivity function $S(s)$ is a factor in the transfer function from v til e for the control system. However, $S(s)$ has an additional meaning related to the compensation of a disturbance, namely it expresses the degree of the reduction of the control error due to using closed loop control. *With* feedback (i.e. closed loop system) the response in the control error due to the disturbance is $e_v(s) = -S(s)H_v(s)v(s)$. *Without* feedback (open loop) this response is $e_v(s) = -H_v(s)v(s)$. The ratio between these responses is

$$\frac{e_v(s)_{\text{with feedback}}}{e_v(s)_{\text{without feedback}}} = \frac{-S(s)H_v(s)v(s)}{-H_v(s)v(s)} = S(s) \tag{6.87}$$

Assuming that the disturbance is sinusoidal with frequency ω rad/s, (6.87) with $s = j\omega$, that is $S(j\omega)$, expresses the ratio between sinusoidal responses.

Again, effective control, which here means effective disturbance compensation, corresponds to a small value of $|S|$ (value zero or close to zero), while ineffective control corresponds to $|S|$ close to or greater than 1. We can define the *bandwidth* of the control system with respect to its compensation property. Here are two alternate bandwidth definitions:

- The bandwidth ω_s – the sensitivity bandwidth – is the upper limit of the frequency range of effective compensation. One possible definition is
$$|S(j\omega_s)| \approx 0.29 \approx -11 \text{ dB} \tag{6.88}$$
which means that the amplitude of the error *with* feedback control is less than 29% of amplitude *without* feedback control. The number 0.29 is chosen to have the same bandwidth definition regarding disturbance compensation as regarding setpoint tracking, cf. page 163.

- The bandwidth ω_c is the crossover frequency of the loop transfer functions ω_c, that is,
$$|L(j\omega_c)| = 0 \text{ dB} \approx 1 \tag{6.89}$$

Note: The feedback does not reduce the control error due to a sinusoidal disturbance if its frequency is above the bandwidth. But still the disturbance may be well attenuated through the (control) system. This attenuation is due to the typical inherent lowpass filtering characteristic of physical systems (processes). Imagine a liquid tank, which attenuates high-frequent temperature variations existing in the inflow fluid temperature or in the environmental temperature. This inherent lowpass filtering is *self regulation.*

Example 6.6 ***Frequency response analysis of disturbance compensation***

This example is based on the control system described in Example 6.5 (page 164).

Figure 6.11 shows simulated responses in the process output y due to a sinusoidal disturbance v of amplitude 10% (with bias 10%) and frequency

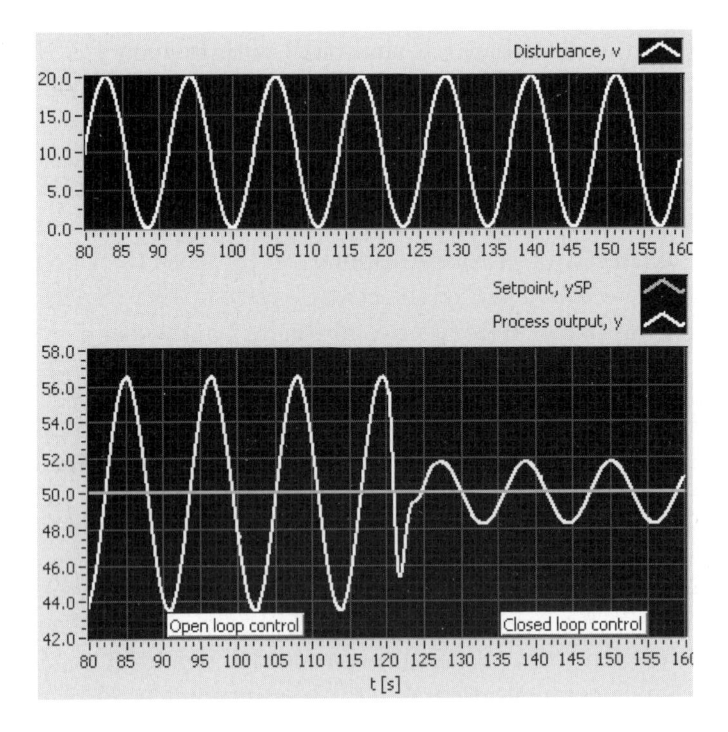

Figure 6.11: Example 6.6: Simulated responses of the control system. The disturbance v is sinusoidal with frequency $\omega_1 = 0.55$ rad/s. The PID-controller is in manual mode (i.e. open loop control) the first 40 seconds, and in automatic mode (closed loop control) thereafter.

$\omega_1 = 0.55$rad/s. This frequency is for illustration purpose chosen equal to the sensitivity bandwidth of the control system, cf. Figure 6.9. The setpoint y_{SP} is 50%. The control error can be read off as the difference between y_{SP} and y. In the first 40 seconds of the the simulation the PID controller is in manual mode, so the control loop is open. In the following 40 seconds the PID controller is in automatic mode, so the control loop is closed. We clearly see that the feedback control is effective to compensate for the disturbance at this frequency (0.55rad/s). The amplitude of the control error is 6.6 without feedback and 1.9 with feedback. Thus, the ratio between the closed loop error and the open loop error is $1.9/6.6 = 0.29$, which is in accordance with the amplitude of the sensitivity function at this frequency, cf. Figure 6.9.

Figure 6.12 shows the same kind of simulation, but with disturbance frequency $\omega_1 = 1.7$rad/s, which is higher than the sensitivity bandwidth, which is 0.55rad/s. From the simulations we see that closed loop control at this relatively high frequency, 1.7rad/s, does not compensate for the disturbance — actually the open loop works better. This is in accordance

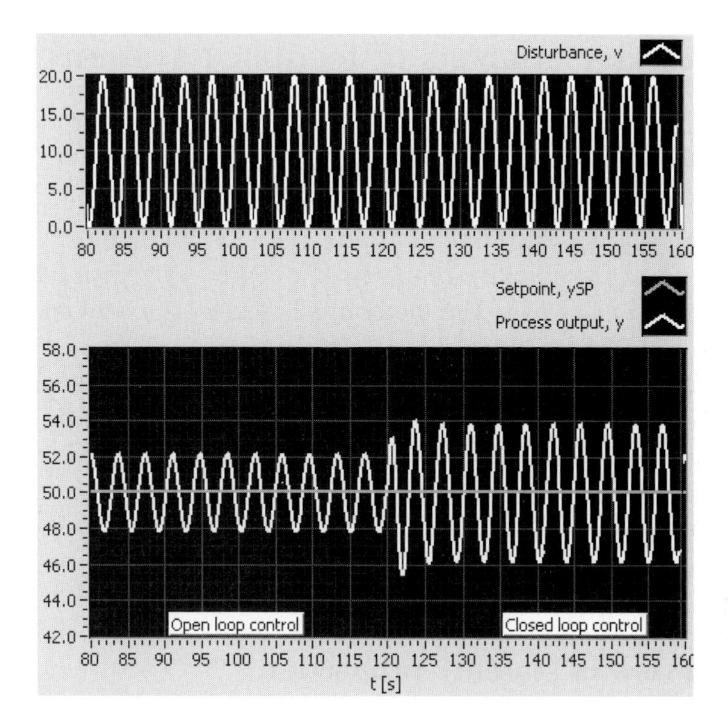

Figure 6.12: Example 6.6: Simulated responses of the control system. The disturbance v is sinusoidal with frequency $\omega_1 = 1.7$ rad/s. The PID-controller is in manual mode (i.e. open loop control) the first 40 seconds, and in automatic mode (closed loop control) thereafter.

with the fact that $|S(j\omega)|$ is greater than 1 at $\omega = 1.7$rad/s, cf. the Bode plot in Figure 6.9.

Finally, let us compare the simulated responses shown in Figure 6.12 and in Figure 6.8. The amplitude of the control error is less in Figure 6.12, despite the fact that the closed loop or feedback control is not efficient (at frequency 1.7 rad/s). The relatively small amplitude of the control error is due to the self regulation of the process, which means that the disturbance is attenuated through the process, whether the process is controlled or not.

[End of Example 6.6]

In Example 6.6 I did not choose the disturbance frequency, 1.7rad/s, by random. 1.7rad/s is actually the loop transfer function crossover frequency of the control system. Thus, the example demonstrates that the crossover frequency may give a poor measure of the performance of the control system. The sensitivity bandwidth is a better measure of the performance.

6.4 Stability analysis of feedback systems

6.4.1 Introduction

A control system must be *asymptotically stable.* A method for determination of the stability property of a control system will be described in the following. The method is based on the frequency response of the loop transfer function[6], $L(j\omega)$, and it is denoted *Nyquist's stability criterion.* This is a graphical analysis method. There are also algebraic analysis methods, as the Routh's stability criterion [4] which is based on the coefficients of the characteristic polynomial of the control system. Such stability analysis methods are not described in this book since they have limited practical importance (the mathematical operations become quite complicated except for the simplest models).

6.4.2 Review of stability properties

Let us review some basic facts from the stability theory (cf. e.g. [7]). Figure 6.13 shows the relation between the stability properties and the *impulse response* $h(t)$ of the system.

Figure 6.14 shows the relation between the stability properties and the placement of the transfer function *poles* in the complex plane. The relations shown in Figure 6.13 and Figure 6.14 are summarized below.

- **Asymptotically stable system**: The stationary impulse response is zero:

$$\lim_{t\to\infty} h(t) = 0 \tag{6.90}$$

  All the poles of system's transfer function lies in the left half plane, that is, all the poles have strictly negative real parts.

- **Marginally stable system**: The stationary impulse response is different from zero, but limited:

$$0 < \lim_{t\to\infty} h(t) < \infty \tag{6.91}$$

  One or more poles lie on the imaginary axis, that is, these poles have real part equal to zero, and none of these poles are multiple. No poles lies in the right half plane.

[6] the product of all the transfer functions in the control loop, cf. (6.25)

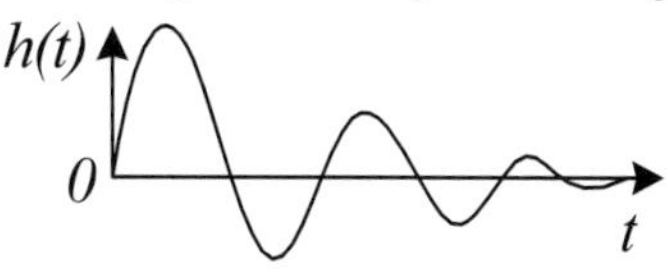

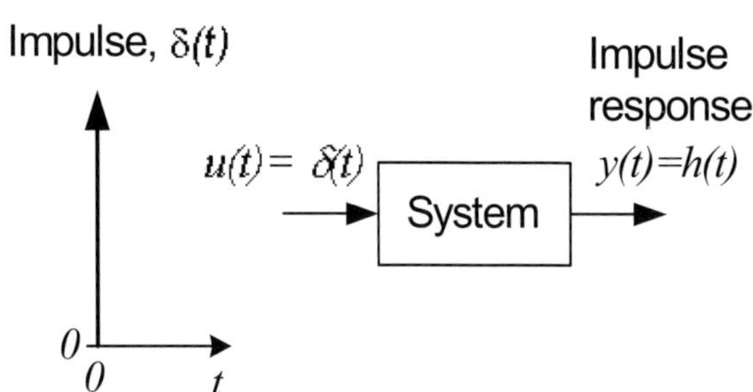

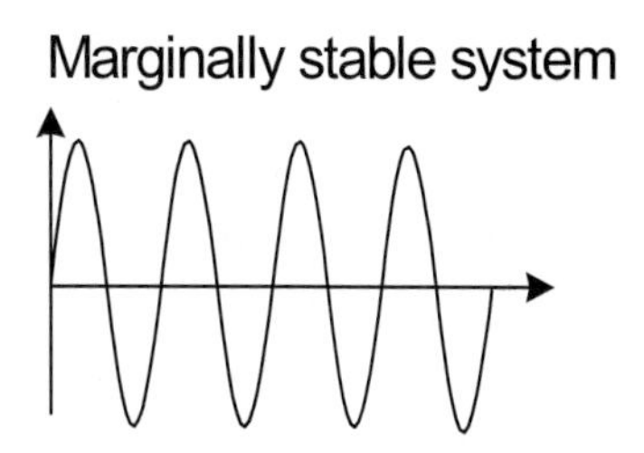

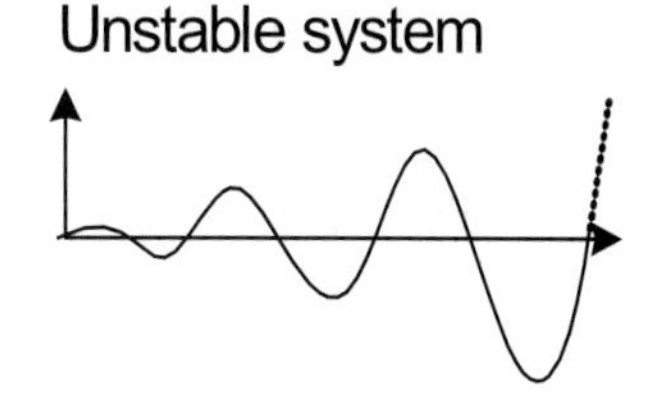

Figure 6.13: Illustration of the different stability properties expressed by the impulse response of the system

- **Unstable system**: The stationary impulse response is unlimited:

$$\lim_{t \to \infty} h(t) = \infty \tag{6.92}$$

At least one of the poles lies in the right half plane, that is, has real part larger than zero. Or: There are multiple poles on the imaginary axis.

Which is the transfer function to be used to determine the stability analysis of *control systems*? See Figure 6.3 (page 152). We must select a transfer function from one of the input signals to the closed loop to one of the output signals from the loop. Let us select the transfer function from the setpoint $y_{m_{SP}}$ to the process measurement y_m. This transfer function is

$$\frac{y_m(s)}{y_{m_{SP}}(s)} = \frac{H_c(s)H_u(s)H_m(s)}{1 + H_c(s)H_u(s)H_m(s)} = \frac{L(s)}{1 + L(s)} = T(s) \tag{6.93}$$

which is the *tracking transfer function* of the control system. (If we had selected some other transfer function, for example the transfer function

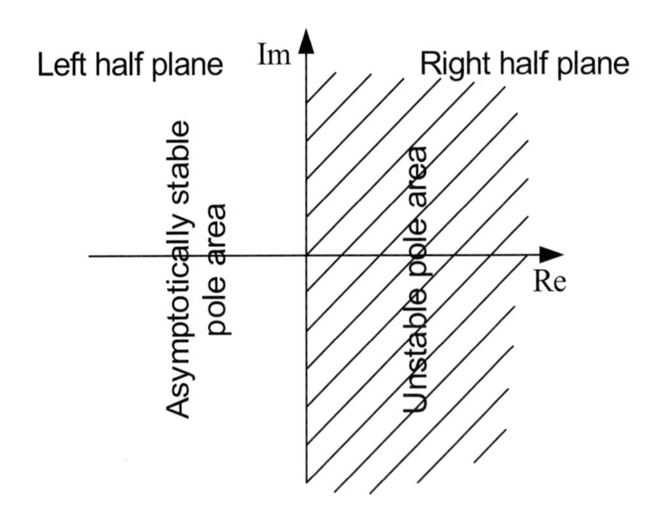

Figure 6.14: Relation between stability properties and pole placement in the complex plane

from the disturbance to the process output variable, the result of the analysis would have been the same.) $L(s)$ is the loop transfer function of the control system:

$$L(s) = H_c(s)H_u(s)H_m(s) \tag{6.94}$$

Figure 6.15 shows a compact block diagram of a control system. The transfer function from $y_{m_{SP}}$ to y_m is the tracking function:

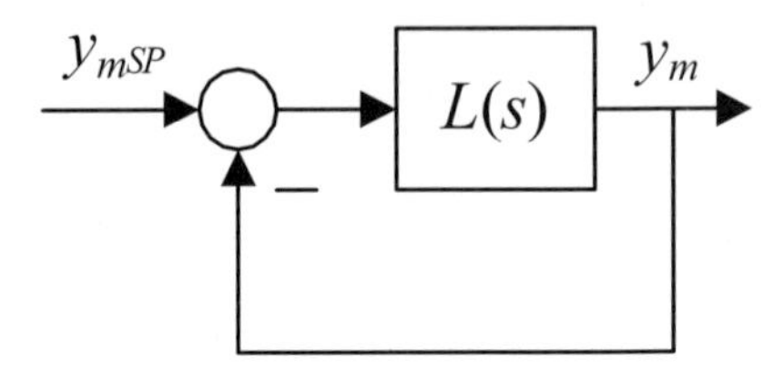

Figure 6.15: Compact block diagram of a control system with setpoint $y_{m_{SP}}$ as input variable and process measurement Y_m as output variable

$$T(s) = \frac{L(s)}{1 + L(s)} = \frac{\frac{n_L(s)}{d_L(s)}}{1 + \frac{n_L(s)}{d_L(s)}} = \frac{n_L(s)}{d_L(s) + n_L(s)} \tag{6.95}$$

where $n_L(s)$ and $d_L(s)$ are the numerator and denominator polynomials of $L(s)$, respectively. The *characteristic polynomial* of the tracking function is

$$c(s) = d_L(s) + n_L(s) \tag{6.96}$$

The stability of the control system is determined by the placement of the roots of (6.96) in the complex plan.

6.4.3 Nyquist's stability criterion

The Nyquist's stability criterion will now be derived. We start with a little rewriting: The roots of (6.96) are the same as the roots of

$$\frac{d_L(s) + n_L(s)}{d_L(s)} = 1 + \frac{n_L(s)}{d_L(s)} = 1 + L(s) = 0 \tag{6.97}$$

which, therefore, too can be denoted the characteristic equation of the control system. (6.97) is the equation from which the Nyquist's stability criterion will be derived. In the derivation we will use the Argument variation principle:

Argument variation principle: Given a function $f(s)$ where s is a complex number. Then $f(s)$ is a complex number, too. As with all complex numbers, $f(s)$ has an angle or argument. If s follows a closed contour Γ (gamma) in the complex s-plane which encircles a number of poles and a number of zeros of $f(s)$, see Figure 6.16, then the following applies:

$$\arg_{\Gamma} f(s) = 360^{\circ} \cdot (\text{number of zeros minus number of poles of } f(s) \text{ inside } \Gamma) \tag{6.98}$$

where $\arg_{\Gamma} f(s)$ means the change of the angle of $f(s)$ when s has followed Γ once in positive direction of circulation (i.e. clockwise).

For our purpose, we let the function $f(s)$ in the Argument variation principle be

$$f(s) = 1 + L(s) \tag{6.99}$$

The Γ contour must encircle the entire right half s-plane, so that we are certain that all poles and zeros of $1 + L(s)$ are encircled. From the Argument Variation Principle we have:

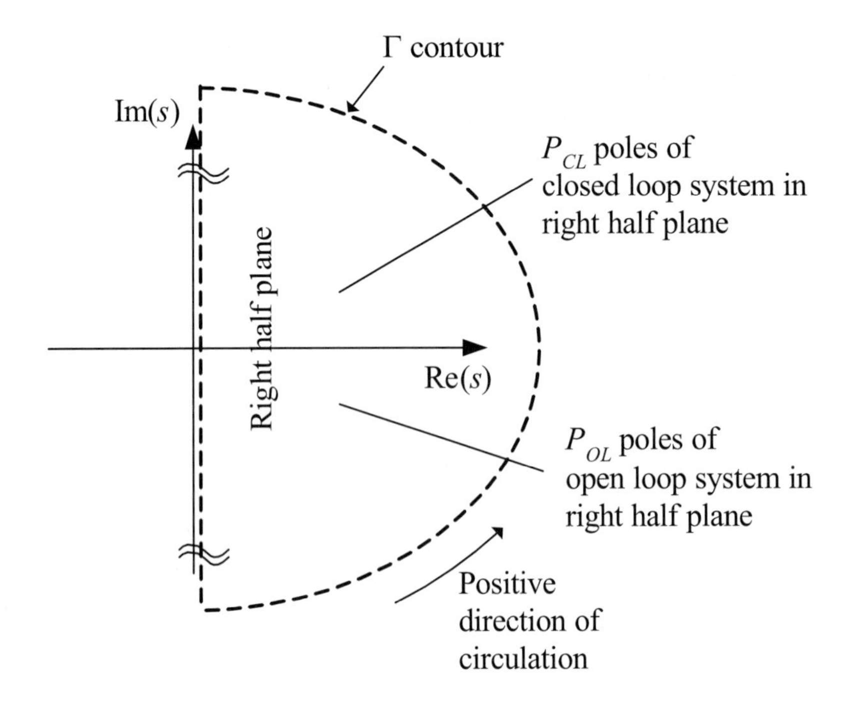

Figure 6.16: s shall follow the Γ contour once in positive direction (counter clockwise).

$$\begin{aligned}
\arg_{\Gamma}[1+L(s)] &= \arg_{\Gamma}\frac{d_L(s)+n_L(s)}{d_L(s)} && (6.100)\\
&= 360^\circ \cdot (\text{number of roots of } (d_L+n_L) \text{ in RHP} \\
&\quad \text{minus number roots of } d_L \text{ in RHP}) && (6.101)\\
&= 360^\circ \cdot (\text{number poles of closed loop system in RHP} \\
&\quad \text{minus number poles of open system in RHP}) \\
&= 360^\circ \cdot (P_{CL} - P_{OL}) && (6.102)
\end{aligned}$$

where RHP means right half plane. By "open system" we mean the (imaginary) system having transfer function $L(s) = n_L(s)/d_L(s)$, i.e., the original feedback system with the feedback broken. The poles of the open system are the roots of $d_L(s) = 0$.

Finally, we can formulate the Nyquist's stability criterion. But before we do that, we should remind ourselves what we are after, namely to be able to determine the number poles P_{CL} of the closed loop system in RHP. It those poles which determines whether the closed loop system (the control system) is asymptotically stable or not. *If $P_{CL} = 0$ the closed loop system is asymptotically stable.*

Nyquist's stability criterion: Let P_{OL} be the number of poles of the open system in the right half plane, and let $\arg_\Gamma L(s)$ be the angular change of the vector $L(s)$ as s have followed the Γ contour once in positive direction of circulation. Then, the number poles P_{CL} of the closed loop system in the right half plane, is

$$P_{CL} = \frac{\arg_\Gamma L(s)}{360°} + P_{OL} \tag{6.103}$$

If $P_{CL} = 0$, the closed loop system is asymptotically stable.

Let us take a closer look at the terms on the right side of (6.103): P_{OL} are the roots of $d_L(s)$, and there should not be any problem calculating these roots. To determine the angular change of the vector $1 + L(s)$. Figure 6.17 shows how the vector (or complex number) $1 + L(s)$ appears in a *Nyquist diagram* for a typical plot of $L(s)$. A Nyquist diagram is simply a Cartesian diagram of the complex plane in which L is plotted. $1 + L(s)$ is the vector from the point $(-1,\ 0j)$, which is denoted the *critical point*, to the Nyquist curve of $L(s)$.

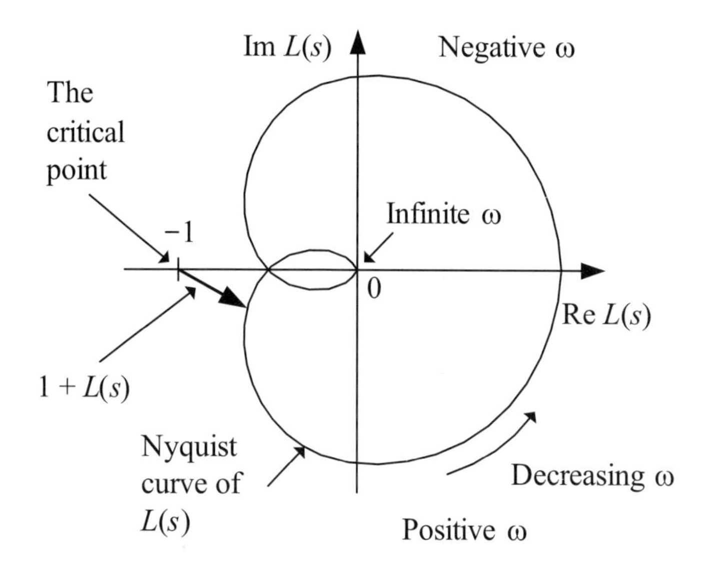

Figure 6.17: Typical Nyquist curve of $L(s)$. The vector $1 + L(s)$ is drawn.

More about the Nyquist curve of $L(j\omega)$

Let us take a more detailed look at the Nyquist curve of L as s follows the Γ contour in the s-plane, see Figure 6.16. In practice, the denominator polynomial of $L(s)$ has higher order than the numerator polynomial. This implies that $L(s)$ is mapped to the origin of the Nyquist diagram when $|s| = \infty$. Thus, the whole semicircular part of the Γ contour is mapped to the origin.

The imaginary axis constitutes the rest of the Γ contour. How is the mapping of $L(s)$ as s runs along the imaginary axis? On the imaginary axis $s = j\omega$, which implies that $L(s) = L(j\omega)$, which *is the frequency response of* $L(s)$. A consequence of this is that we can in principle determine the stability property of a feedback system by just looking at the frequency response of the open system, $L(j\omega)$.

ω has negative values when $s = j\omega$ is on the negative imaginary axis. For $\omega < 0$ the frequency response has a mathematical meaning. From general properties of complex functions,

$$|L(-j\omega)| = |L(j\omega)| \tag{6.104}$$

and

$$\angle L(-j\omega) = -\angle L(j\omega) \tag{6.105}$$

Therefore the Nyquist curve of $L(s)$ for $\omega < 0$ will be identical to the Nyquist curve of $\omega > 0$, but mirrored about the real axis. Thus, we only need to know how $L(j\omega)$ is mapped for $\omega \geq 0$. The rest of the Nyquist curve then comes by itself! Actually we need not draw more of the Nyquist curve (for $\omega > 0$) than what is sufficient for determining if the critical point is encircled or not.

We must do some extra considerations if some of the poles in $L(s)$, which are the poles of the open loop system, lie in the origin. This corresponds to pure integrators in control loop, which is a common situation in feedback control systems because the controller usually has integral action, as in a PI or PID controller. If $L(s)$ contains integrators, the Γ contour must go *outside* the origo. But to the left or to the right? We choose to the right, see Figure 6.18. (We have thereby decided that the origin belongs to the left half plane. This implies that P_{OL} does not count these poles.) The radius of the semicircle around origin is arbitrarily small. The Nyquist curve then becomes as shown in the diagram to the right in the same figure. The arbitrarily small semicircle in the s-plane is mapped to an infinitely large semicircle in the L-plane. The is because as $s \to 0$, the loop transfer function is approximately

$$L(s) \approx \frac{K}{s}$$

(if we assume one pole in the origin). On the small semicircle,

$$s = re^{j\theta} \tag{6.106}$$

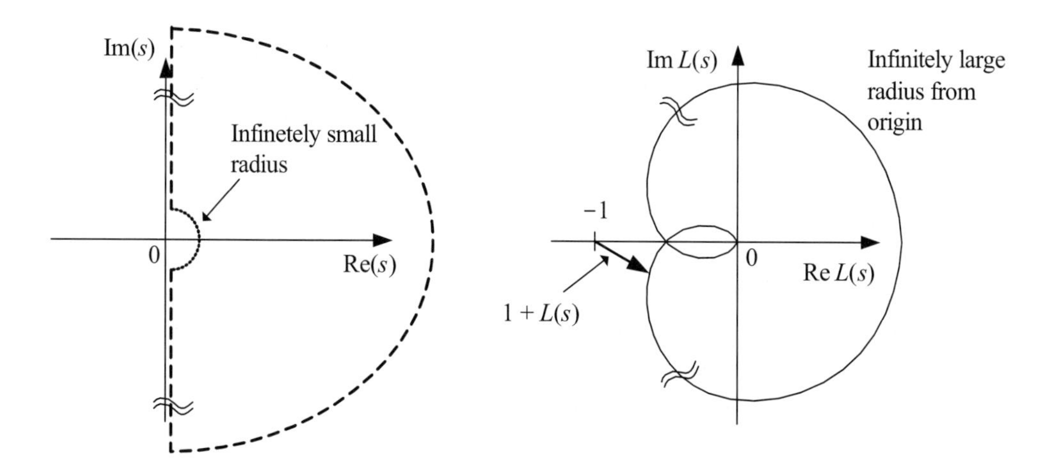

Figure 6.18: Left diagram: If $L(s)$ has a pole in origin, the Γ contour must pass the origin along an arbitrarily small semicircle to the right. Right diagram: A typical Nyquist curve of L.

which gives

$$L(s) \approx \frac{K}{r} e^{-j\theta} \tag{6.107}$$

When $r \to 0$ and when simultaneously θ goes from $+90°$ via $0°$ to $-90°$, the Nyquist curve becomes an infinitely large semicircle, as shown.

The Nyquist's stability criterion for non-rational transfer functions

The Nyquist's stability criterion gives information about the poles of feedback systems. So far it has been assumed that the loop transfer function $L(s)$ is a rational transfer function. What if $L(s)$ is irrational? Here is one example:

$$L(s) = \frac{1}{s} e^{-\tau s} \tag{6.108}$$

where $e^{-\tau s}$ represents time delay. In such cases the tracking ratio $T(s)$ will also be irrational, and the definition of poles does not apply to such irrational transfer functions. Actually, the Nyquist's stability criterion can be used as a graphical method for determining the stability property on basis of the frequency response $L(j\omega)$.

Nyquist's special stability criterion

In most cases the open system is stable, that is, $P_{OL} = 0$. (6.103) then becomes

$$P_{CL} = \frac{\arg_{\Gamma}[L(s)]}{360°} \tag{6.109}$$

This implies that the feedback system is asymptotically stable if the Nyquist curve does not encircle the critical point. This is the *Nyquist's special stability criterion* or the *Nyquist's stability criterion for open stable systems.*

The Nyquist's special stability criterion can also be formulated as follows: *The feedback system is asymptotically stable if the Nyquist curve of L has the critical point on its left side for increasing ω.*

Another way to formulate Nyquist's special stability criterion involves the *amplitude crossover frequency* ω_c and the *phase crossover frequency* ω_{180}. ω_c is the frequency at which the $L(j\omega)$ curve crosses the unit circle, while ω_{180} is the frequency at which the $L(j\omega)$ curve crosses the negative real axis. In other words:

$$|L(j\omega_c)| = 1 \tag{6.110}$$

and

$$\arg L(j\omega_{180}) = -180° \tag{6.111}$$

See Figure 6.19. Note: The Nyquist diagram contains no explicit frequency axis. We can now determine the stability properties from the relation between these two crossover frequencies:

- Asymptotically stable closed loop system: $\omega_c < \omega_{180}$
- Marginally stable closed loop system: $\omega_c = \omega_{180}$
- Unstable closed loop system: $\omega_c > \omega_{180}$

Stability margins in terms of gain margin GM and phase margin PM

An asymptotically stable feedback system may become marginally stable if the loop transfer function changes. The *gain margin* GM and the *phase margin* PM [radians or degrees] are *stability margins* which in their own ways expresses how large parameter changes can be tolerated before an asymptotically stable system becomes marginally stable. Figure 6.20 shows the stability margins defined in the Nyquist diagram. GM is the

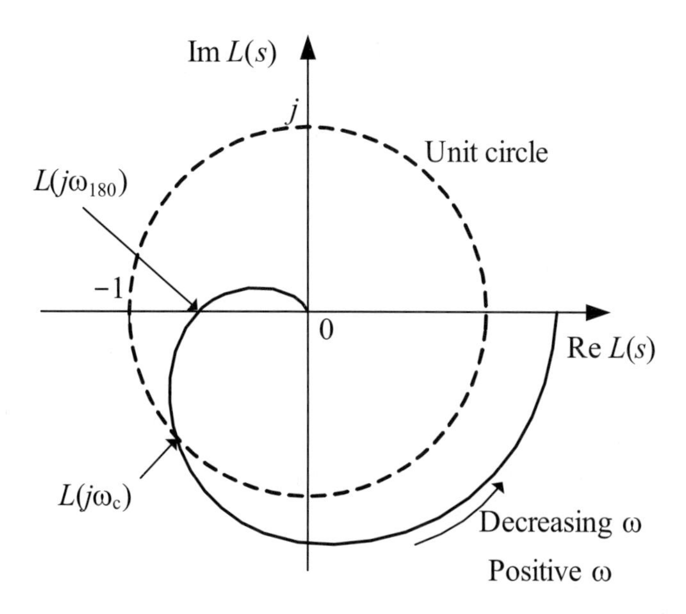

Figure 6.19: Definition of amplitude crossover frequency ω_c and phase crossover frequency ω_{180}

(multiplicative, not additive) increase of the gain that L can tolerate at ω_{180} before the L curve (in the Nyquist diagram) passes through the critical point. Thus,

$$|L(j\omega_{180})| \cdot GM = 1 \tag{6.112}$$

which gives

$$GM = \frac{1}{|L(j\omega_{180})|} = \frac{1}{|\mathrm{Re}\, L(j\omega_{180})|} \tag{6.113}$$

(The latter expression in (6.113) is because at ω_{180}, $\mathrm{Im}\, L = 0$ so that the amplitude is equal to the absolute value of the real part.)

If we use decibel as the unit (like in the Bode diagram which we will soon encounter), then

$$GM \text{ [dB]} = -\,|L(j\omega_{180})| \text{ [dB]} \tag{6.114}$$

The phase margin PM is the phase reduction that the L curve can tolerate at ω_c before the L curve passes through the critical point. Thus,

$$\arg L(j\omega_c) - PM = -180^\circ \tag{6.115}$$

which gives

$$PM = 180^\circ + \arg L(j\omega_c) \tag{6.116}$$

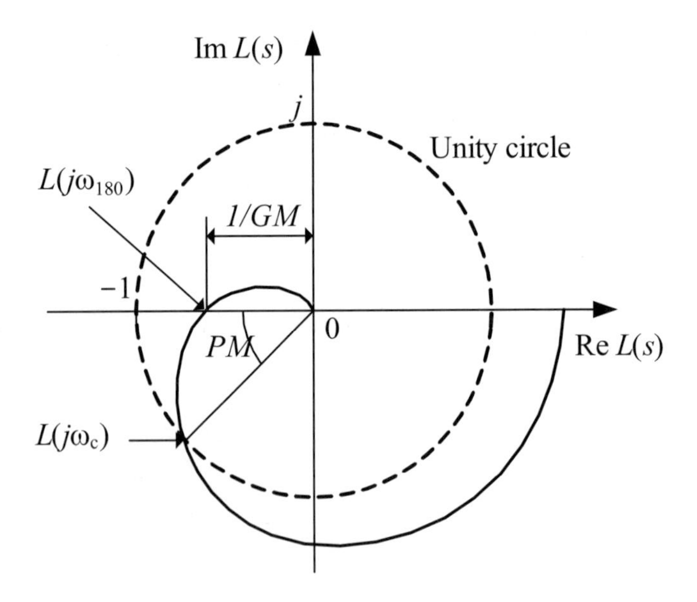

Figure 6.20: Gain margin GM and phase margin PM defined in the Nyquist diagram

We can now state as follows: The feedback (closed) system is asymptotically stable if

$$GM > 0\text{dB} \ = 1 \text{ and } PM > 0^\circ \tag{6.117}$$

This criterion is often denoted the *Bode-Nyquist stability criterion.*

Reasonable ranges of the stability margins are

$$2 \approx 6\text{dB} \leq GM \leq 4 \approx 12\text{dB} \tag{6.118}$$

and

$$30^\circ \leq PM \leq 60^\circ \tag{6.119}$$

The larger values, the better stability, but at the same time the system becomes more sluggish, dynamically. If you are to use the stability margins as design criterias, you can use the following values (unless you have reasons for specifying other values):

$$GM \geq 2.5 \approx 8\text{dB and } PM \geq 45^\circ \tag{6.120}$$

For example, the controller gain, K_p, can be adjusted until one of the inequalities becomes an equality.[7]

[7]But you should definitely check the behaviour of the control system by simulation, if possible.

It can be shown[8] that for $PM \leq 70°$, the damping of the feedback system approximately corresponds to that of a second order system with relative damping factor

$$\zeta \approx \frac{PM}{100°} \tag{6.121}$$

For example, $PM = 50° \sim \zeta = 0.5$.

Stability margins in terms of maximum sensitivity amplitude, $|S(j\omega)|_{\max}$

An alternative quantity of a stability margin, is the minimum distance from the $L(j\omega)$ curve to the critical point. This distance is $|1 + L(j\omega)|$, see Figure 6.21. So, we can use the minimal value of $|1 + L(j\omega)|$ as a stability

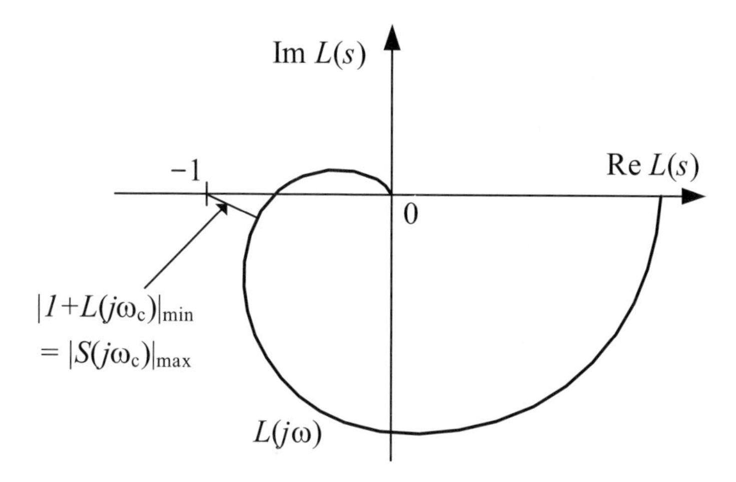

Figure 6.21: The minimum distance between the $L(j\omega)$ curve and the critical point can be interpreted as a stability margin. This distance is $|1 + L|_{\min} = |S|_{\max}$.

margin. However, it is more common to take the inverse of the distance: Thus, a stability margin is the *maximum* value of $1/|1 + L(j\omega)|$. And since $1/[1 + L(s)]$ is the sensitivity function $S(s)$, then $|S(j\omega)|_{\max}$ represents a stability margin. Reasonable values are in the range

$$1.5 \approx 3.5\text{dB} \leq |S(j\omega)|_{\max} \leq 3.0 \approx 9.5\text{dB} \tag{6.122}$$

If you use $|S(j\omega)|_{\max}$ as a criterion for adjusting controller parameters, you can use the following criterion (unless you have reasons for some other

[8] The result is based on the assumption that the loop transfer function is $L(s) = \omega_0^2/[(s + 2\zeta\omega_0)s]$ which gives tracking transfer function $T(s) = L(s)/[1 + L(s)] = \omega_0^2/\left[s^2 + 2\zeta\omega_0 s + \omega_0^2\right]$. The phase margin PM can be calculated from $L(s)$.

specification):

$$|S(j\omega)|_{\max} = 2.0 \approx 6\text{dB} \tag{6.123}$$

Frequency of the sustained oscillations

There are sustained oscillations in a marginally stable system. *The frequency of these oscillations is $\omega_c = \omega_{180}$.*This can be explained as follows: In a marginally stable system, $L(\pm j\omega_{180}) = L(\pm j\omega_c) = -1$. Therefore, $d_L(\pm j\omega_{180}) + n_L(\pm j\omega_{180}) = 0$, which is the characteristic equation of the closed loop system with $\pm j\omega_{180}$ inserted for s. Therefore, the system has $\pm j\omega_{180}$ among its poles. The system usually have additional poles, but they lie in the left half plane. The poles $\pm j\omega_{180}$ leads to sustained sinusoidal oscillations. Thus, ω_{180} (or ω_c) is the frequency of the sustained oscillations in a marginally stable system.

6.4.4 Stability analysis in a Bode diagram

It is most common to use a Bode diagram for frequency response based stability analysis of closed loop systems. The Nyquist's Stability Criterion says: The closed loop system is marginally stable if the Nyquist curve (of L) goes through the critical point, which is the point $(-1, 0)$. But where is the critical point in the Bode diagram? The critical point has phase (angle) $-180°$ and amplitude $1 = 0$dB. The critical point therefore constitutes two lines in a Bode diagram: The 0dB line in the amplitude diagram and the $-180°$ line in the phase diagram. Figure 6.22 shows typical L curves for an asymptotically stable closed loop system. In the figure, GM, PM, ω_c and ω_{180} are indicated.

Example 6.7 ***Stability analysis of a feedback control system***

Given a feedback control system with structure as shown in Figure 6.23. The loop transfer function is

$$L(s) = H_c(s)H_p(s) = \underbrace{K_p}_{H_c(s)} \underbrace{\frac{1}{(s+1)^2 s}}_{H_p(s)} = \frac{K_p}{(s+1)^2 s} = \frac{n_L(s)}{d_L(s)} \tag{6.124}$$

We will determine the stability property of the control system for different values of the controller gain K_p in three ways: Pole placement, Nyquist's

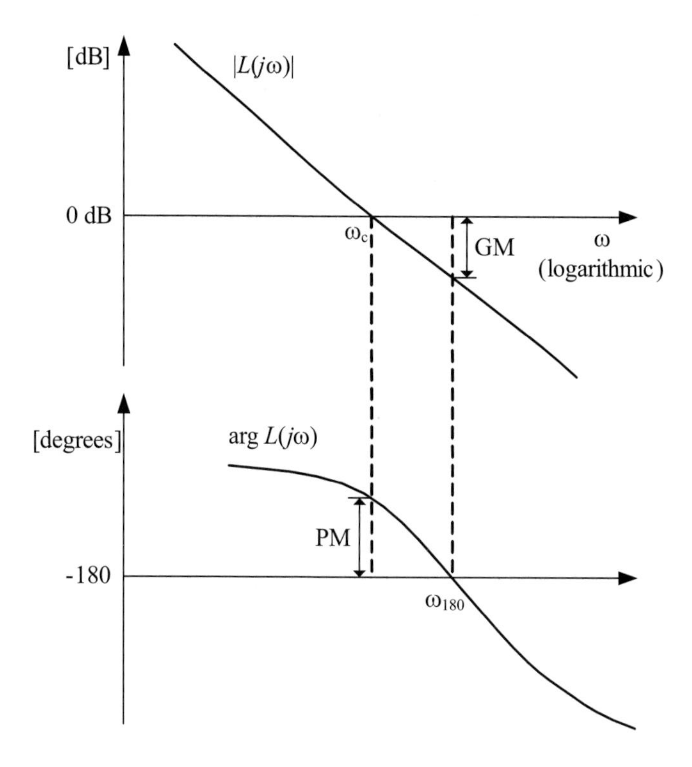

Figure 6.22: Typical L curves of an asymptotically stable closed loop system with GM, PM, ω_c and ω_{180} indicated

Stability Criterion, and simulation. The tracking transfer function is

$$T(s) = \frac{y_m(s)}{y_{m_{SP}}(s)} = \frac{L(s)}{1 + L(s)} = \frac{n_L(s)}{d_L(s) + n_L(s)} = \frac{K_p}{s^3 + 2s^2 + s + K_p} \tag{6.125}$$

The characteristic polynomial is

$$c(s) = s^3 + 2s^2 + s + K_p \tag{6.126}$$

Figures 6.24 – 6.26 show the step response after a step in the setpoint, the poles, the Bode diagram and Nyquist diagram for three K_p values which result in different stability properties. The detailed results are shown below.

- $K_p = 1$: Asymptotically stable system, see Figure 6.24. From the Bode diagram we read off stability margins $GM = 6.0\text{dB} = 2.0$ and $PM = 21°$. we see also that $|S(j\omega)|_{\max} = 11$ dB $= 3.5$ (a large value, but it corresponds with the small the phase margin of $PM = 20°$).
- $K_p = 2$: Marginally stable system, see Figure 6.25. From the Bode diagram, $\omega_c = \omega_{180}$. The L curve goes through the critical point in

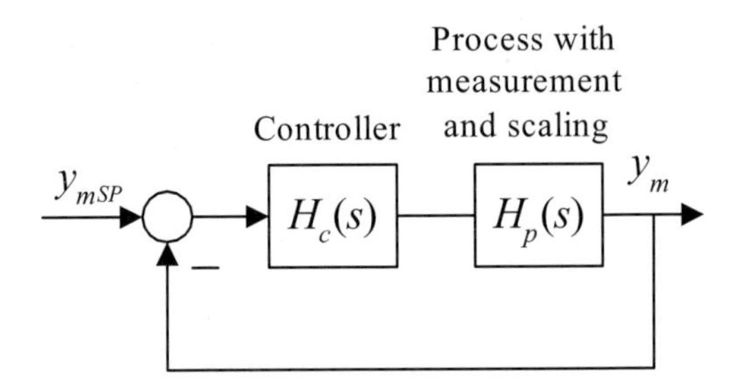

Figure 6.23: Example 6.7: Block diagram of feedback control system

the Nyquist diagram. $|S|_{\max}$ has infinitely large value (since the minimum distance, $1/|S|_{\max}$, between $|L|$ and the critical point is zero).

Let us calculate the period T_p of the undamped oscillations: Since $\omega_{180} = 1.0$rad/s, the period is $T_p = 2\pi/\omega_{180} = 6.28$s, which fits well with the simulation shown in Figure 6.25.

- $K_p = 4$: Unstable system, see Figure 6.26. From the Bode diagram, $\omega_c > \omega_{180}$. From the Nyquist diagram we see that the L curve passes outside the critical point. (The frequency response curves of M and N have no physical meaning in this the case.)

[End of Example 6.7]

6.4.5 Stability margins and robustness

Per definition the stability margins expresses the robustness of the feedback control system against certain parameter changes in the loop transfer function:

- *The gain margin GM* is how much the loop gain, K, can increase before the system becomes unstable. For example, is $GM = 2$ when $K = 1.5$, the control system becomes unstable for K larger than $1.5 \cdot 2 = 3.0$.

- *The phase margin PM* is how much the phase lag function of the loop can be reduced before the loop becomes unstable. One reason of reduced phase is that the time delay in control loop is increased. A change of the time delay by $\Delta\tau$ introduces the factor $e^{-\Delta\tau s}$ in $L(s)$ and contributes to $\arg L$ with $-\Delta\tau \cdot \omega$ [rad] or $-\Delta\tau \cdot \omega\frac{180^{\circ}}{\pi}$ [deg]. $|L|$

is however not influenced because the amplitude function of $e^{-\tau s}$ is 1, independent of the value of τ. The system becomes unstable if the time delay have increased by $\Delta\tau_{\max}$ such that[9]

$$PM = \Delta\tau_{\max} \cdot \omega_c \frac{180^\circ}{\pi} \text{ [deg]} \tag{6.127}$$

which gives the following maximum change of the time delay:

$$\Delta\tau_{\max} = \frac{PM}{\omega_c} \frac{\pi}{180^\circ} \tag{6.128}$$

If you want to calculate how much the phase margin PM is reduced if the time delay is increased by $\Delta\tau$, you can use the following formula which stems from (6.127):

$$\Delta PM = \Delta\tau \cdot \omega_c \frac{180^\circ}{\pi} \text{ [deg]} \tag{6.129}$$

For example, assume that a given control system has $\omega_c = 0.2\text{rad/min}$ and $PM = 50^\circ$. If the time delay increases by 1min, the phase margin is reduced by $\Delta PM = 1 \cdot 0.2 \frac{180^\circ}{\pi} = 11.4^\circ$, i.e. from 50° to 38.6°.

[9]Remember that PM is found at ω_c.

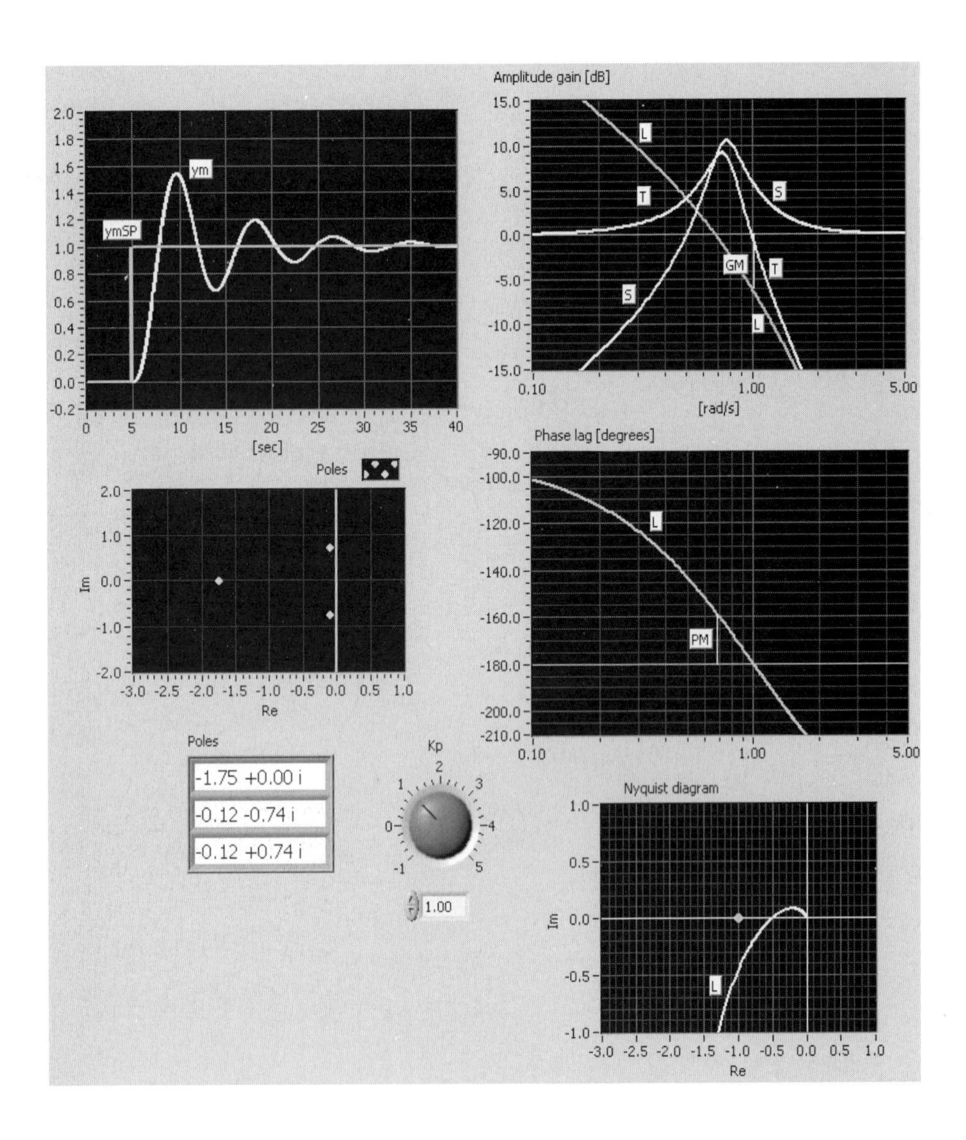

Figure 6.24: Example 6.7: Step response (step in setpoint), poles, Bode diagram and Nyquist diagram with $K_p = 1$. The control system is asymptotically stable.

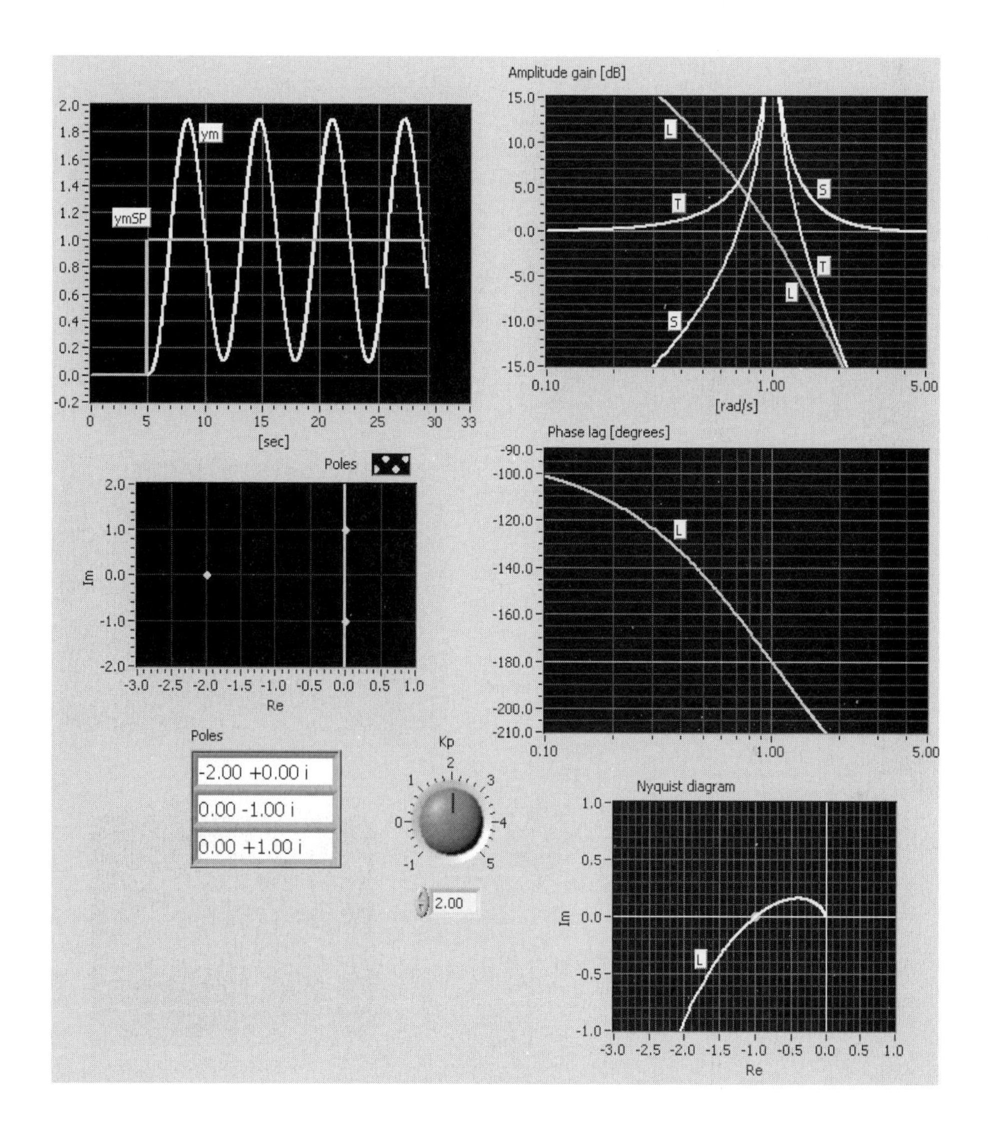

Figure 6.25: Example 6.7: Step response (step in setpoint), poles, Bode diagram and Nyquist diagram with $K_p = 2$. The control system is marginally stable.

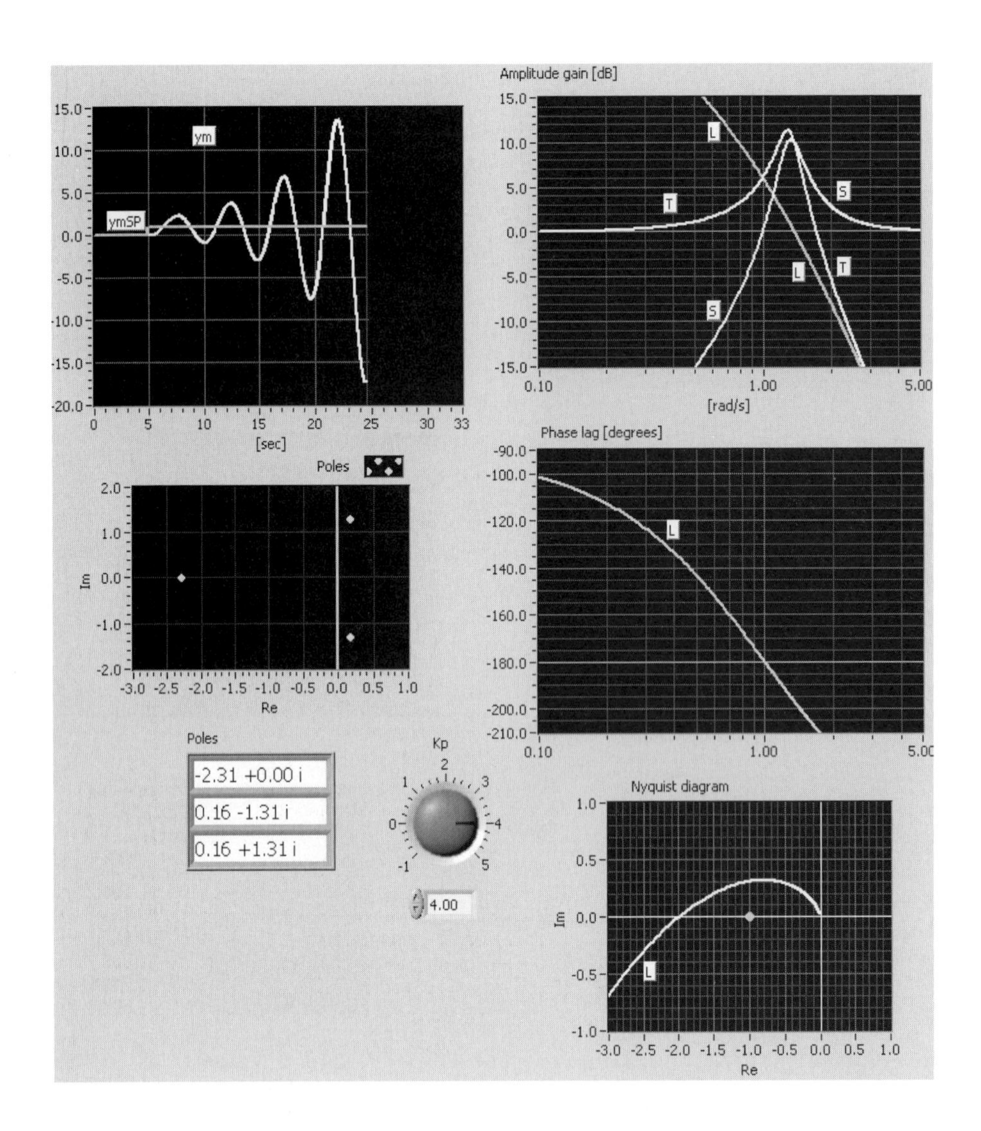

Figure 6.26: Example 6.7: Step response (step in setpoint), poles, Bode diagram and Nyquist diagram with $K_p = 4$. The control system is unstable.

Chapter 7

Transfer function based PID tuning

7.1 Introduction

This chapter describes PID tuning methods based on *transfer function models* of the process to be controlled. It is assumed that the control system has a transfer function block diagram as shown in Figure 7.1. As a

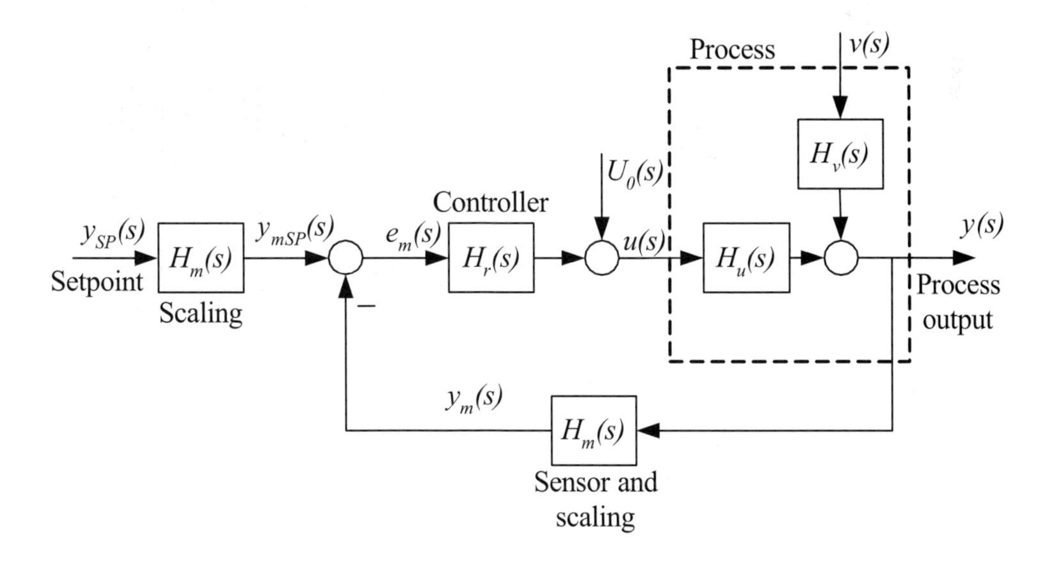

Figure 7.1: Transfer function based block diagram of the control system

basis for the design of the controllers we will work with the combined, equivalent block diagram shown in Figure 7.2. The sensor and scaling block is combined with the process blocks. The two combined process transfer functions then becomes

$$\frac{y_m(s)}{u(s)} = \underbrace{H_u(s) \cdot H_m(s)}_{H_p(s)} \tag{7.1}$$

and

$$\frac{y_m(s)}{v(s)} = \underbrace{H_v(s) \cdot H_m(s)}_{H_{vm}(s)} \tag{7.2}$$

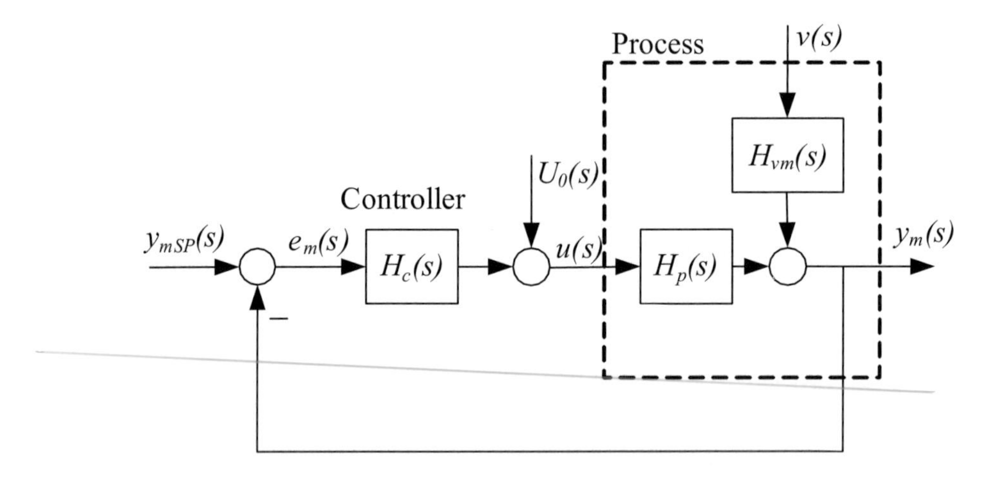

Figure 7.2: Combined, equivalent block diagram of the control system showing the process measurement as the variable to be controlled (cf. Figure 7.1)

Some comments about the disturbance acting on the process: In most processes the dominating disturbance influences the process in the same way, dynamically, as the control variable. Such a disturbance is called an *input disturbance* . Here are a few examples:

- Liquid tank: The control variable controls the inflow. The outflow is a disturbance.
- Motor: The control variable controls the motor torque. The load torque is a disturbance.
- Thermal process: The control variable controls the power supply via an heating element. The power loss via heat transfer through the walls and heat outflow through the outlet are disturbances.

In the cases above (with input disturbances) the dynamic properties represented by $H_p(s)$ and $H_{vm}(s)$ are similar. For example, the following transfer functions represents the same dynamics if the time constants T and T_{vm} are equal:

$$H_p(s) = \frac{K}{Ts+1} \tag{7.3}$$

$$H_{vm}(s) = \frac{K_{vm}}{T_{vm}s+1} = \frac{K_{vm}}{Ts+1} \tag{7.4}$$

In the simulations shown in this chapter, it is assumed that the disturbance is an input disturbance type, with as described above.

Let us repeat some basic relations: For a control system as shown in Figure 7.2, the Laplace transform of the control error is, cf. Section 6.3.3,

$$e_m(s) = \underbrace{S(s)y_{mSP}(s)}_{e_{y_{mSP}}(s)} \underbrace{-S(s)H_{vm}(s)v(s)}_{e_{vm}(s)} \tag{7.5}$$

where $S(s)$ is the sensitivity function given by

$$S(s) = \frac{1}{1+L(s)} \tag{7.6}$$

where $L(s)$ is the loop transfer function, here given by

$$L(s) = H_c(s)H_p(s) \tag{7.7}$$

The tracking transfer function is

$$T(s) = \frac{L(s)}{1+L(s)} \tag{7.8}$$

In most control systems the controller should have *integral action*, since it brings the static (steady-state) control error to zero for a constant setpoint and a constant disturbance, cf. Section 6.3.3. Therefore we will consider only PI and PID controllers (not P and PD controllers) in this chapter.

In this chapter formulas for the controller parameters will be presented. Remember to always check that the control system behaves as expected by running *simulations*. The model used in the simulator should include nonlinearities, as saturation limits, which have been neglected in the controller design.

7.2 Controller tuning from specified characteristic polynomial

7.2.1 Introduction

The subsequent sections explain controller tuning based on specifications of the characteristic polynomial of the control system. Using this method you can shape the dynamic properties of the control system quite freely. However, the method is in practice applicable only to processes of low order due to the mathematical operations involved, and here only integrator processes and first order (time constant) process will be considered.

If the process order is high, or if the process contains time delay, you should consider using the Ziegler-Nichols' tuning methods, cf. Chapter 4 (Ziegler-Nichols' tuning methods actually can not be used for integrators or first order processes since the parameters needed in the methods, as the ultimate gain, can not be found or is infinitely large for these processes).

For all the processes that we soon will encounter (integrator and first order system), Skogestad's method, cf. Section 7.5, can be used. Although this tuning method certainly works fine, the method is based on some model approximations. In some cases it is useful to be able to perform an exact controller design. One important example is the level controller design for a liquid tank, cf. Example 7.2.

7.2.2 Tuning a controller for an integrator process

The process transfer function is

$$H_p(s) = \frac{K}{s} \tag{7.9}$$

and the disturbance transfer function is

$$H_{vm}(s) = \frac{K_{vm}}{s} \tag{7.10}$$

One example of such a process is a liquid tank where the level h is to be controlled by controlling the outflow w_{out} from the tank. The transfer function from w_{out} to level measurement h_m is on the form (7.9). (This example is described in detail in Example 7.1 (page 194).)

We will use a PI controller (the derivative term in the PID controller serves no purpose for this process), which has transfer function

$$H_c(s) = K_p \frac{T_i s + 1}{T_i s} \tag{7.11}$$

The controller parameters K_p and T_i will be calculated from a specified *bandwidth*, which represents the speed of the control system. In addition we must require that the control system has acceptable *stability*. We start by finding the tracking transfer function $T(s)$, which is given by (7.8) where the loop transfer function is

$$L(s) = H_c(s)H_p(s) = K_p \frac{T_i s + 1}{T_i s} \cdot \frac{K}{s} \tag{7.12}$$

From (7.8) we get

$$T(s) = \frac{L(s)}{1 + L(s)} = \frac{K_p K \left(s + \frac{1}{T_i}\right)}{\underbrace{s^2 + K_p K s + \frac{K_p K}{T_i}}_{c(s)}} \tag{7.13}$$

where $c(s)$ is the characteristic polynomial of the control system. We write it as a standard second order polynomial:

$$c(s) = s^2 + K_p K s + \frac{K_p K}{T_i} = s^2 + 2\zeta\omega_0 s + {\omega_0}^2 \tag{7.14}$$

where ω_0 is the undamped resonance frequency and ζ is the relative damping factor [7]. Comparison of coefficients between the two polynomials in (7.14) gives the following identities:

$$K_p K \equiv 2\zeta\omega_0 \text{ and } \frac{K_p K}{T_i} \equiv {\omega_0}^2 \tag{7.15}$$

Solving for K_p and T_i gives the following formulas for the controller parameters:

$$K_p = \frac{2\zeta\omega_0}{K} \tag{7.16}$$

$$T_i = \frac{2\zeta}{\omega_0} \tag{7.17}$$

Using (7.16) and (7.17), $T(s)$ can be written as

$$T(s) = \frac{K_p K \left(s + \frac{1}{T_i}\right)}{s^2 + K_p K s + \frac{K_p K}{T_i}} = \frac{2\zeta\omega_0 s + {\omega_0}^2}{s^2 + 2\zeta\omega_0 s + {\omega_0}^2} \tag{7.18}$$

ω_0 can be interpreted as the bandwidth of the tracking function (7.18). A rough estimate of the response time[1] of the control system is

$$T_r \approx \frac{1}{\omega_0} \tag{7.19}$$

A reasonable choice of ζ is

$$\zeta = 0.5 \tag{7.20}$$

which gives step responses with well damped oscillations. If larger damping of the time responses is desired, ζ can be given a larger value (closer to 1).

Example 7.1 ***PI control of an integrator process***

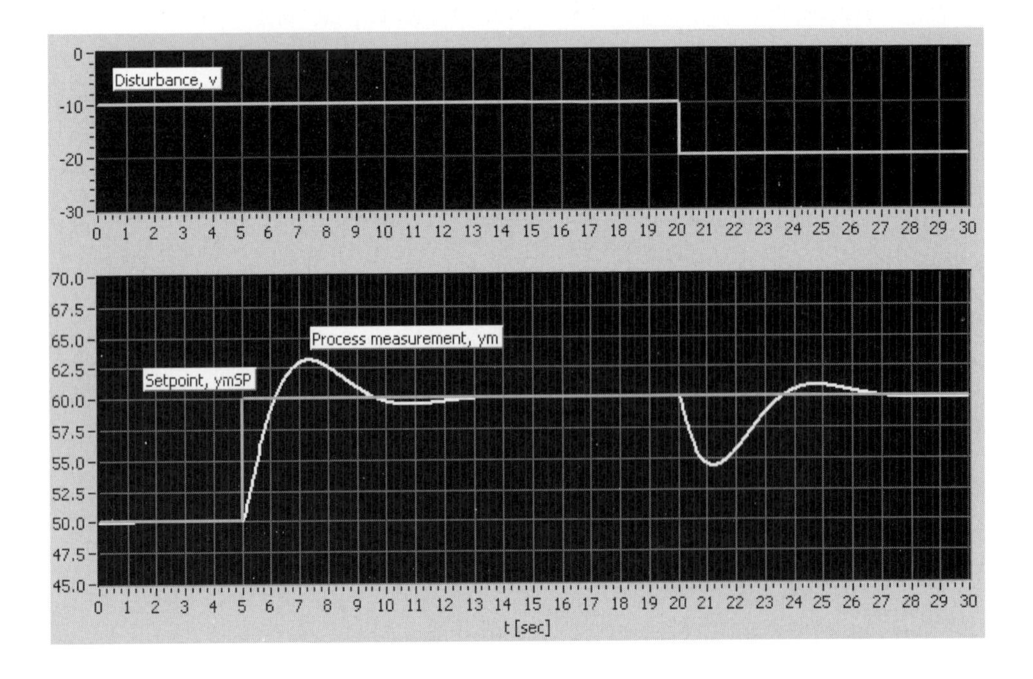

Figure 7.3: Example 7.1: Simulated responses in the control system

Assume that $K = 1$ and $K_{vm} = 1$ in (7.9) and (7.10). We specify $\omega_0 = 1$ and $\zeta = 0.5$. (7.16) and (7.17) gives

$$K_p = 1; T_i = 1 \tag{7.21}$$

Figure 7.3 shows simulated responses in a control system with transfer functions (7.9) and (7.10). There is a setpoint step and a disturbance step. The simulations indicates that the stability of the control system is

[1] The response time can be regarded as an approximate time constant.

acceptable. The response time is read off as $T_r \approx 0.9$s which is quite similar to the estimate $T_r = 1/\omega_0 = 1/1 = 1$s according to (7.19).

[End of Example 7.1]

Tuning the controller for *sluggish* control

The aim of controller tuning is not always fast control, but in stead sluggish control! This is the case for a level controlled liquid tank in a process line. The tank is an integrator, dynamically. The level control system ensures the mass balance. In addition the control system behaves like a lowpass filter between the (free) inflow w_{in} and the outflow w_{out}. To obtain enough attenuation of inflow variations through the system, the level control system must be sluggish! Example 7.2 goes into the details.

Example 7.2 *Level control of buffer tank*

Figure 7.4 shows the front panel of a simulator for buffer tank with level control system.[2] (The simulated responses are explained later in this example.)The control system has two aims:

- To keep the level on or close to a level setpoint.
- To attenuate variations in the outflow so that it becomes smoother than the inflow.

We need a mathematical process model: Mass balance is

$$\rho A\dot{h} = w_{in} - \underbrace{w_{out}}_{K_u u} \tag{7.22}$$

Laplace transformation of (7.22) is

$$\rho A h(s) = w_{in}(s) - K_u u(s) \tag{7.23}$$

Solving for $h(s)$ gives the following transfer function model:

$$h(s) = \frac{1}{\rho A s} w_{in}(s) - \frac{K_u}{\rho A s} u(s) \tag{7.24}$$

[2] The system may be in e.g. a production line in a factory.

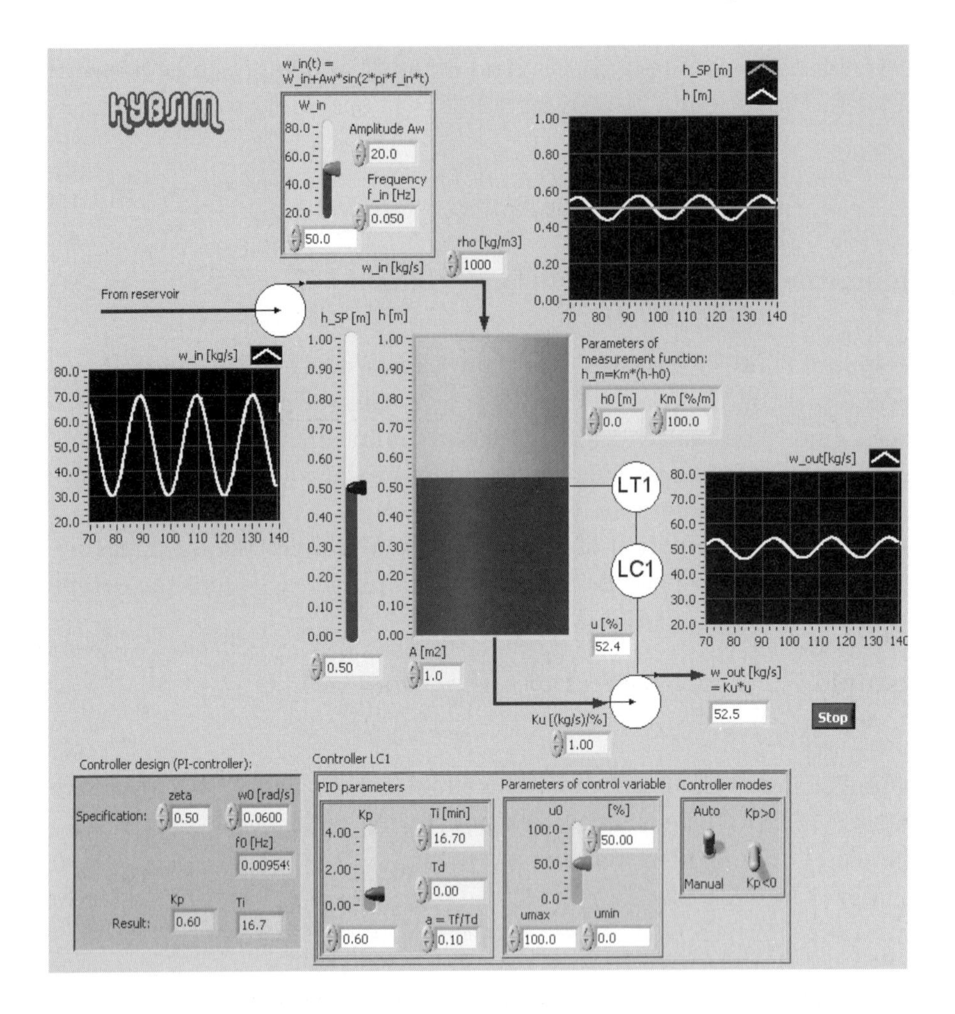

Figure 7.4: Example 7.2: Front panel of simulator for level control system

The transfer function from level h to level measurement h_m is

$$h_m(s) = K_m h(s) \tag{7.25}$$

Combining (7.24) and (7.25) gives the following model:

$$h_m(s) = \frac{K_m}{\rho A s} w_{in}(s) - \frac{K_m K_u}{\rho A s} u(s) \tag{7.26}$$

The transfer function from u to h_m is

$$H_p(s) = \frac{h_m(s)}{u(s)} = -\frac{K_u}{\rho A s} = -\frac{K}{s} \tag{7.27}$$

where

$$K = \frac{K_u K_m}{\rho A} \tag{7.28}$$

is the process gain. Figure 7.5 shows a block diagram of the level control system.

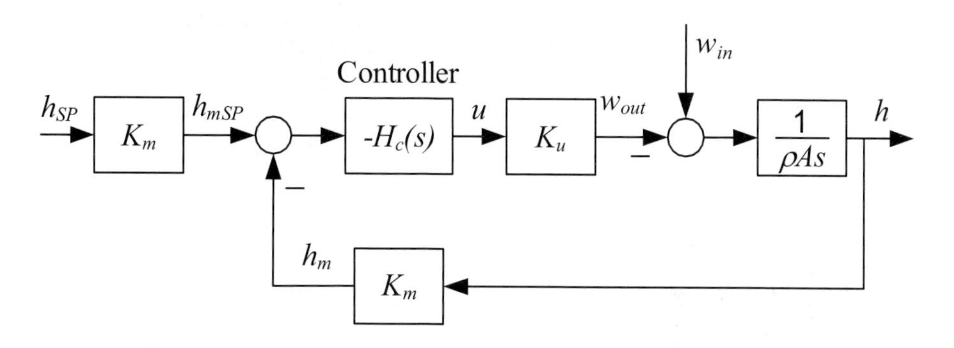

Figure 7.5: Example 7.2: Block diagram of the level control system

The level controller is a PI controller. The integral term ensures zero static control error. The transfer function of the PI controller is

$$H_c(s) = K_p \frac{T_i s + 1}{T_i s} \tag{7.29}$$

What is the reason for the negative sign ahead of the controller transfer function $H_c(s)$ in the block diagram in Figure (7.5)? The negative sign means that the controller in effect has negative gain. Negative controller gain is here necessary since the process gain is negative, cf. Section 2.6.8.

The tracking transfer function of the control system is given by (7.18), which is repeated here:

$$\frac{h_m(s)}{h_{m_{SP}}(s)} = T(s) \tag{7.30}$$

$$= \frac{L(s)}{1 + L(s)} \tag{7.31}$$

$$= \frac{H_c(s)H_p(s)}{1 + H_c(s)H_p(s)} \tag{7.32}$$

$$= \frac{K_p K \left(s + \frac{1}{T_i}\right)}{s^2 + K_p K s + \frac{K_p K}{T_i}} \tag{7.33}$$

$$= \frac{2\zeta\omega_0 s + \omega_0{}^2}{s^2 + 2\zeta\omega_0 s + \omega_0{}^2} \tag{7.34}$$

Once ζ and ω_0 is specified, the controller parameters are given by (7.16) and (7.17). Below we will specify ζ and ω_0 from a specification to the attenuation of the mass flow through the tank.

The relation between the inflow w_{in} and the outflow w_{ut} can be expressed by the transfer function from w_{in} to w_{out}. From the block diagram in Figure 7.5 we see that the relation between w_{in} and w_{out} is identical to the relation between the level setpoint h_{SP} and the level h, which implies that the transfer function from w_{in} to w_{out} is the same as the tracking transfer function! Thus,

$$\frac{w_{out}(s)}{w_{in}(s)} = \frac{h(s)}{h_m(s)} = T(s) \tag{7.35}$$

$$= \frac{K_p K\left(s+\frac{1}{T_i}\right)}{s^2 + K_p K s + \frac{K_p K}{T_i}} \tag{7.36}$$

$$= \frac{2\zeta\omega_0 s + \omega_0{}^2}{s^2 + 2\zeta\omega_0 s + \omega_0{}^2} \tag{7.37}$$

By giving values to ζ and ω_0 we determine the dynamic properties of the controlled tank. Let us set

$$\zeta = 0.5 \tag{7.38}$$

What about ω_0? It can roughly be regarded as the bandwidth of the lowpass filter (7.35). A Bode plot of the amplitude function $|T(j\omega)|$ gives a good picture of the filtering properties, see Figure 7.6 which shows $|T(j\omega)|$ with controller parameters as calculated later in this example. In the figure the frequency unit is Hz. The relation between a frequency f_1 in Hz and the corresponding frequency ω_1 in rad/s is

$$2\pi f_1 = \omega_1 \tag{7.39}$$

Let us specify that a frequency component in w_{in} of frequency $f_{in} = 0.05$Hz is attenuated by a factor of 5 – or in other words: amplified by factor 0.2 which is approximately -14dB. This means that the amplitude gain of T at this frequency must be

$$|T(s)|_{s=j2\pi f_{in}} \tag{7.40}$$

$$= |T(j2\pi f_{in})| \tag{7.41}$$

$$= \left| \left. \frac{2\zeta\omega_0 s + \omega_0{}^2}{s^2 + 2\zeta\omega_0 s + \omega_0{}^2} \right|_{s=j2\pi f_{in}=j2\pi\cdot 0.05} \right| = 0.2 = -14\text{dB} \tag{7.42}$$

Here we use (7.38). In principle we can now solve (7.42) for ω_0 (to be used in (7.16) and (7.17) for calculating the PI parameters). Although it is possible to solve (7.42) for ω_0, it is a bit difficult operation. If we have a computer tool for plotting Bode diagrams, it is easier to iterate on plotting $|T|$ for varying ω_0 until $|T| = 0.2$. The result is

$$\omega_0 = 0.06\text{rad/s} \,\hat{=}\, 0.0095\text{Hz} \tag{7.43}$$

Using (7.38) and (7.43) in (7.16) and (7.17) gives

$$K_p = 0.60;\ T_i = 16.7\text{s} \tag{7.44}$$

Figure 7.6 shows a Bode plot of $|T(j\omega)|$ with the controller parameters (7.44). We read off $|T| = -14.0\text{dB} = 0.20$.

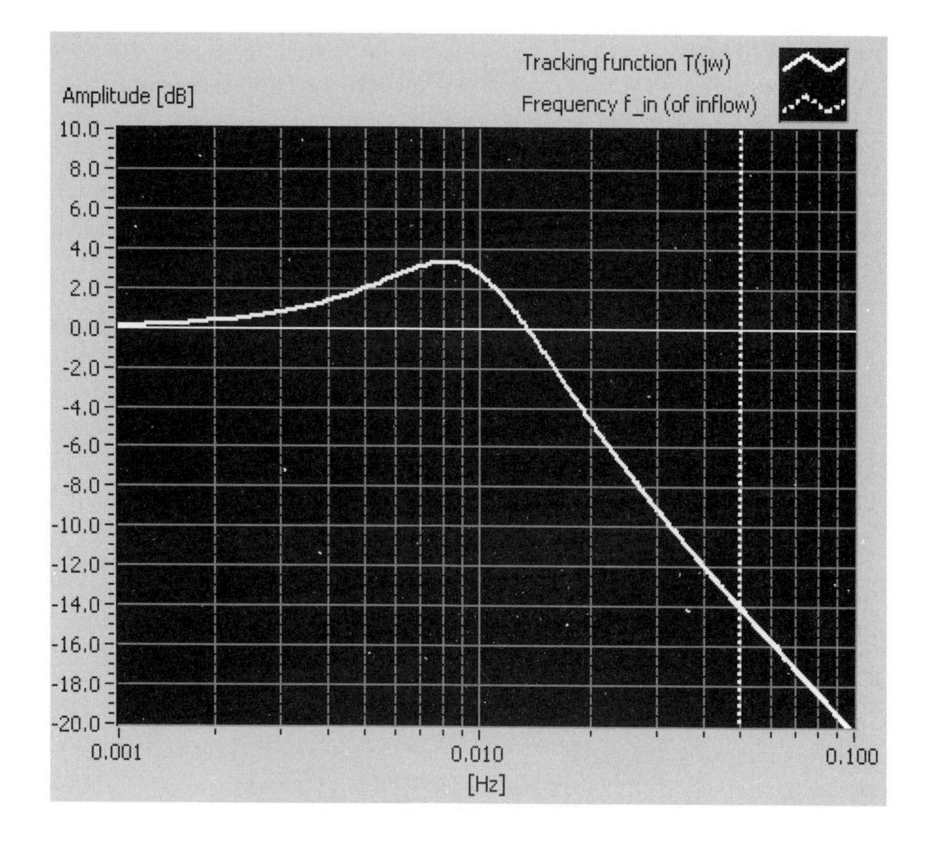

Figure 7.6: Example 7.2: Bode plot of $|T(\omega)|$ with the calculated PI parameters

Figure 7.4 shows simulated responses in the control system with parameter values defined above. An accurate reading from the simulations shows that the amplitude of w_{out} is 4.0kg/s (in steady-state). The amplitude of w_{in} is 20kg/s. Thus, the amplitude ratio is $4.0/20 = 0.20 = -14.0\text{dB}$, which is in accordance with the Bode plot, see Figure 7.6.

[End of Example 7.2]

Strange phenomenon with PI control of integrator process

From (7.15) we get

$$\omega_0 = \sqrt{\frac{K_p K}{T_i}} \tag{7.45}$$

and

$$\zeta = \frac{1}{2}\sqrt{K_p K T_i} \tag{7.46}$$

(7.46) shows that the stability of the control system is *reduced* if the controller gain, K_p, for some reason is reduced (smaller ζ implies less damping and hence less stability). This explains a somewhat surprising observation in control systems where an integrator process is controlled by a PI controller, since in general the stability of control systems is increased when the gain is decreased. One example of such a control system is a level control system of a level tank as shown in Figure 7.4. In addition to the reduced stability, the responses in the control system are more sluggish since ω_0 is reduced with decreased K_p.

Let us look at a few simulations which will illustrate the above. Assume given a process with the following transfer function:

$$H_p(s) = \frac{y(s)}{u(s)} = \frac{K}{s} \tag{7.47}$$

with $K = 1$. The process is controlled by a PI controller. We specify $\omega_0 = 1$ and $\zeta = 0.5$. (7.16) and (7.17) gives $K_p = 1$ and $T_i = 1$. Figure 7.7 shows the response in the process measurement y_m due to a step in the setpoint y_{mSP}. Figure 7.8 shows the response with K_p reduced to 0.1.

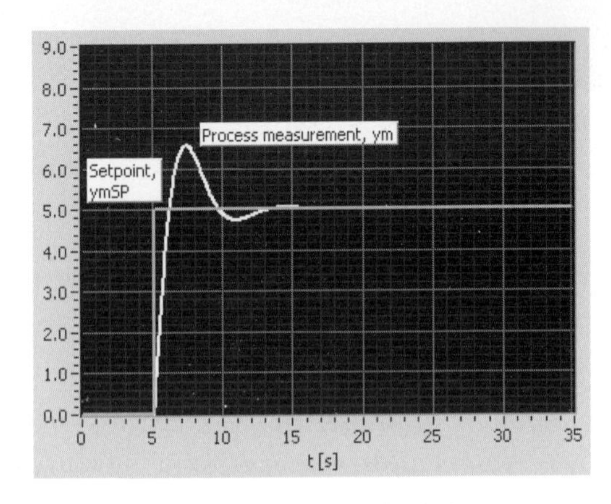

Figure 7.7: Response in process measurement y_m due to a step in the setpoint y_{mSP} with PI parameters $K_p = 1$ and $T_i = 1$ (normal design)

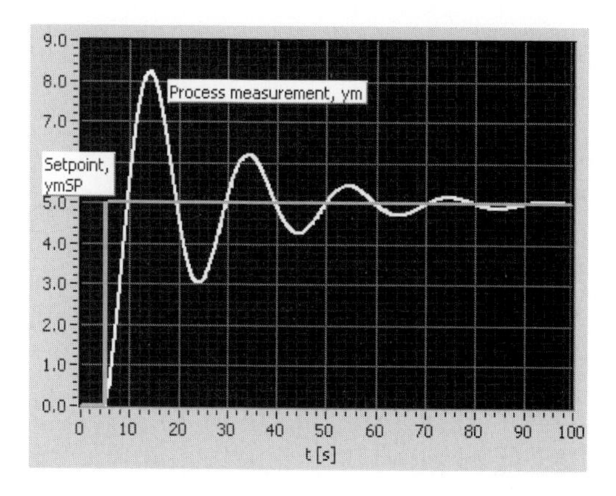

Figure 7.8: Response in y_m with $K_p = 0.1$ (reduced value compared to normal design) and $T_i = 1$

Compared to the response shown in Figure 7.7 it is clear that the reduced K_p value implies reduced stability of the control system. The oscillations are less damped, and they have smaller frequency. (Note the difference in the time scalings between the figures.)

7.2.3 Tuning a controller for a first order process

See Figure 7.1. The process transfer function is now

$$H_p(s) = \frac{K}{Ts+1} \tag{7.48}$$

and the disturbance transfer function is

$$H_{vm}(s) = \frac{K_{vm}}{T_{vm}s+1} = \frac{K_{vm}}{Ts+1} \tag{7.49}$$

Here are examples of processes having a first order model (approximately):

- A stirred tank (with homogenous contents) with continuous flow through the tank where the concentration, say c_A, of a material, A, is to be controlled by controlling the inflow, q_A, to the tank.
- A liquid tank (with homogenous contents) continuous flow through the tank where the temperature T_t is to be controlled by controlling the supplied power P.
- An electrical motor where the rotational speed n is to be controlled by controlling the input motor voltage v.

A PI controller is proper for controlling a first order process. The controller transfer function is

$$H_c(s) = K_p \frac{T_i s + 1}{T_i s} \tag{7.50}$$

The tracking transfer function becomes

$$T(s) = \frac{L(s)}{1+L(s)} = \frac{H_c(s)H_p(s)}{1+H_c(s)H_p(s)} \tag{7.51}$$

$$= \frac{\frac{K_p K}{T_i T}(T_i s + 1)}{s^2 + \frac{K_p K + 1}{T}s + \frac{K_p K}{T_i T}} = \frac{\frac{K_p K}{T_i T}(T_i s + 1)}{s^2 + 2\zeta\omega_0 s + \omega_0^2} \tag{7.52}$$

The characteristic polynomial $c(s)$ of the control system is a second order polynomial:

$$c(s) = s^2 + \frac{K_p K + 1}{T}s + \frac{K_p K}{T_i T} = s^2 + 2\zeta\omega_0 s + \omega_0^2 \tag{7.53}$$

where ω_0 is the undamped resonance frequency and ζ is the relative damping factor [7]. You have to specify ω_0 and ζ for the control system. Roughly, ω_0 is the bandwidth of $T(s)$. A reasonable value of ζ may be

$$\zeta = 0.5 \tag{7.54}$$

which gives step responses with well damped oscillations. If larger damping of the time responses is desired, ζ can be given a larger value (closer to 1).

Comparison of coefficients between the two polynomials in (7.53) gives the following identities:

$$\frac{K_p K + 1}{T} \equiv 2\zeta\omega_0 \text{ and } \frac{K_p K}{T_i T} \equiv \omega_0^2 \tag{7.55}$$

Solving for K_p and T_i gives the following formulas for the controller parameters:

$$K_p = \frac{2\zeta\omega_0 T - 1}{K} \tag{7.56}$$

$$T_i = \frac{2\zeta\omega_0 T - 1}{\omega_0^2 T} \tag{7.57}$$

An estimate of the response time[3] of the control system

$$T_r \approx \frac{1}{\omega_0} \tag{7.58}$$

Example 7.3 ***PI control of a first order process***

[3] The response time can be regarded as an approximate time constant.

Assume given a process model (7.48) and (7.49) with the following parameters: $K = 1$, $K_{vm} = 1$, $T = T_{vm} = 5$. Let us specify $\omega_0 = 1$ and $\zeta = 0.5$. Then (7.56) and (7.57) gives the following PI parameters:

$$K_p = 4;\ T_i = 0.8 \tag{7.59}$$

Figure 7.9 shows the simulated responses in the control system. There is a setpoint step and a disturbance step. The responses are ok.

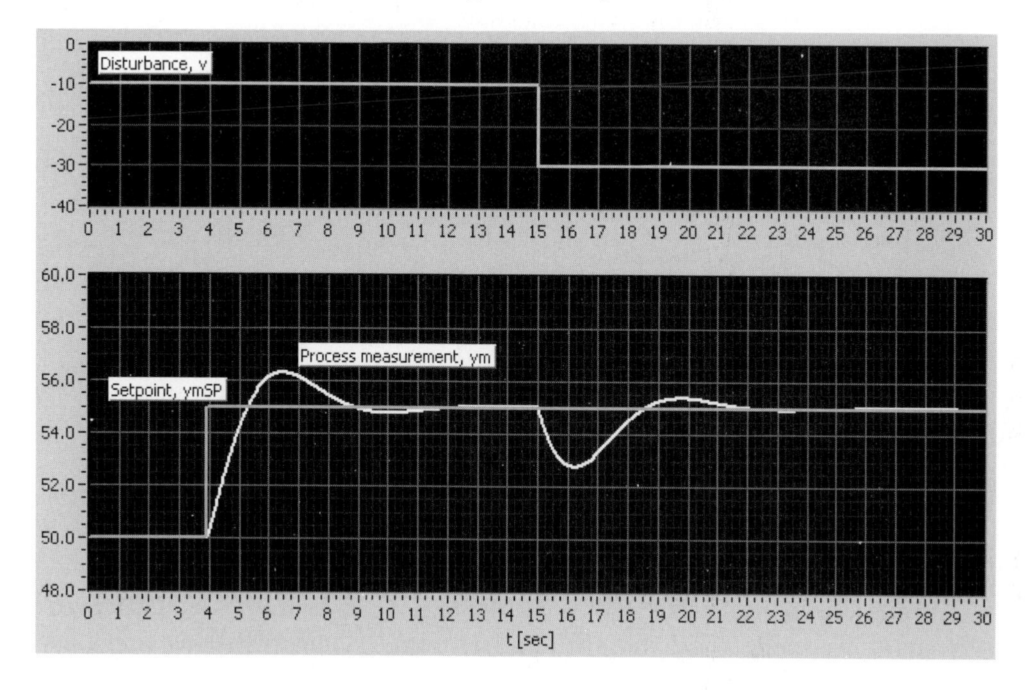

Figure 7.9: Example 7.3: Simulated responses in the control system

[End of Example 7.3]

Pole-zero cancellation as an alternative controller tuning method

An alternative way to calculate the PI controller parameters for a first order process, is pole-zero cancellation. Look at the loop transfer function:

$$L(s) = H_c(s)H_p(s) = K_p \frac{T_i s + 1}{T_i s} \cdot \frac{K}{Ts + 1} \tag{7.60}$$

Here we set the *integral time equal to the process time constant*:

$$T_i = T \tag{7.61}$$

The factor $(T_i s + 1)$ can now be cancelled against the factor $(Ts + 1)$ in $L(s)$, which gives

$$L(s) = \frac{K_p K}{Ts} \tag{7.62}$$

The tracking transfer function becomes

$$T(s) = \frac{L(s)}{1 + L(s)} = \frac{1}{\frac{T}{K_p K}s + 1} = \frac{1}{T_M s + 1} \tag{7.63}$$

which is just a first order transfer function with time constant (response time)

$$T_M = \frac{T}{K_p K} \tag{7.64}$$

We have to specify T_M. Then the controller gain is given by

$$K_p = \frac{T}{T_M K} \tag{7.65}$$

To sum up: The PI parameter formulas are (7.65) and (7.61).

One possible drawback with this simple PI controller tuning method is that T_i becomes large if T is large (since they are equal). For a sluggish process, having large T, the integral action can be slow, resulting in slow disturbance compensation. However, the setpoint tracking may be fast since this response is given by time constant T_M which we can specify. In general it is safer to calculate K_p and T_i from specification of ω_0 and ζ in the second order characteristic polynomial since this may give fast disturbance compensation (in addition to fast setpoint tracking.

Example 7.4 ***PI controller tuning using pole-zero cancellation***

Given the same process as in Example 7.3. We specify $T_M = 1$. (7.65) and (7.61) gives

$$K_p = 5; \; T_i = 5 \tag{7.66}$$

Figure 7.10 shows simulated responses in the control system. The setpoint tracking is as expected: Like the step response of a first order system with time constant $T_M = 1$. The disturbance compensation is however *much slower* than in Example 7.3. This demonstrates that the pole-zero cancelling may be a bad tuning method if the process time constant is large.

[End of Example 7.4]

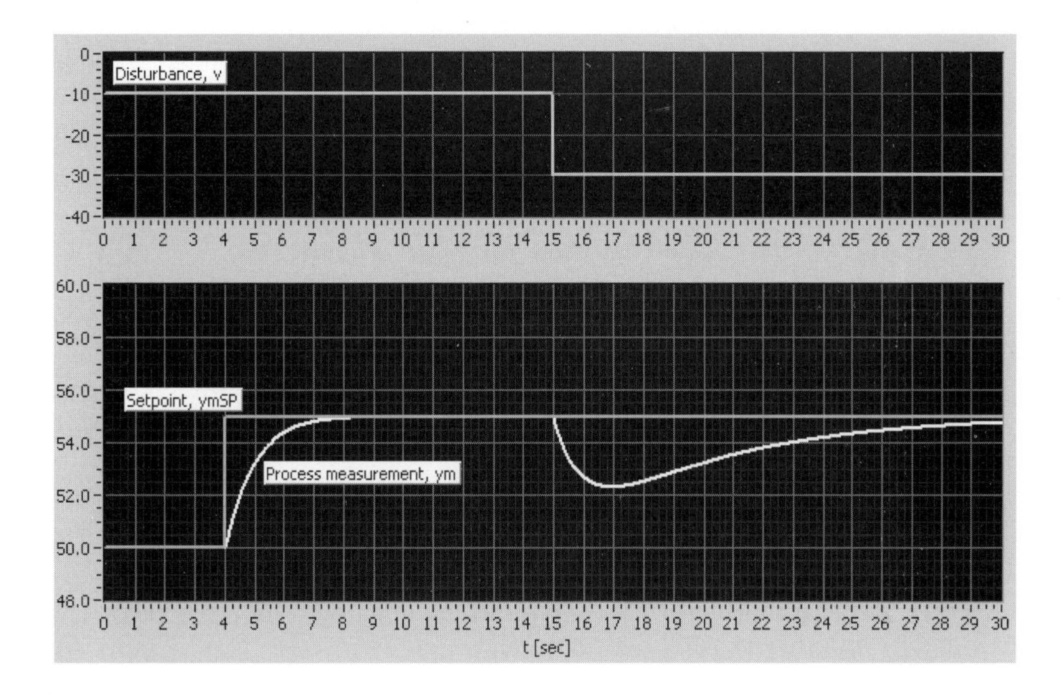

Figure 7.10: Example 7.4: Simulated responses in the control system. The controller is tuned using pole-zero cancellation.

7.3 Controller design using the direct method

The *direct method* for designing a controller is in principle a very simple method, but there is no guarantee that the controller becomes a PID controller. Non PID controllers may be difficult to implement in standard commercial control equipment, but in flexible and mathematical tools as MATLAB, SIMULINK and LabVIEW such controllers are straightforward to implement.

The direct method is based on a specified tracking function $T(s)$, or a specified sensitivity function $S(s)$. (The direct method is the basis of the dead-time compensator described in Section 7.4 and Skogestad's method described in Section 7.5.)

Assume given the tracking function $T(s)$. From (7.8) we get

$$L(s) = H_c(s)H_p(s) = \frac{T(s)}{1 - T(s)} \tag{7.67}$$

which gives the following controller transfer function:

$$H_c(s) = \frac{1}{H_p(s)} \cdot \frac{T(s)}{1 - T(s)} \tag{7.68}$$

If in stead the sensitivity function $S(s)$ is specified, the controller is found from (7.6) as

$$H_c(s) = \frac{1}{H_p(s)} \cdot \frac{1 - S(s)}{S(s)} \tag{7.69}$$

Example 7.5 ***The direct method for controller design***

Given the process

$$H_p(s) = \frac{K}{T_p s + 1} \tag{7.70}$$

We specify the tracking function as

$$T(s) = \frac{1}{T_t s + 1} \tag{7.71}$$

where T_t is the time constant of the tracking function or of the control system. Inserting (7.71) into (7.68) gives

$$H_c(s) = \frac{1}{H_p(s)} \cdot \frac{T(s)}{1 - T(s)} = \frac{1}{\frac{K}{T_p s+1}} \cdot \frac{\frac{1}{T_t s+1}}{1 - \frac{1}{T_t s+1}} = \frac{T_p s + 1}{K T_t s} \tag{7.72}$$

which happens to be a PI controller with

$$K_p = \frac{T_p}{K T_t}; \; T_i = T_p \tag{7.73}$$

[End of Example 7.5]

Why feedback?

Above we specified the tracking transfer function $T(s)$ and calculated the controller transfer function $H_c(s)$, assuming that the control system is a *feedback* system. Suppose we drop the assumption about feedback, and that we just want to achieve $T(s)$ by placing a compensating transfer function, $H_{sc}(s)$, in front and in series with the process, as shown in Figure 7.11.This is an open loop control strategy (or feedforward control). No sensor is needed. We can calculate $H_{sc}(s)$ by requiring

$$H_{sc}(s) H_u(s) = T(s) \tag{7.74}$$

Solving for $H_{sc}(s)$ gives

$$H_{sc}(s) = \frac{T(s)}{H_u(s)} \tag{7.75}$$

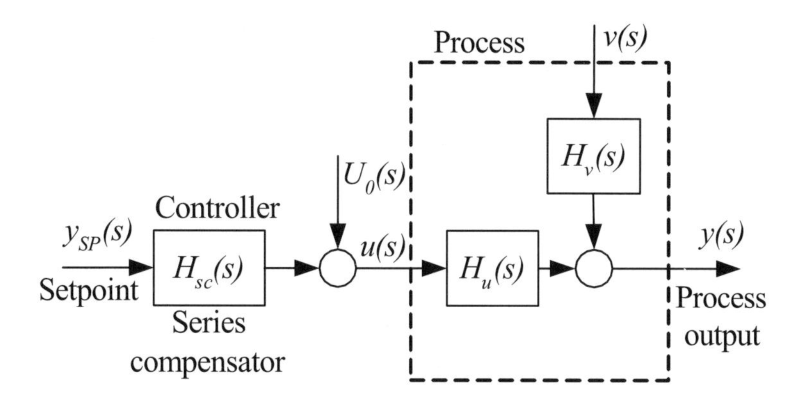

Figure 7.11: Compensating transfer function, $H_{sc}(s)$, in series with the process

Assume as an example that

$$H_u(s) = \frac{K_u}{T_u s + 1} \tag{7.76}$$

We specify the tracking function as

$$T(s) = \frac{1}{T_t s + 1} \tag{7.77}$$

Now (7.75) gives

$$H_{sc}(s) = \frac{T(s)}{H_u(s)} = \frac{\frac{1}{T_t s+1}}{\frac{K_u}{T_u s+1}} = \frac{1}{K_u} \frac{T_u s + 1}{T_t s + 1} \tag{7.78}$$

which will give the same $T(s)$ as in Example 7.5 where feedback was assumed. So, why not choose the open loop solution, since we can then drop the sensor? Due to *model errors and disturbances* the control error may be different from zero, and since there is no feedback from the actual process output signal, no automatic adjustment of the control signal (as a function of the control error) can take place, so the control error may be (too) large.

The open loop control shown in Figure 7.11 is actually *feedforward* control. Feedforward control may be used in combination with feedback control, cf. Section 9.1.

7.4 Dead-time compensator (Smith predictor)

The dead-time compensator – also called the Smith predictor[4] – is a control method particularly designed for processes with dead-time (time delay). Compared to ordinary feedback control with a PI(D) controller, dead-time compensation gives improved setpoint tracking in all cases, and it may under certain conditions give improved disturbance compensation.

It is assumed that the process to be controlled has a mathematical model on the following transfer function form:

$$H_p(s) = H_u(s)e^{-\tau s} \tag{7.79}$$

where $H_u(s)$ is a partial transfer function without time delay, and $e^{-\tau s}$ is the transfer function of the time delay.

Simply stated, with dead-time compensation the bandwidth (quickness) of the control system is independent of the dead-time, and relatively high bandwidth can be achieved. However, the controller function is more complicated than the ordinary PID controller since it contains a transfer function model of the process. A dead-time compensator are implemented in several controller products.[5]

Figure 7.12 shows the structure of the control system based on dead-time compensation. In the figure y_{mp} is a predicted value of y_m – therefore the name Smith *predictor*. y_{m1p} is a predicted value of the non time delayed internal process variable y_{m1}. There is a feedback from the predicted or calculated value y_{m1}. The PID controller is the controller for the non delayed process, $H_u(s)$, and *it is tuned for this process.*[6] The bandwidth of this loop can be made (much) larger compared to the bandwidth if the time delay were included in the loop. The latter corresponds to the ordinary feedback control structure, cf. Figure 7.1.

As long as the model predicts a correct value of y_m, the prediction error e_p is zero, and the signal in the outer feedback is zero. But if e_p is different from zero (due to modeling errors), there will be a compensation for this error via the outer feedback.

What is the tracking transfer function, $T(s)$, of the control system? To make it simple, we will assume that there are no modeling errors. From

[4] After the originator, O. J. M. Smith.

[5] E.g. in the Provox system by Fisher.

[6] The controller tuning can be made using any standard method, e.g. the Skogestad's method described in Section 7.5.

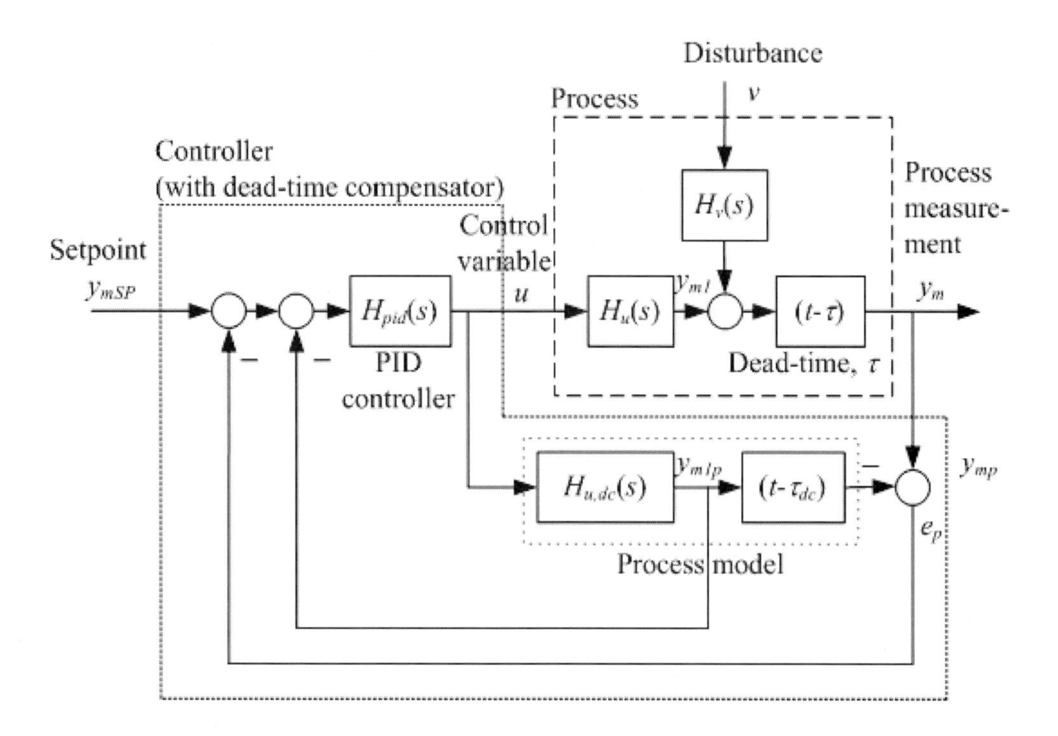

Figure 7.12: Structure of a control system based on dead-time compensation

the block diagram in Figure 7.12 the following can be found:

$$T(s) = \frac{y_m(s)}{y_{m_{SP}}(s)} = \frac{H_{pid}(s)H_u(s)}{1 + H_{pid}(s)H_u(s)}e^{-\tau s} = \frac{L(s)}{1 + L(s)}e^{-\tau s} \tag{7.80}$$

where

$$L(s) = H_{pid}(s)H_u(s) \tag{7.81}$$

is the loop transfer function of the loop consisting of the PID controller and the partial *non time delayed* process $H_u(s)$.[7]

How is the setpoint tracking and the disturbance compensation performance of the control system?

- **Setpoint tracking** is as if the feedback loop did not have time delay, and therefore *faster setpoint tracking* can be achieved with a dead-time compensator than with ordinary feedback control (with

[7] Actually, the controller having $y_{m_{SP}}$ as setpoint, y_m as measurement signal fed back to the conroller, and u as control signal (shown in the dashed frame in Figure 7.12) can be derived by the direct controller design method described in Section 7.3 with (7.80) as the specified tracking function.

PID controller). However, the time delay of the response in the process measurement can not be avoided with a dead-time compensator.

- **Disturbance compensation**: In [13] the disturbance compensation with a dead-time compensating control system is investigated for a first order with time delay process. It was found that dead-time compensation gave better disturbance compensation (assuming a step in the disturbance) compared to ordinary feedback control only if the time delay (dead-time) τ is larger than the time constant T of the process.

Example 7.6 ***Dead-time compensator***

Given a process with the following transfer functions (cf. Figure 7.12):

$$H_p(s) = \underbrace{\frac{K_u}{T_u s + 1}}_{H_u(s)} e^{-\tau s} \tag{7.82}$$

$$H_v(s) = \frac{K_v}{T_v s + 1} \tag{7.83}$$

where

$$K_u = 1;\ T_u = 0.5;\ K_v = 1;\ T_v = 0.5;\ \tau = 2 \tag{7.84}$$

The following two control systems are simulated:

- *Dead-time compensator* for the process defined above. The internal controller, $H_{pid}(s)$, is a PI controller with the following parameter values:

$$K_p = 2.0;\ T_i = 0.36 \tag{7.85}$$

  These PI parameters are calculated using Skogestad's method, cf. Table 7.2, with $T_C = 0.25\ (= T/2)$ and $k_1 = 1.44$.

- *Ordinary feedback control* with PI controller for the process defined above. The control system has a structure as in Figure 7.1. The PI controller, $H_{pid}(s)$, is a PI controller with the following parameter values:

$$K_p = 0.12;\ T_i = 0.5 \tag{7.86}$$

  These PI parameters are calculated using Skogestad's method, cf. Table 7.1, with $T_C = 2\ (= \tau)$ and $k_1 = 1.44$.

Figure 7.13 shows the simulated responses for the two control systems due to a setpoint step and a disturbance step. The dead-time compensator gives better setpoint tracking and better disturbance compensation than ordinary feedback control does.

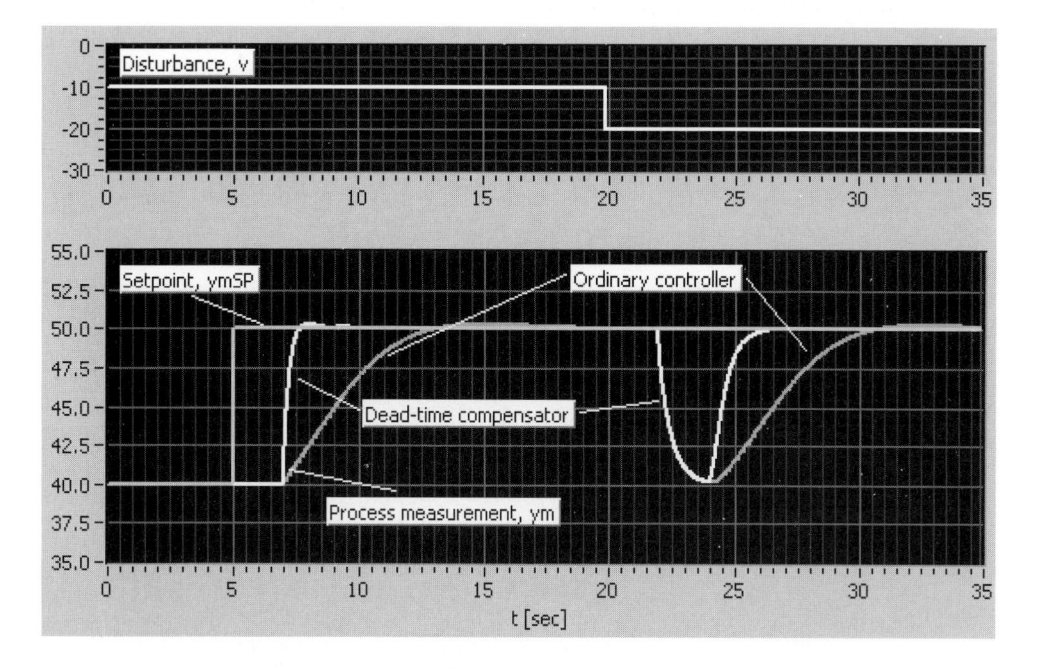

Figure 7.13: Example 7.6: Simulated responses for the two control systems due to a setpoint step and a disturbance step

[End of Example 7.6]

The dead-time compensator is model-based since the controller includes a model of the process. Consequently, the stability and performance robustness of the control system depend on the accuracy of the model. Running a sequence of simulations with a varied process model (changed model parameters) in each run is one way to investigate the robustness.

7.5 Skogestad's method

7.5.1 Introduction

[17] describes controller tuning for several types of transfer function processes – with and without time delay (dead-time). It is assumed that

the block diagram of the control system is as shown in Figure 7.2. The method, which can be denoted Skogestad's method after the originator[8], is based on the direct method described in Section 7.3: The control system tracking function $T(s)$ is specified as a first order transfer function with time delay:

$$T(s) = \frac{y_m(s)}{y_{m_{SP}}(s)} = \frac{1}{T_C s + 1} e^{-\tau s} \tag{7.87}$$

where T_C is the time constant of the control system which the user must specify, and τ is the process time delay which is given by the process model (the method can however be used for processes without time delay, too). Figure 7.14 shows the step response for (7.87).

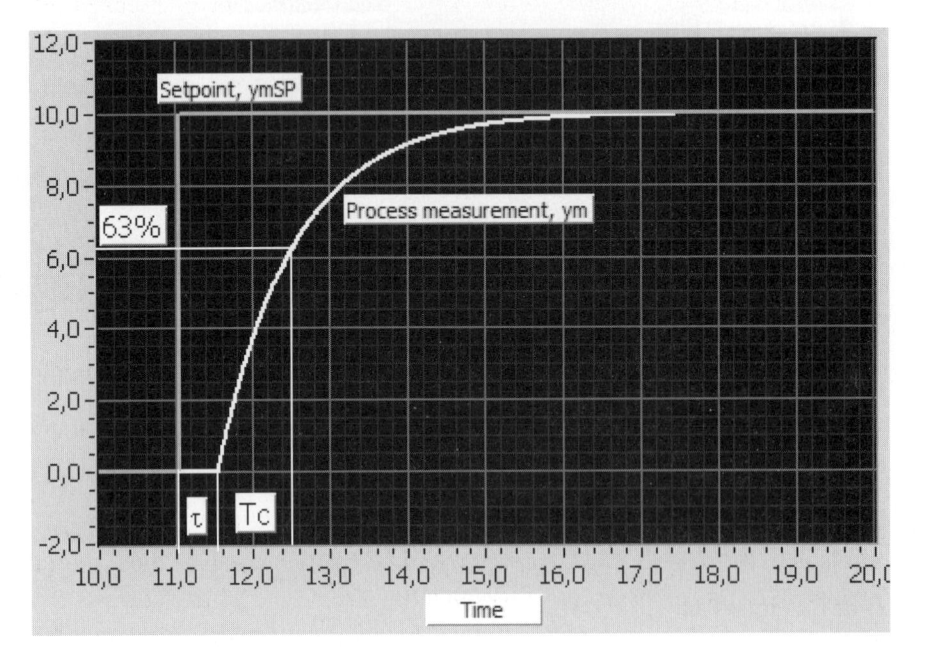

Figure 7.14: Step response of the specified tracking transfer function (7.87) in Skogestad's PID tuning method

The method is based on initially calculating the controller transfer function, $H_c(s)$, by (7.68) which is repeated here:

$$H_c(s) = \frac{1}{H_p(s)} \cdot \frac{T(s)}{1 - T(s)} \tag{7.88}$$

The process transfer function $H_p(s)$ may be of higher order than $T(s)$. Therefore, the specification (7.87) implies pole-zero cancellations in the control system loop transfer function, $L(s) = H_c(s)H_p(s)$. It is assumed

[8] Prof. Sigurd Skogestad

that the process $H_p(s)$ contains a time delay, $e^{-\tau s}$. The controller $H_c(s)$ according to (7.88) will contain the term $e^{-\tau s}$. This term is in $H_c(s)$ approximated by a first order Taylor series expansion which is $1 - \tau s$, and it turns out that the controller is a PI controller or a PID controller (depending on the process to be controlled).

Skogestad's method is in principle the same as dead-time compensation, which is described in Section 7.4, but in the latter there is no approximation of the time delay term. As with dead-time compensation Skogestad's method gives good setpoint tracking. The method gives formulas for the integral time, T_i, which are supposed to avoid slow disturbance compensation. In other controller design methods based on pole-zero cancellations there is a danger of slow disturbance compensation if the cancelled pole is close to zero (corresponding to cancellation of a large process time constant using a large T_i). This problem was demonstrated in Section 7.2.3.

The PID controller is assumed to be on serial form:

$$H_c(s) = K_p \frac{(T_i s + 1)(T_d s + 1)}{T_i s (T_f s + 1)} \tag{7.89}$$

If the PID controller you are going to apply is actually on parallel form,

$$H_c(s) = K_p + \frac{K_p}{T_i s} + \frac{K_p T_d s}{T_f s + 1} \tag{7.90}$$

you should consider *transforming* the PID parameters from serial form to parallel form to be sure that your parallel controller behaves like a serial controller. The transformation formulas are (2.51) – (2.53). (If the controller is a P or a PI controller, the transformation formulas need not be applied since in that case the serial and the parallel form are identical.)

7.5.2 Skogestad's tuning formulas

Skogestad's tuning formulas for several processes are shown in Table 7.1.[9] According to [17] the factor k_1 in Table 7.1 is 4, but there may be reasons to give it a different value, as argued on page 216. For the second order the process in Table 7.1 T_1 is the largest and T_2 is the smallest time constant.[10]

[9] [17] describes controller tuning for one additional process, namely a pure time delay, and the resulting controller is an I controller (Integral controller). However, a pure time delay can be approximated by a first order system with a small time constant (compared to the time delay), and this process is one of the processes in Table 7.1.

[10] [17] also describes methods for model reduction so that more complicated models can be approximated with one of the models shown in Table 7.1.

$H_p(s)$ (process)	K_p	T_i	T_d
$\frac{K}{s}e^{-\tau s}$	$\frac{1}{K(T_C+\tau)}$	$k_1\left(T_C+\tau\right)$	0
$\frac{K}{Ts+1}e^{-\tau s}$	$\frac{T}{K(T_C+\tau)}$	$\min\left[T,\, k_1\left(T_C+\tau\right)\right]$	0
$\frac{K}{(Ts+1)s}e^{-\tau s}$	$\frac{1}{K(T_C+\tau)}$	$k_1\left(T_C+\tau\right)$	T
$\frac{K}{(T_1s+1)(T_2s+1)}e^{-\tau s}$	$\frac{T_1}{K(T_C+\tau)}$	$\min\left[T_1,\, k_1\left(T_C+\tau\right)\right]$	T_2
$\frac{K}{s^2}e^{-\tau s}$	$\frac{1}{4K(T_C+\tau)^2}$	$4\left(T_C+\tau\right)$	$4\left(T_C+\tau\right)$

Table 7.1: Skogestad's formulas for PI(D) tuning. Standard value of k_1 is 4, but a smaller value, e.g. $k_1 = 1.44$ can give faster disturbance compensation. For the second order the process T_1 is the largest and and T_2 is the smallest time constant. (min means the minimum value.)

Unless you have reasons for a different specification, [17] suggests

$$T_C = \tau \tag{7.91}$$

to be used for T_C in Table 7.1.

The Ziegler-Nichols' closed loop method may be applied to most of the processes in Table 7.1 (since the processes have time delay). Generally, Skogestad's method results in better tracking property of the control system (without the quite large overshoot in the response after a step in the setpoint which is typical with Ziegler-Nichols' method), but the disturbance compensation may for some processes become more sluggish than with the Ziegler-Nichols' method. This sluggish compensation can however be speeded up by selecting a smaller value of k_1, cf. the discussion on page 216. It is here assumed that the disturbance is an input disturbance as explained on page 190.

Example 7.7 ***Control of first order system with time delay***

Let us try Skogestad's method and Ziegler-Nichols' closed loop method for tuning a PI controller for the process

$$H_p(s) = \frac{K}{Ts+1}e^{-\tau s} \tag{7.92}$$

where

$$K = 1;\, T = 0.5;\, \tau = 1 \tag{7.93}$$

(The time delay is relatively large compared to the time constant.) The controller parameters are as follows:

- Skogestad's method, cf. Table 7.1 with (7.91) and $k = 4$:

$$K_p = 0.25;\ T_i = 0.5 \tag{7.94}$$

- Ziegler-Nichols' closed loop method:

$$K_p = 0.68;\ T_i = 2.43 \tag{7.95}$$

Figure 7.15 shows control system responses for the two controller tunings. Skogestad's method works clearly better than Ziegler-Nichols' method, both with respect to setpoint tracking and disturbance compensation.

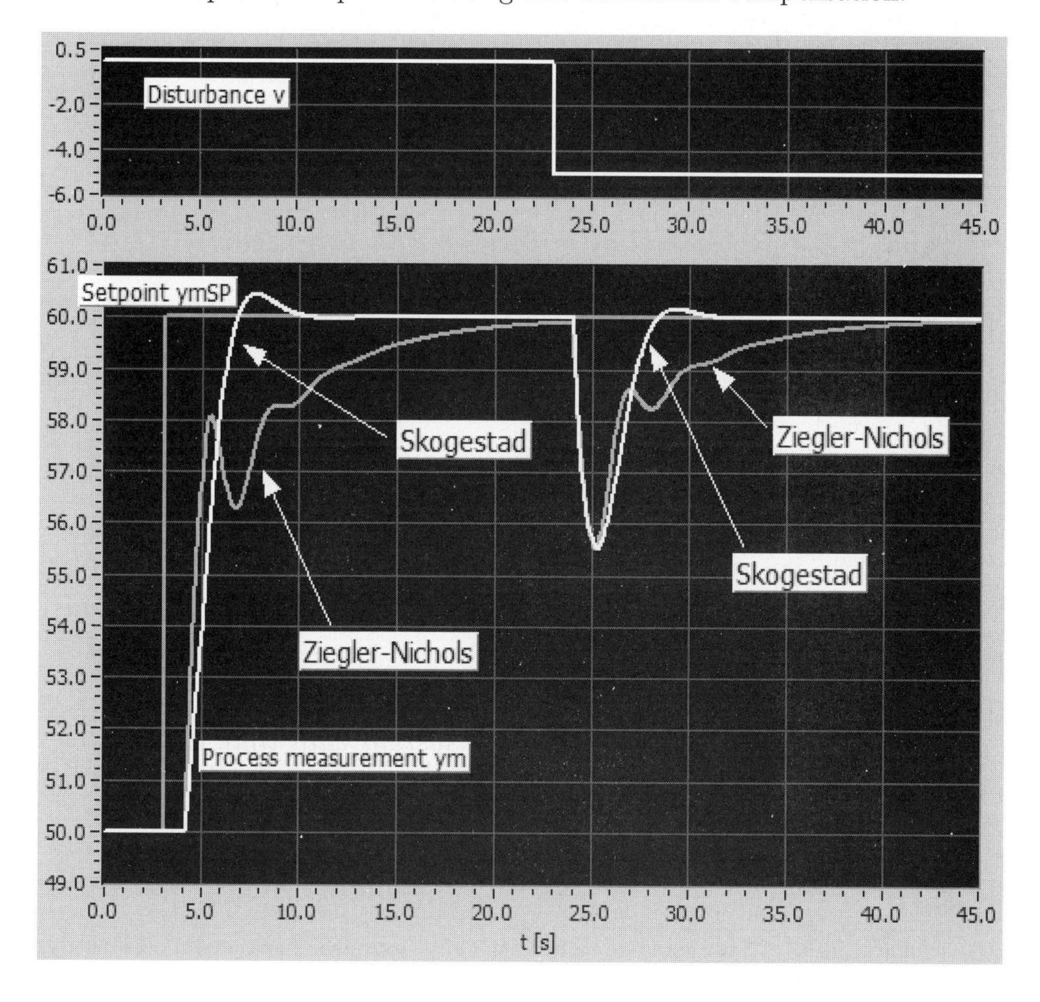

Figure 7.15: Example 7.7: Simulated responses in the control system for two different controller tunings

[End of Example 7.7]

7.5.3 Skogestad's method with faster disturbance compensation

According to [17], k_1 is 4 in Table 7.1. However, through simulations I have observed that $k_1 = 4$ in several cases gives quite sluggish disturbance compensation, although the parameter formulas in Table 7.1 are developed to avoid unnecessary sluggish compensation. A reduced k_1 value, as $k_1 = 1.44$, can give considerably faster disturbance compensation (since the integral time T_i is reduced).[11] A drawback of this modification of Skogestad's method is that there will be somewhat larger overshoot in the response after setpoint step, but in most cases such an increased overshoot is acceptable (if the setpoint is constant, which is typical, there is no overshoot, of course). Another drawback of the modification is that the stability robustness of the loop is somewhat reduced because of the reduced T_i.

Example 7.8 ***PI control of integrator with time delay***

The process

$$H_p(s) = \frac{K}{s}e^{-\tau s} \tag{7.96}$$

where

$$K = 1;\ \tau = 0.5 \tag{7.97}$$

will be controlled by a PI controller. (The wood-chip tank described in Example 2.3 has such a transfer function model.) Below are the PI parameters according to various tuning methods:

- Skogestad's method, cf. Table 7.1, with (7.91) and $k_1 = 4$:

$$K_p = 1;\ T_i = 4 \tag{7.98}$$

- Skogestad's method, cf. Table 7.1, with (7.91) and $k_1 = 1.44$:

$$K_p = 1;\ T_i = 1.44 \tag{7.99}$$

[11] According to [17] the standard value $k_1 = 4$ gives a transfer function from disturbance v to process measurement y_m in the control system with characteristic polynomial as of a critically damped second order system, i.e. the relative damping factor is $\zeta = 1$. This is quite a conservative choice. Faster but less damped dynamics is obtained with $\zeta < 1$. Simulations shows that $\zeta = 0.6$ is a reasonable value. It gives almost 3 times smaller T_i and therefore faster disturbance compensation. $\zeta = 0.6$ is obtained with $k_1 = 1.44$. It can be shown that the phase margin, PM, of a loop having second order characteristic polynomial is approximately equal to $100° \cdot \zeta$. With $\zeta = 0.6$ this equals $60°$ – a reasonable value in most cases.

- Ziegler-Nichols' closed loop method:

$$K_p = 1.3;\ T_i = 1.78 \tag{7.100}$$

Figure 7.16 shows simulated responses in the control system for the three different sets of PI parameter values. Skogestad's method with $k_1 = 4$ seems to give the best set point tracking, but there are no oscillations, indicating good (too good?) stability. The disturbance compensation with Skogestad's method with $k_1 = 4$ is clearly the slowest of the three alternatives.

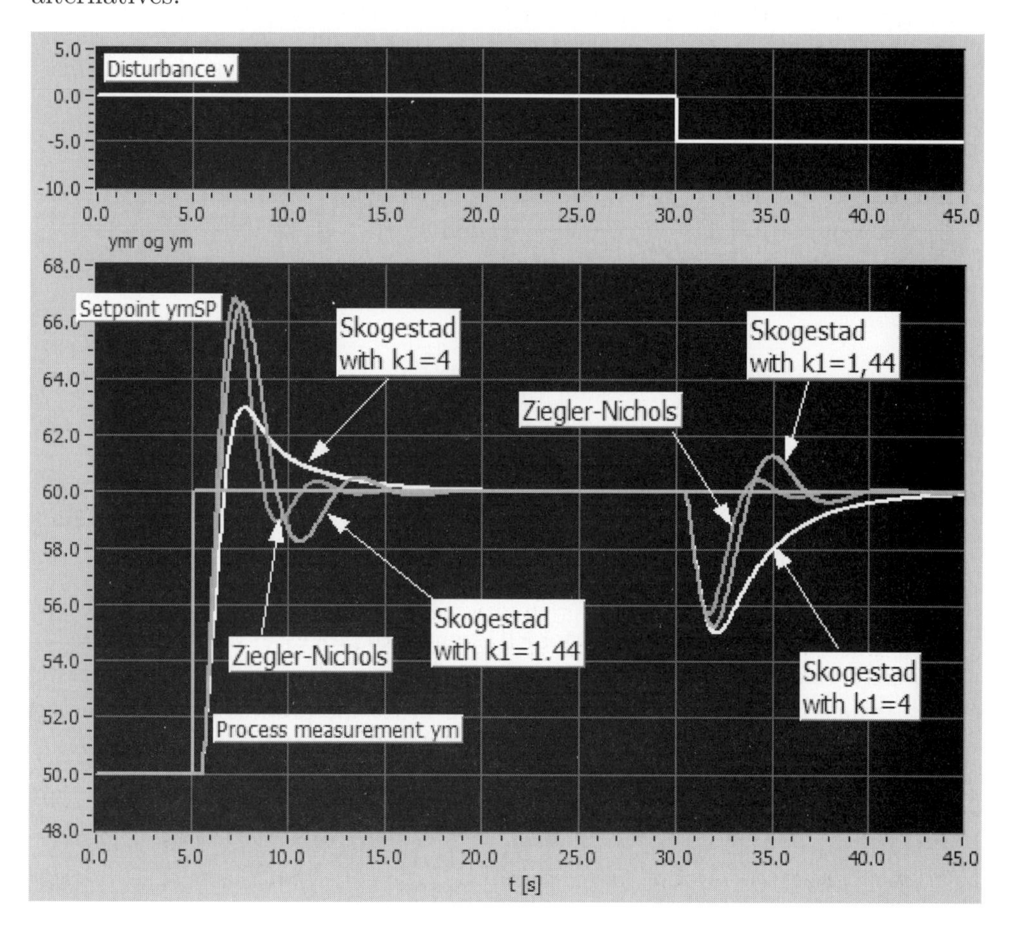

Figure 7.16: Example 7.8: Simulated responses in the control system for various PI tunings

[End of Example 7.8]

Example 7.7 demonstrated that it may be beneficial to set $k_1 = 1.44$ in stead of the standard value $k_1 = 4$ because faster disturbance is then

obtained. Let us review Example 7.7 which demonstrated that $k_1 = 4$ gave fast and properly damped disturbance compensation. Since $k_1 = 4$ worked well in that example, will the disturbance compensation in that example be worse with $k_1 = 1.44$ than with $k_1 = 4$? The answer is no, because: K_p is in any case independent of k_1, so it has value 0.25. However, T_i is dependent of k_1. According to Table 7.1,
$T_i = \min[T,\, k_1(T_C + \tau)] = \min[T,\, 2k_1\tau]$, but this minimum value is 0.5 no matter if k_1 is 4 or 1.44. So, this example has indicated that even if $k_1 = 4$ works fine, the suggestion $k_1 = 1.44$ makes no harm in this case.

7.5.4 Skogestad's method for processes without time delay

Each of the processes in Table 7.1 has time delay ($\tau > 0$). Can Skogestad's method be applied to processes *without* time delay? Yes, but in such cases we can not specify T_C according to (7.91) since τ is zero. We must specify T_C larger than zero. The controller parameter formulas are as shown in Table 7.2 (which is equal to Table 7.1 with $\tau = 0$).

$H_p(s)$ (process)	K_p	T_i	T_d
$\frac{K}{s}$	$\frac{1}{KT_C}$	$k_1 T_C$	0
$\frac{K}{Ts+1}$	$\frac{T}{KT_C}$	$\min[T,\, k_1 T_C]$	0
$\frac{K}{(Ts+1)s}$	$\frac{1}{KT_C}$	$k_1 T_C$	T
$\frac{K}{(T_1 s+1)(T_2 s+1)}$	$\frac{T_1}{KT_C}$	$\min[T_1,\, k_1 T_C]$	T_2
$\frac{K}{s^2}$	$\frac{1}{4K(T_C)^2}$	$4T_C$	$4T_C$

Table 7.2: Skogestad's formulas for PID tuning for processes without time delay. Standard value of k_1 is 4, but a smaller value, e.g. $k_1 = 1.44$ can give faster disturbance compensation. For the second order the process T_1 is the largest and and T_2 is the smallest time constant. (min means the minimum value.)

Example 7.9 ***PI control of first order system without time delay***

Given the following process:

$$H_p(s) = \frac{K}{Ts + 1} \tag{7.101}$$

where

$$K = 1;\, T = 5 \tag{7.102}$$

Let us specify $T_C = 1$. We try both $k_1 = 4$ (the standard value) and $k_1 = 1.44$ (which may give faster disturbance compensation). According to Table 7.2 the controller parameters (of a PI controller) are as follows:

- Skogestad's method, cf. Table 7.2, with $T_C = 1$ and $k_1 = 4$:

$$K_p = 5;\ T_i = 4 \tag{7.103}$$

- Skogestad's method, cf. Table 7.2, with $T_C = 1$ and $k_1 = 1.44$:

$$K_p = 1;\ T_i = 1.44 \tag{7.104}$$

Figure 7.17 shows simulated responses in the control system with the PI parameters values given above. We see that $k_1 = 1.44$ gives somewhat faster setpoint tracking, but with some overshoot, and in addition better disturbance compensation than with $k_1 = 4$.

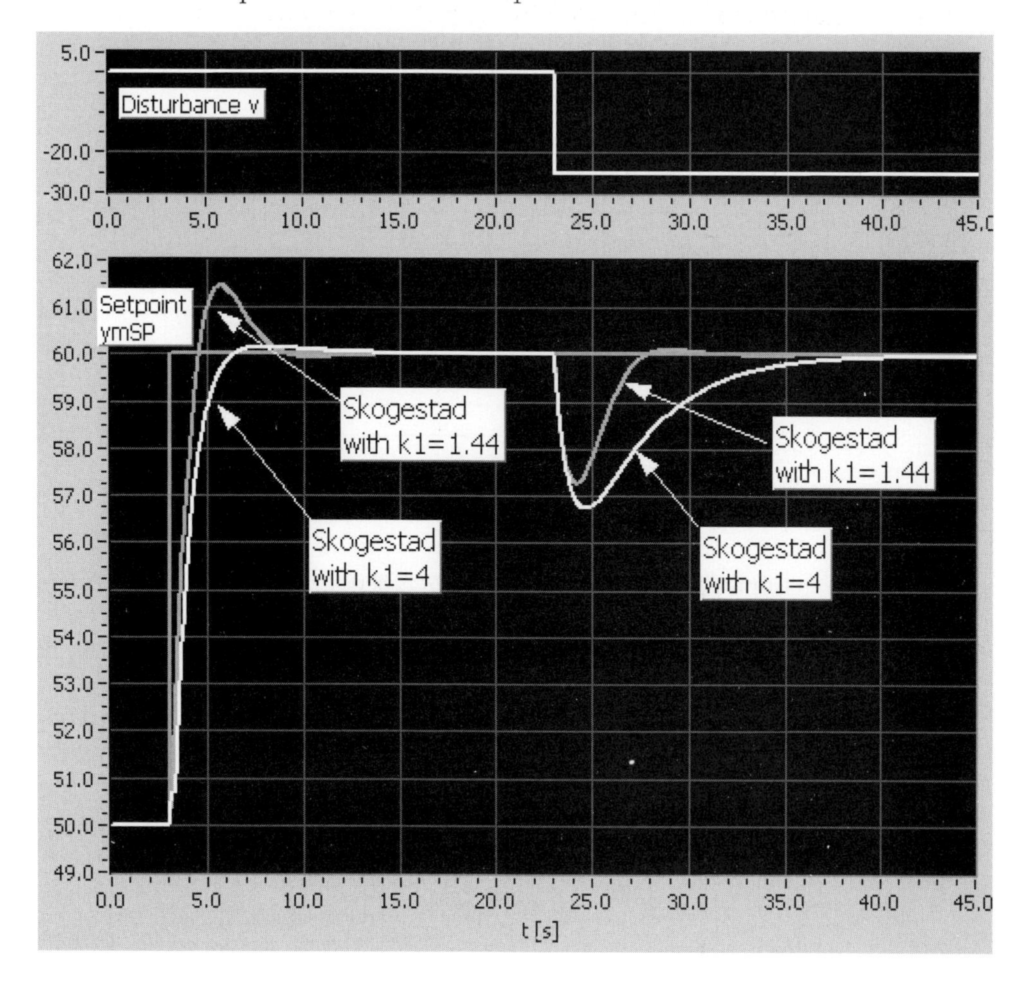

Figure 7.17: Example 7.9: Simulated responses in the control system for two different PI tunings

[End of Example 7.9]

7.6 Controllers with two degrees of freedom

Earlier in this chapter we have used one controller function to obtain (hopefully) both satisfactory setpoint tracking and disturbance compensation. In general we can not just expect one controller to satisfy independent requirements. In cases where this is a problem a solution is to use two controller functions, or a controller with two degrees of freedom . Figure 7.18 shows a block diagram for such a controller. The partial PID controller may be tuned so that the control loop gives optimal disturbance compensation, while the setpoint signal filter is designed so that the setpoint tracking property for the combined system consisting of the filter in series with the control loop becomes optimal. Such a structure is implemented in some commercial controllers.

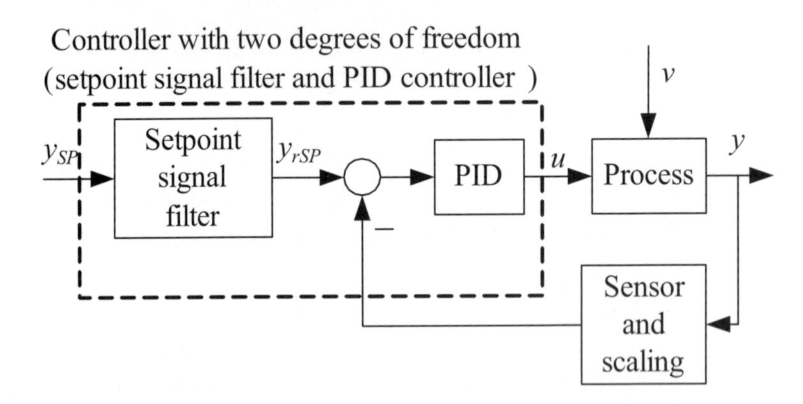

Figure 7.18: Controller with two degrees of freedom (consisting of setpoint signal filter and PID controller)

Figure 7.19 shows an alternative structure which also have two degrees of freedom. Now a feedforward controller couples directly from the setpoint to the control variable. Feedforward control is described in detail in Section 9.1.

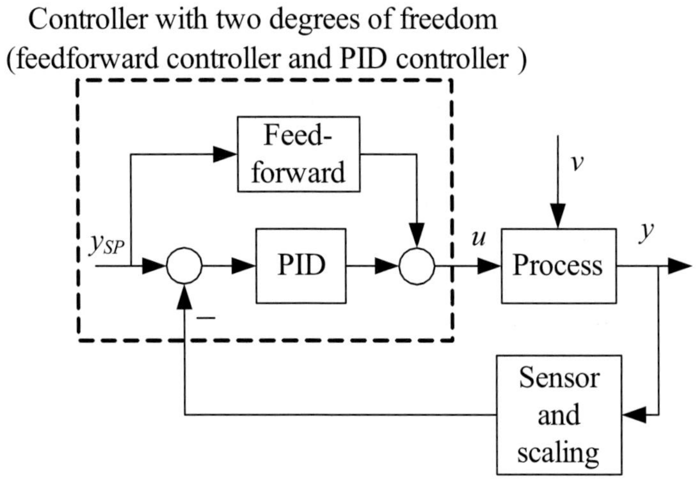

Figure 7.19: Controller with two degrees of freedom (including feedforward controller and PID controller)

Chapter 8

Frequency response based PID tuning

8.1 Introduction

In this chapter you will see how PID parameters can be tuned from frequency response plots in Bode diagram. The basic method is the Ziegler-Nichols' closed loop method interpreted in the frequency domain. The resulting method – the Ziegler-Nichols' *frequency response method* – can be used to tune PID controllers from the *frequency response*, $H_p(j\omega)$, of the process to be controlled ($H_p(s)$ is the transfer function from control variable u to process measurement y_m).

You will also see how the PID controller parameters can be *adjusted* or fine tuned from requirements about the frequency response, $L(j\omega)$, of the loop transfer function. The starting point for the adjustment is a set of PID parameters already tuned with some tuning method, for example one of the Ziegler-Nichols' methods, cf. Chapter 4, or Skogestad's method or some other transfer function based tuning method, cf. Section 7.5.

In this chapter it is assumed that the PID controller is on *serial form*, cf. Section 2.6.7. The serial form is more convenient in frequency response design of control systems than is the parallel form. This is due to the factorized form of the serial controllers' transfer function. If the PID controller you actually use has parallel form, you can transform the serial parameters to corresponding parallel parameters using the transformation formulas (2.51)–(2.53) so that your parallel controller behaves like a serial controller.

Frequency response based controller design is computational demanding, so you should use software tools[1].

8.2 Useful facts about frequency response of a control loop

8.2.1 Frequency response of the loop transfer function

In this chapter we will assume that the mathematical model of the control system is represented by the transfer function based block diagram shown in Figure 8.1. In Chapter 6 we saw that the loop transfer function $L(s)$ is

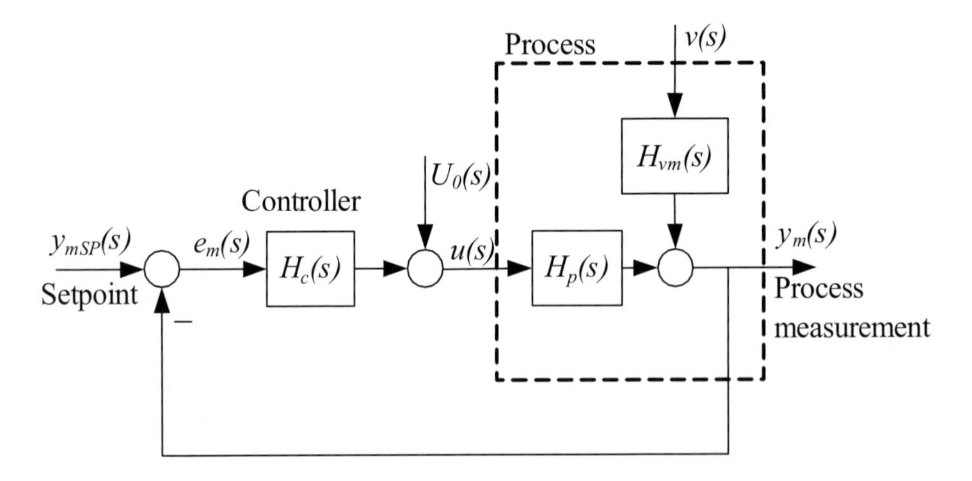

Figure 8.1: Block diagram of control system

crucial in stability analysis and in analysis of setpoint tracking and disturbance compensation of a control function. $L(s)$ is

$$L(s) = H_c(s)H_p(s) \tag{8.1}$$

In the following sections the frequency response of $L(s)$ is used. (8.1) implies the following relation for the amplitude gain function of $L(s)$:

$$|L(j\omega)| = |H_c(j\omega)||H_p(j\omega)| \tag{8.2}$$

or, in decibel:

$$|L(j\omega)|[\mathrm{dB}] = |H_c(j\omega)|[\mathrm{dB}] + |H_p(j\omega)|[\mathrm{dB}] \tag{8.3}$$

[1] For example Control System Toolbox in MATLAB or Control Design Toolkit in LabVIEW.

And for the phase lag function we have:

$$\arg L(j\omega) = \arg H_c(j\omega) + \arg H_p(j\omega) \tag{8.4}$$

How do we get $H_c(j\omega)$ and $H_p(j\omega)$? $H_c(j\omega)$ can easily be found from the controller transfer function. For example, a PI controller has the following transfer function:

$$H_c(s) = K_p \frac{T_i s + 1}{T_i s} \tag{8.5}$$

Its frequency response becomes

$$H_c(j\omega) = K_p \frac{T_i j\omega + 1}{T_i j\omega} \tag{8.6}$$

The process frequency response, $H_p(j\omega)$, can be found in one of the following ways:

- The transfer function $H_p(s)$ may be derived from the mathematical model of the process – either directly from the model if the model is linear, or from a linearized model if the model is nonlinear. Then,

$$H_p(j\omega) = H_p(s) \text{ with } s = j\omega \tag{8.7}$$

- $H_p(j\omega)$ may stem from a discrete-time transfer function, $H_{p_d}(z)$[2] which may have been found by discretizing the continuous-time transfer function $H_p(s)$ or from model identification based on experimental time series of control signal u and process measurement y_m.[3] Once $H_p(s)$ exists, $H_p(j\omega)$ can be found by

$$H_p(j\omega) = H_{p_d}(z) \text{ with } z = e^{-j\omega T_s} \tag{8.8}$$

  where T_s is the sampling interval.[4]

- $H_p(j\omega)$ can be found from experimental time series of u and y_m directly using Fourier Transform based methods.

8.2.2 Frequency response of PID and PI controller

In frequency response based analysis and design of control systems, it is useful to know the shape of the asymptotical and exact frequency response

[2] z is here the z-variable in discrete-time systems theory.

[3] Model identification may be performed with for example System Identification Toolbox in MATLAB or System Identification Toolkit in LabVIEW.

[4] Discrete-time systems theory is described in documents available on http://techteach.no.

of a PID controller and a PI controller. These frequency responses are described in the following. In Section 8.3.3 the PID controller and the PI controller are compared with respect to frequency response characteristics of a control system.

Frequency response of PID controller

As mentioned in the introduction to this chapter, it is assumed that the PID controller is on *serial form* with the following transfer function:

$$H_c(s) = K_p \frac{(T_i s + 1)(T_d s + 1)}{T_i s(T_f s + 1)} \tag{8.9}$$

A serial PID controller has corner frequencies in a Bode diagram which are easily determined. The corner frequencies are $1/T_i$, $1/T_d$ and $1/T_f$.

From (8.9) we find

$$|H_c(j\omega)| = |H_c(s)_{s=j\omega}| = K_p \frac{\sqrt{(T_i\omega)^2 + 1}\sqrt{(T_d\omega)^2 + 1}}{T_i\omega\sqrt{(T_f\omega)^2 + 1}} \tag{8.10}$$

and

$$\arg H_c(j\omega) = \arg\left[H_c(s)_{s=j\omega}\right] \tag{8.11}$$

$$= \arctan(T_i\omega) + \arctan(T_d\omega) - \arctan(T_f\omega) - 90° \tag{8.12}$$

These expressions can be used for manual calculation (or for programming of the calculation) of the controller frequency response.[5]

I most cases – as when using one of the Ziegler-Nichols' methods – the order of the corner frequencies from small to high frequency is $1/T_i$, $1/T_d$ and $1/T_f$. Figure 8.2 shows asymptotic and exact amplitude and phase curves of the PID controller (8.9). (The PID parameters are $K_p = 1$, $T_i = 1$, $T_d = 0.1$, $T_f = 0.01$.) Below are comments to the frequency response:

- At low frequencies, that is, below the corner frequency $1/T_i$, the PID controller has dominating integral action. This ensures *zero static control error*.
- The integral term gives negative phase contribution to $|L(j\omega)|$, which in itself reduces the phase margin. This is counteracted by the positive phase of the derivative term, see below.

[5]However, it may be time efficient to use functions in Control System Toolbox in MATLAB or in Control Design Toolkit in LabVIEW.

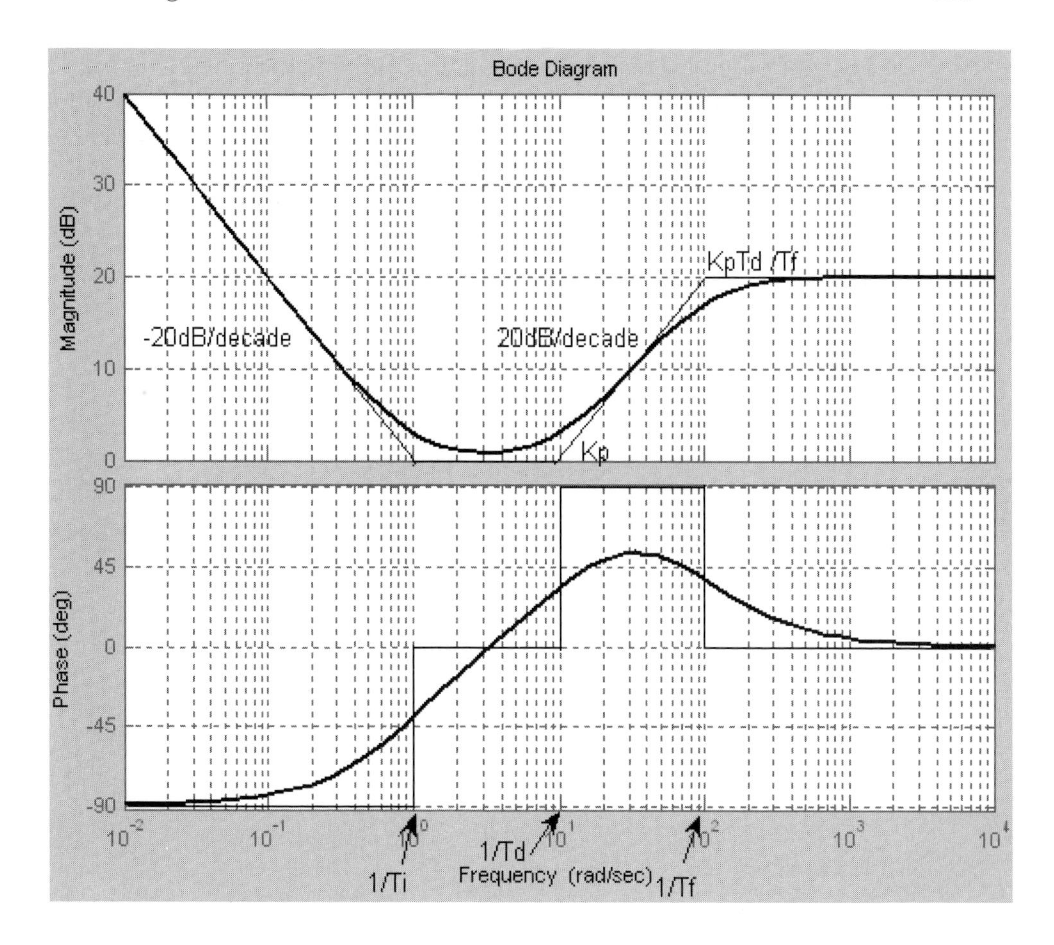

Figure 8.2: Asymptotic and exact Bode plots of the amplitude gain and the phase lag of the PID controller (8.9). (The PID parameters are $K_p = 1$, $T_i = 1$, $T_d = 0.1$, $T_f = 0.01$.)

- At frequencies between the corner frequencies $1/T_d$ and $1/T_f$ the PID controller has a dominating derivative action. This gives the possibility to obtain improved stability or higher bandwidth, see below.

- The maximum phase, $\arg H_c(j\omega)$, is positive. The positive angle is due to the derivative term. Think about the transfer function of a derivator, which is s. Its frequency response is $j\omega$ which have angle $+90°$.

- Typically $\arg H_c(j\omega)$ is between $40°$ and $50°$, but it depends on the ratio of the corner frequencies. $\arg H_c(j\omega)$ occurs near the logarithmic mean of $1/T_d$ and $1/T_f$, which is $1/\sqrt{T_d T_f}$.

- The positive phase gives positive phase contribution to $\arg L(j\omega)$, which can be used to *stabilize* the control loop (since positive phase contribution increases the phase margin) or to increase *the bandwidth* (crossover frequency ω_c of L) without a reduction of the stability. But the derivative term also increases the amplitude of L (observe the positive slope of the amplitude curve), and if T_d becomes too large, $|L|$ may become too large at ω_c, and the stability margins may become small.

Frequency response of PI controller

The PI controller is a special case of the PID controller: We set $T_d = 0$ and $T_f = 0$ in (8.9), which gives the PI controller transfer function:

$$H_c(s) = K_p \frac{T_i s + 1}{T_i s} \tag{8.13}$$

From (8.13) we find

$$|H_c(j\omega)| = |H_c(s)_{s=j\omega}| = K_p \frac{\sqrt{(T_i\omega)^2 + 1}}{T_i\omega} \tag{8.14}$$

and

$$\arg H_c(j\omega) = \arg H_c(s)_{s=j\omega} = \arctan(T_i\omega) - 90^\circ \tag{8.15}$$

Figure 8.3 shows asymptotic and exact amplitude and phase curves of the PI controller. (The PI parameters are $K_p = 1$ and $T_i = 1$.)

Below are comments to the frequency response:

- At low frequencies, that is, below the corner frequency $1/T_i$, the PI controller has dominating integral action, as for the PID controller. This ensures *zero static control error.*

- The integral term gives negative phase contribution to $|L(j\omega)|$, which reduces the phase margin. Compared to using a PID controller: To maintain the stability margins, the decrease of the phase margin the must be counteracted by reducing the loop gain, causing reduction of the crossover frequency ω_c and hence a reduction of the bandwidth of the control system.

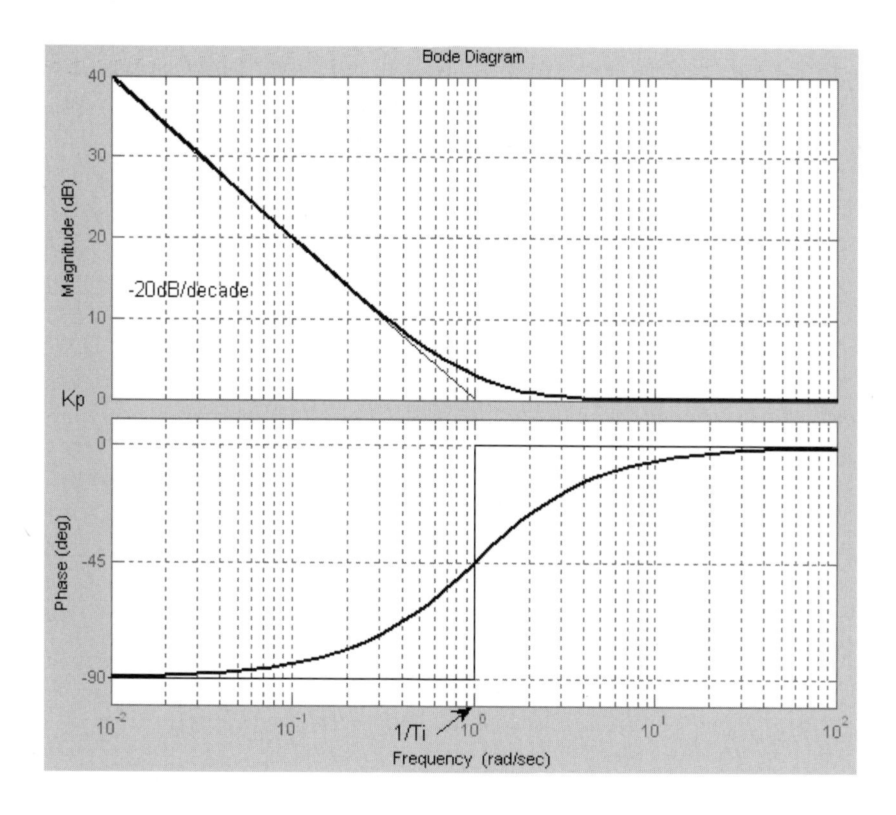

Figure 8.3: Asymptotic and exact Bode plots of the amplitude gain and the phase lag of the PI controller (8.13). (The PI parameters are $K_p = 1$, $T_i = 1$.)

8.3 Ziegler-Nichols' frequency response method

8.3.1 Introduction

Ziegler-Nichols' closed loop method, which is described in Section 4.4, can be interpreted in the frequency domain, and it is convenient to use the name Ziegler-Nichols' frequency response method. The point is that the ultimate controller gain, K_{p_u}, and the ultimate period, T_u, are found from frequency response plots, typically in a Bode diagram. Once K_{p_u} and T_u are known, the controller parameters can be calculated using the Ziegler-Nichols' formulas shown in Table 4.1 (page 95).

The Ziegler-Nichols' frequency response method can be used only if the control system can be brought to the stability limit with a P controller. In other words: It is necessary that $\arg L(j\omega)$, which is equal to $\arg H_p(j\omega)$ when the controller is a P controller, crosses the $-180°$ line at some finite frequency. This requirement is not satisfied neither for integrators, first order processes, nor second order processes. For such processes transfer function based tuning may be used, cf. Chapter 7.

8.3.2 Tuning procedure

The Ziegler-Nichols' frequency response method is as follows:

1. Given a plot of the process frequency response $H_p(j\omega)$ in a Bode diagram.

2. Let the controller be a P controller. The loop transfer function frequency response $L(j\omega)$ is in this case

$$L(j\omega) = H_c(j\omega)H_p(j\omega) = K_pH_p(j\omega) \tag{8.16}$$

which is $H_p(j\omega)$ multiplied by K_p. This implies that $|L(j\omega)| = K_p\,|H_p(j\omega)|$ or

$$|L(j\omega)|\,[\text{dB}] = K_p[\text{dB}] + |H_p(j\omega)|\,[\text{dB}] \tag{8.17}$$

and

$$\arg L(j\omega) = \arg H_p(j\omega) \tag{8.18}$$

Initially, set

$$K_p = 1 = 0\text{dB} \tag{8.19}$$

Find the ultimate gain K_{p_u} as the K_p value which makes the closed loop system marginally stable[6]*. In other words: Find K_p so that $\omega_c = \omega_{180}$*, cf. Chapter 6.4.

A few words about how to do the above operations in a Bode diagram: An increase/decrease of K_p from the initial value of $1 = 0$dB implies that $|L(j\omega)|$ is raised/lowered the amount of the change of K_p in dB. But, in stead of drawing a new $|L(j\omega)|$ plot for a new K_p value, it is easier to lower/raise the 0dB line of the Bode plot of $|H_p(j\omega)|$ to get new $|L(j\omega)|$ plots.

3. The ultimate period T_u is the period of the sustained oscillations in the closed loop method. In Chapter 6.4 it was shown that T_u is equal to $T_u = 2\pi/\omega_{180}$ where ω_{180} is the phase crossover frequency of $\arg L$. But since the controller is a P controller, ω_{180} of $L(j\omega)$ is identical to ω_{180} of $\arg H_p(j\omega)$. So,

$$T_u = \frac{2\pi}{\omega_{180}} \tag{8.20}$$

where ω_{180} is the phase crossover frequency of $\arg H_p(j\omega)$.

[6] This corresponds to the sustained oscillations in the closed loop method.

4. Calculate the controller parameters from Table 4.1. The derivative filter time constant T_f of the PID controller can be set to

$$T_f = aT_d \tag{8.21}$$

where a may be set to 0.1, cf. page 44.

5. Draw a Bode plot of $L(j\omega)$ and observe if the stability margins are acceptable. GM should be 6 and 12dB, PM between 30 and 60 degrees, alternatively $|S(j\omega)|_{\max}$ between 3dB = 1.4 and 6dB = 2.0. If the stability margins are too small (large), try decreasing (increasing) K_p. Adjustment of T_i and T_d may also be tried (such adjustments are described in Section 8.4).

6. If possible: Simulate the control system to verify that the theoretical design gives acceptable results.

Example 8.1 ***Ziegler-Nichols' frequency response method for PID tuning***

See the block diagram shown in Figure 8.1. Assume that the transfer functions in the block diagram are as follows:

$$H_p(s) = \frac{K}{(T_1 s + 1)(T_2 s + 1)} e^{-\tau s} \tag{8.22}$$

$$H_{vm}(s) = \frac{K_{vm}}{(T_1 s + 1)(T_2 s + 1)} e^{-\tau s} \tag{8.23}$$

where

$$K = 1;\ K_{vm} = 1;\ T_1 = 1\text{s};\ T_2 = 0.5\text{s};\ \tau = 0.5\text{s} \tag{8.24}$$

Figure 8.4 shows a Bode plot of $H_p(j\omega)$ which is $L(j\omega)$ with P controller with $K_p = 1$. From the Bode plot we read off $\omega_{180} = 2.27$rad/s and $K_{p_u} = 11.5\text{dB} = 3.76$. According to (8.20) $T_u = 2.78$s.[7] Inserting K_{p_u} and T_u into Table 4.1 gives the following PID parameters (the filter time constant T_f has been given the typical value of $0.1T_d$):

$$K_p = 0.6K_{p_u} = 0.6 \cdot 3.76 = 2.26 \tag{8.25}$$

$$T_i = \frac{T_u}{2} = \frac{2.78}{2} = 1.39\text{s} \tag{8.26}$$

$$T_d = \frac{T_u}{8} = \frac{2.78}{8} = 0.35\text{s} \tag{8.27}$$

$$T_f = 0.1T_d = 0.035\text{s} \tag{8.28}$$

Figure 8.5 shows several Bode plots together with simulated responses for the control system with PID controller tuned above. You can check if (8.3)

[7] I tried the Ziegler-Nichols' closed loop method on the system – with the same values of K_{p_u} and T_u.

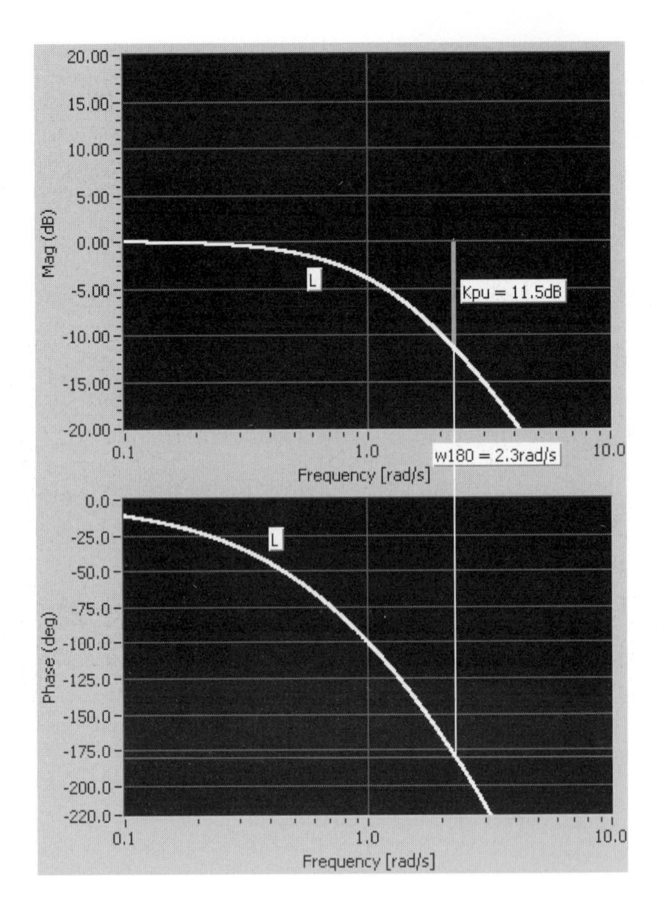

Figure 8.4: Bode plot of $H_p(j\omega)$, which is $L(j\omega)$ with $K_p = 1$. We read off $\omega_{180} = 2.3$rad/s and $K_{p_u} = 11.5$dB.

and (8.4) is confirmed in this example.

From Figure 8.5 we read off the following characteristic frequencies and stability margins:

$$\omega_c = 1.85\text{rad/s} \tag{8.29}$$
$$\omega_s = 0.50\text{rad/s} \tag{8.30}$$
$$\omega_{180} = 2.78\text{rad/s} \tag{8.31}$$
$$GM = 3.9\text{dB} \tag{8.32}$$
$$PM = 30.3^\circ \tag{8.33}$$
$$|S|_{\max} = 10.3\text{dB} \tag{8.34}$$

ω_c (loop transfer function crossover frequency) and ω_s (the sensitivity bandwidth) are two possible bandwidths, cf. Section 6.3.4.

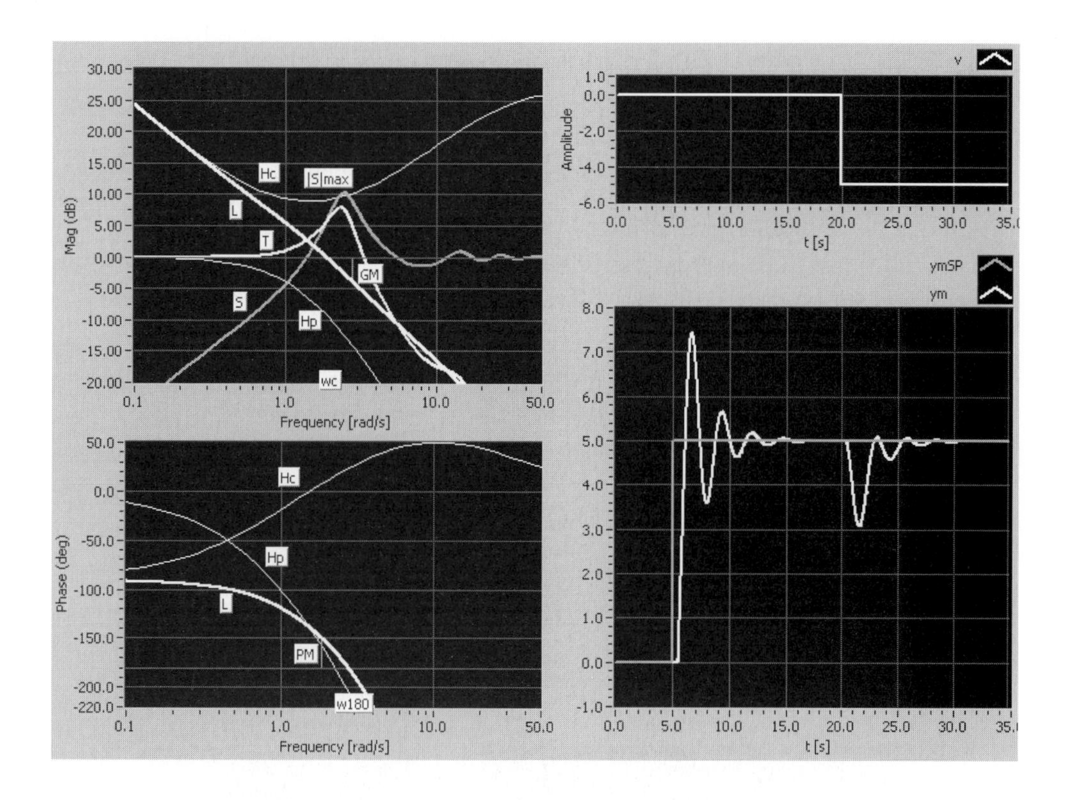

Figure 8.5: Example 8.1: Bode plots and simulated responses of the PID control system

GM has a small value, and $|S|_{\max}$ has a large value – hence the stability margins are small. The simulated responses show an amplitude decay ratio of approximately 1/4 which is a typical value for Ziegler-Nichols' methods.

[End of Example 8.1]

8.3.3 Comparing PID and PI tuning in frequency domain

In several examples in previous chapters we have seen that the PID controller may give faster control than the PI controller. The following example will demonstrate this using frequency domain characteristics. We will see that the PID controller gives *larger bandwidth* of the control system, and hence faster control.

Example 8.2 *Comparing PI and PID tuning in frequency domain*

In Example 8.1 we tuned a PID controller. Tuning a PI controller in stead (using the Ziegler-Nichols' frequency response method) gives the following controller parameters:

$$K_p = 1.69; T_i = 2.31\text{s} \quad (8.35)$$

Figure 8.6 shows several Bode plots for the control system with PI controller tuned above, together with simulated responses (with a setpoint step and a disturbance step). For comparison, Figure 8.7 shows Bode

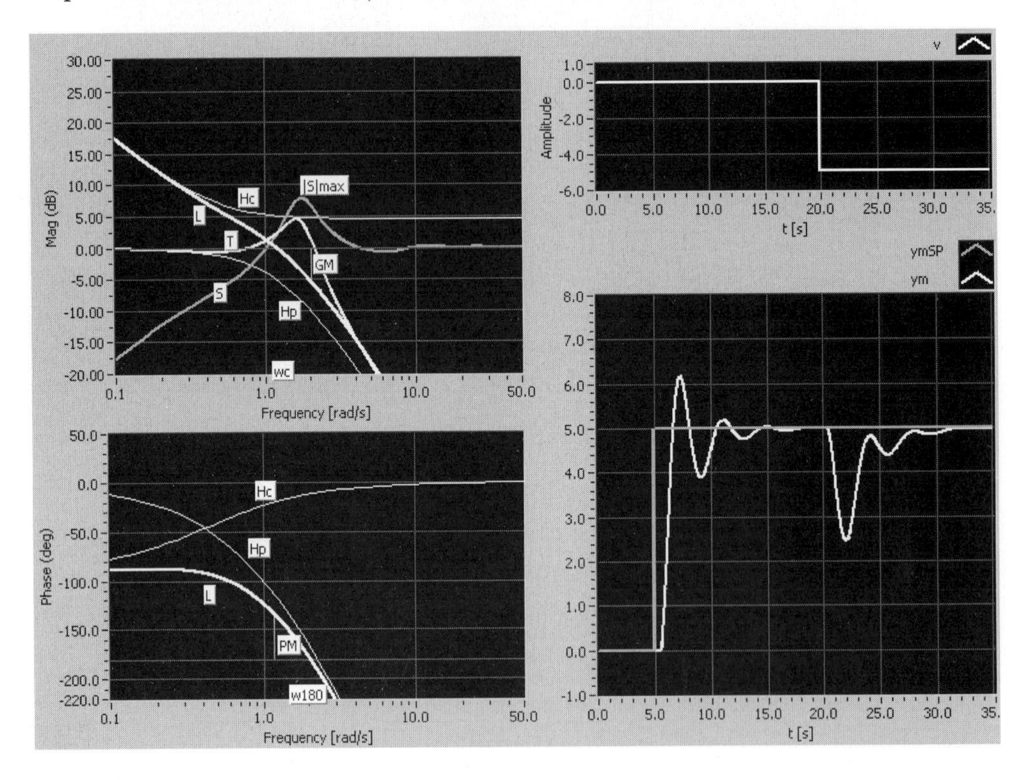

Figure 8.6: Example 8.2: Bode plots and simulated responses of the PI control system

plots of $H_p(j\omega)$, $H_c(j\omega)$, and $L(j\omega)$ with indications of gain margins (GM) and phase margins (PM) for the PI control system and the PID control system.

From Figure 8.5 we read off characteristic frequencies and stability margins as shown in Table 8.1. (Some of the values, namely ω_c, ω_{180}, GM and PM, can alternatively be read off from Figure 8.7.)

From the values in Table 8.1 we can conclude as follows:

- With PID controller, the bandwidth is higher (this applies to both ω_c

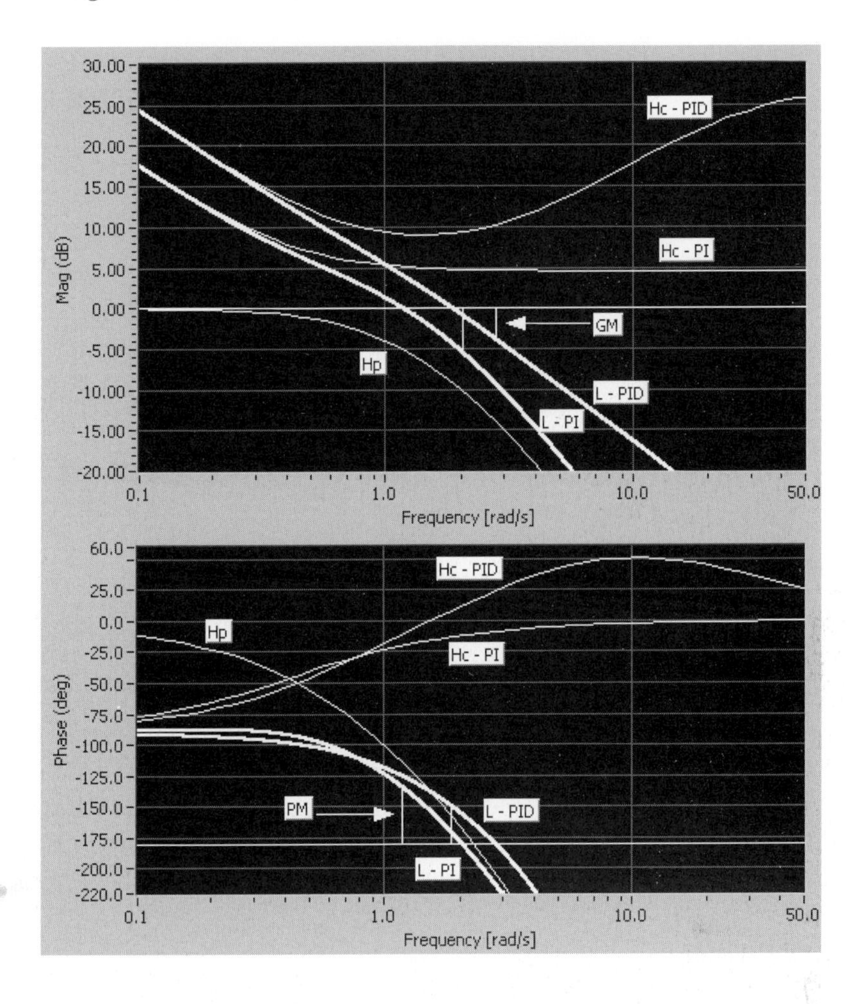

Figure 8.7: Example 8.2: Bode plots with indications of gain margins and phase margins for PI control system and the PID control system

and ω_s), and therefore the control action is quicker.

- With PI controller, the stability is better (GM and PM are larger, and $|S|_{\max}$ is smaller).

The above results are actually quite typical of control systems where Ziegler-Nichols' method has been used. But, it is of course possible to re-tune the controllers to obtain certain specifications. For example, larger stability margins can be obtained by reducing K_p. Controller parameter adjustment is the topic of the following section.

[End of Example 8.2]

	PID	PI
ω_c	1.85rad/s	1.18rad/s
ω_s	0.50rad/s	0.25rad/s
ω_{180}	2.78rad/s	2.04rad/s
GM	3.9dB	5.48dB
PM	30.3°	45.6°
$\lvert S\rvert_{\max}$	10.3dB	7.8dB

Table 8.1: Frequency domain characteristics of PID and PI controller tuned with the Ziegler-Nichols' frequency response method

8.4 Frequency response based adjustment of PID parameters

8.4.1 Introduction

In this section we will observe how the PID parameters K_p, T_i and T_d influence the stability and bandwidth properties of a control system. This knowledge can be used to adjust or re-fine PID parameters from frequency response considerations until specifications regarding bandwidth and stability margins are met. In most cases the first parameter to try to adjust is the proportional gain, K_p, and may be this is the only adjustment needed.

The PID controller may already have been tuned with e.g. Ziegler-Nichols' closed loop method (cf. Chapter 4.4) or Ziegler-Nichols' frequency response method (cf. Section 8.3) or with a transfer function based method (Chapter 7). We will here concentrate on the PID controller. The P, PI and PD controllers are special cases of the PID controller.

A concrete case will demonstrate how the stability and the bandwidth of a control system typically are influenced by changes of the PID parameters K_p, T_i and T_d. This knowledge may be useful for adjusting PID parameters in general. However, there is (of course) no guaranty that the results of the parameter adjustments are as observed in the case. The case is the same as in Example 8.1. The block diagram of the control system is as shown in Figure 8.1. The transfer functions in the block diagram are repeated here for convenience:

$$H_p(s) = \frac{K}{(T_1 s + 1)(T_2 s + 1)} e^{-\tau s} \tag{8.36}$$

$$H_{vm}(s) = \frac{K_{vm}}{(T_1 s + 1)(T_2 s + 1)} e^{-\tau s} \tag{8.37}$$

with parameter values

$$K = 1;\ K_{vm} = 1;\ T_1 = 1\text{s};\ T_2 = 0.5\text{s};\ \tau = 0.5\text{s} \tag{8.38}$$

$H_{vm}(s)$ is not a part of the loop and has therefore no impact on the stability or the bandwidth of the control loop. The Ziegler-Nichols' frequency response method gives the following PID parameters (they were found in Example 8.1):

$$K_p = 2.26;\ T_i = 1.39\text{s};\ T_d = 0.35\text{s};\ T_f = 0.035\text{s} \tag{8.39}$$

Figure 8.8 shows several Bode plots of the control system, including asymptotes of the frequency response of the controller, and simulated responses (due to a setpoint step and a disturbance step). From the Bode

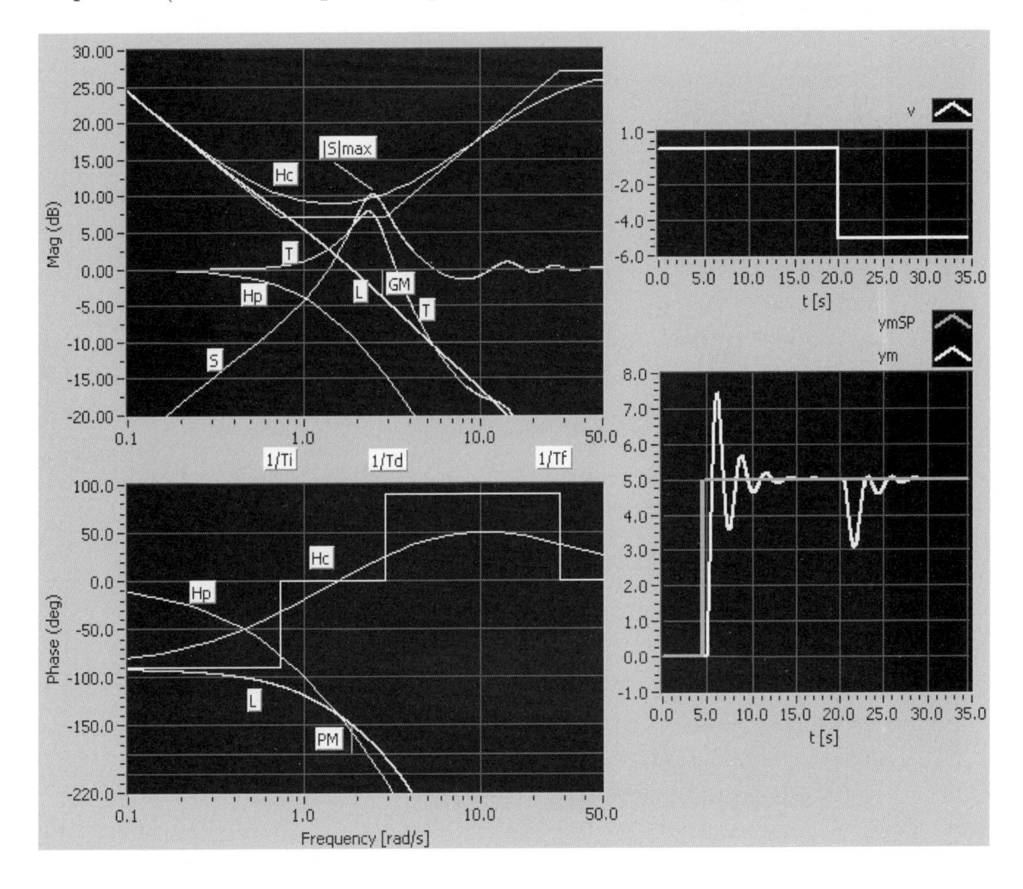

Figure 8.8: Bode plots and simulated responses (due to setpoint step and disturbance step) of the control system with nominal PID parameter values.

plots in Figure 8.8 we read off several characteristic frequencies and stability margins. These values are denoted "original" in the tables shown in the following sections.

8.4.2 Adjusting K_p

The plots shown in Figure indicates that the stability margin of the control system is somewhat small, since $GM = 3.9\text{dB}$ (small) and $|S|_{\max} = 10.3\text{dB}$ (large). If we want to obtain better stability, the first adjustment to try is to *decrease* K_p. Let us here decrease K_p from the original value of 2.26 in (8.39) to its half value: 1.13. Thus, the PID parameters are

$$K_p = 1.13;\ T_i = 1.39\text{s};\ T_d = 0.35\text{s};\ T_f = 0.035\text{s} \tag{8.40}$$

Figure 8.9 shows several Bode plots, including asymptotes of the frequency response of the controller, and simulated responses. From the Bode plots

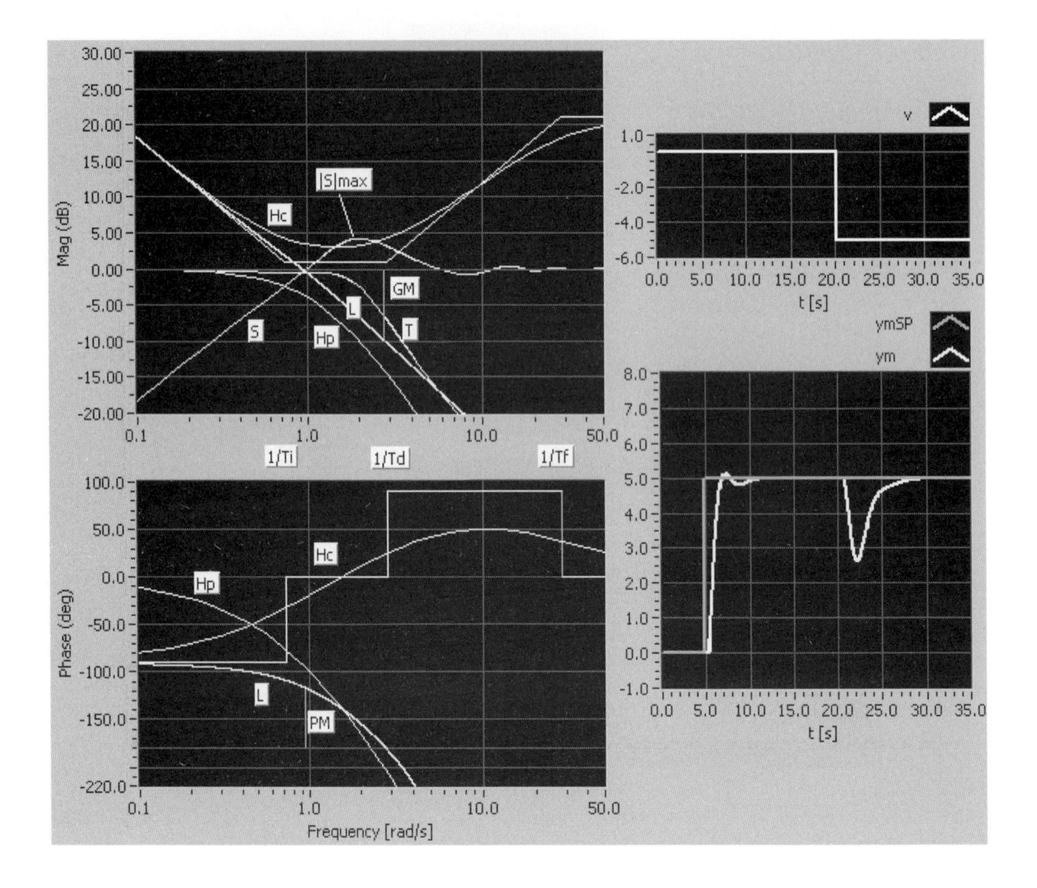

Figure 8.9: Bode plots and simulated responses of the control system with decreased K_p.

in Figure 8.9 we read off the characteristic frequencies and stability margins as shown in the table below:

	Decreased K_p	**Original param.**
ω_c	0.93rad/s	1.85rad/s
ω_s	0.25rad/s	0.50rad/s
ω_{180}	2.74rad/s	2.78rad/s
GM	9.9dB	3.9dB
PM	63.8°	30.3°
$\lvert S\rvert_{\max}$	4.2dB	10.3dB

From the above table we observe that the effect of decreasing K_p is

- better stability (larger stability margins),
- more sluggish responses (smaller bandwidths).

By fine-tuning K_p we can obtain a specified value of e.g. $|S|_{\max}$. For example, it can be shown that $K_p = 1.52$ yields $|S|_{\max} = 6.0\text{dB}$.

Above, we have seen what happens if K_p is *decreased*. The typical consequences of *increasing* K_p are the reverse of the above observations, namely reduced stability and quicker and more oscillating responses.

8.4.3 Adjusting T_i

Let us try *decreasing* T_i from the original value of 1.39 in (8.39) to 0.90. Thus, the PID parameters are

$$K_p = 2.26;\ T_i = 0.90\text{s};\ T_d = 0.35\text{s};\ T_f = 0.035\text{s} \tag{8.41}$$

Figure 8.10 shows several Bode plots of the control system, including asymptotes of the frequency response of the controller, and simulated responses. From the Bode plots in Figure 8.10 we read off the characteristic frequencies and stability margins as shown in the table below:

	Decreased T_i	**Original param.**
ω_c	1.97rad/s	1.85rad/s
ω_s	0.60rad/s	0.50rad/s
ω_{180}	2.51rad/s	2.78rad/s
GM	2.6dB	3.9dB
PM	16.5°	30.3°
$\lvert S\rvert_{\max}$	14.1dB	10.3dB

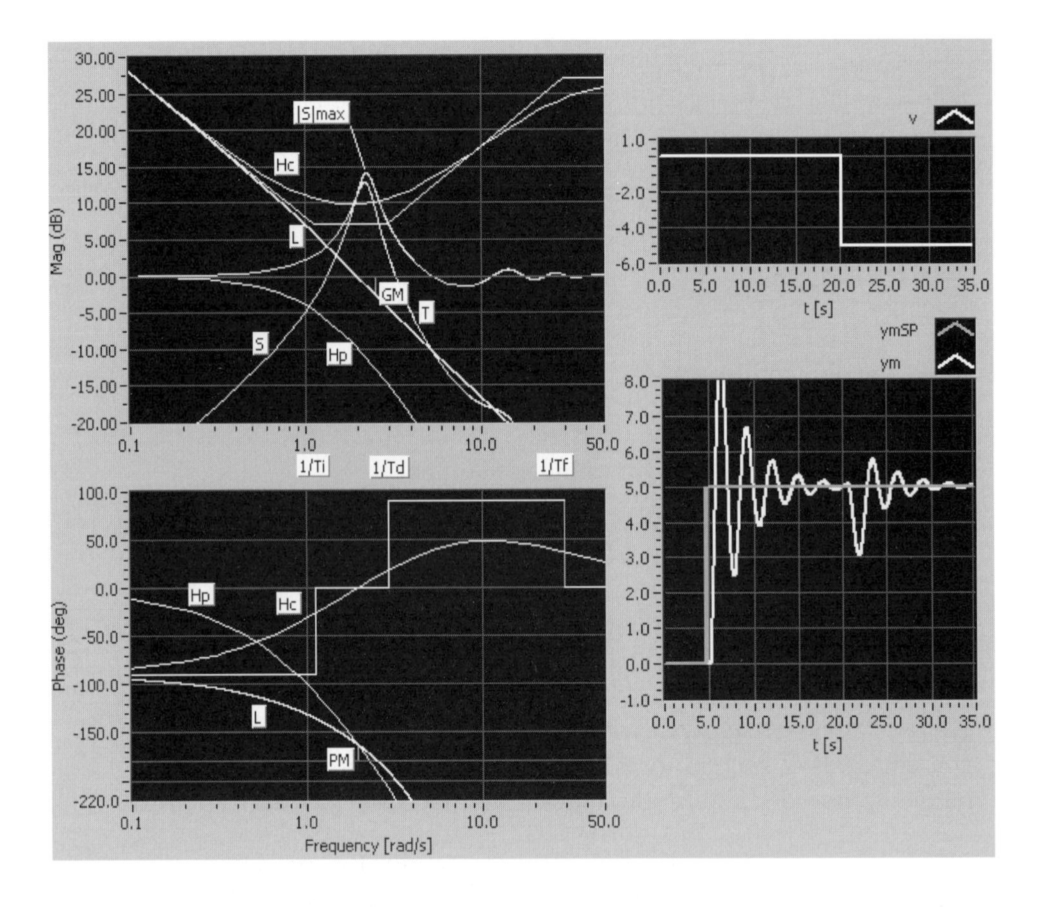

Figure 8.10: Bode plots and simulated responses of the control system with reduced T_i.

From this table we observe that the effect of decreasing T_i is

- reduced stability (smaller stability margins),
- quicker responses (larger bandwidths), but more oscillating responses (due to reduced stability).

One explanation of the reduced stability as T_i is reduced is that the corner frequency $1/T_i$ (of the asymptotic frequency response of the controller) is increased, thereby increasing the frequency range where the PID controller has dominating integral action. An integrator has negative phase (angle). Therefore the PID controller, H_c, contributes with a more negative phase to the phase of the loop transfer function, L, causing decreased phase margin.

One explanation of the increased bandwidth as T_i is reduced is that the more dominating integral action (due to reduced T_i) increases the contribution from H_c to the gain of L, and when the loop gain increases the bandwidth increases.

We have seen what happens if T_i is decreased. The typical consequences of *increasing* T_i are the reverse, namely somewhat better stability and slower responses.

8.4.4 Adjusting T_d

Let us try *increasing* T_d from the original value of 0.35 in (8.39) to 0.70 (the filter time constant T_f is increased proportionally from 0.035 to 0.07). Thus, the PID parameters are

$$K_p = 2.26;\ T_i = 1.39\text{s};\ T_d = 0.70\text{s};\ T_f = 0.070\text{s} \tag{8.42}$$

Figure 8.11 shows Bode plots and simulated responses for the control system. From the Bode plot in Figure 8.11 we read off the characteristic frequencies and stability margins as shown in the table below:

	Increased T_d	**Original param.**
ω_c	2.75rad/s	1.85rad/s
ω_s	0.55rad/s	0.50rad/s
ω_{180}	3.15rad/s	2.78rad/s
GM	1.0dB	3.9dB
PM	14.2°	30.3°
$\lvert S\rvert_{\max}$	19.8dB	10.3dB

From the above table we observe that the effect of increasing T_d is

- reduced stability (smaller stability margins),
- quicker responses (larger bandwidths), but more oscillating responses (due to reduced stability).

One explanation of the reduced stability as T_d is increased is that the corner frequency $1/T_d$ (and $1/T_f$) are decreased so that the amplitude contribution from $|H_c|$ to $|L|$ around the phase crossover frequency ω_{180} is increased, which tends to decrease the gain margin. Simultaneously, $\arg H_c$ contributes positively to $\arg L$, which in itself increases the phase margin,

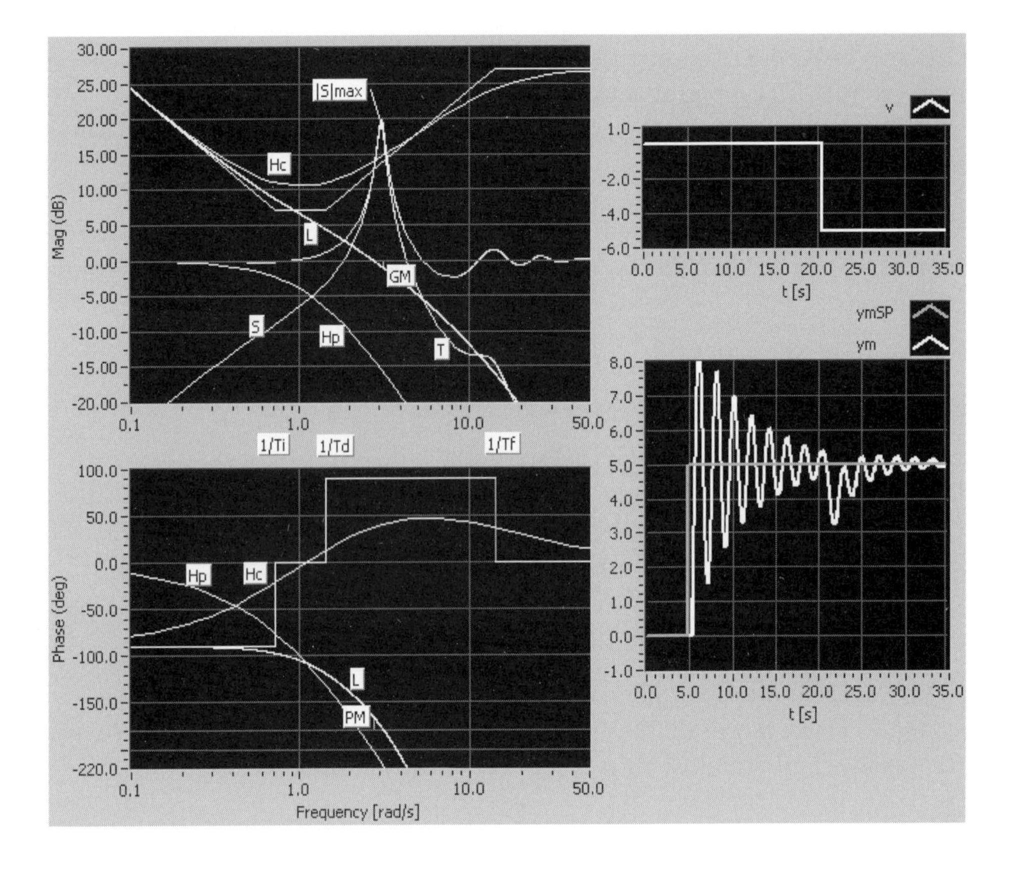

Figure 8.11: Bode plots and simulated responses of the control system with increased T_d.

but in this example, and it is typical at least for systems containing time delay, the decrease of the gain margin is "stronger" than the increase of the phase margin, resulting in overall decreased stability. In systems not including time delay the stability need not be reduced when increasing T_d, but remember from Section 2.6.7 the problem of the controller's increased sensitivity to measurement noise as T_d is increased.

An explanation of the increased bandwidth as T_d is increased is that the loop gain $|L|$ is increase due to the positive amplitude contribution of H_c.

What are the consequences of *reducing* T_d from its original value? The stability of the control system may be reduced. This is because $\arg L$ will decrease, and hence the phase margin will decrease, because the positive phase contribution to $\arg L$ is reduced. In our example, it can be found that setting $T_d = 0$, which means that the derivative term has been removed completely (without re-tuning the controller), causes poor stability as GM is as small as 1.63dB and $|S|_{\max}$ is as large as 17.6dB.

8.4.5 Summary

We can sum up our observations in the examples above as follows (the statements apply to most control systems):

- Decreasing K_p decreases bandwidth and improves stability. (Increasing K_p has the opposite effects.)
- Decreasing T_i increases bandwidth and reduces stability. (Increasing K_p has the opposite effects.)
- Increasing T_d increases bandwidth and may reduce stability. Decreasing T_d (from a nominal value) reduces stability. Consequently, you should be careful about adjusting T_d in either direction once it has been tuned.

Chapter 9

Various control methods and control structures

This chapter describes various control methods and control structures and for industrial applications. Most of the structures involves one or more PID control loops.

9.1 Feedforward control

9.1.1 Introduction

We know from previous chapters that feedback control can bring the process output variable to or close to the setpoint. Feedback control is in most cases a sufficiently good control method. But improvements can be made, if required. A problem with feedback is that it is no adjustment of the control variable before the control error is different from zero, since the control variable is adjusted as a function of the control error. This problem does not exist in *feedforward control*, which may be used as the only control method, or, more typically, as a supplement to feedback control.

In feedforward control there is a *coupling from the setpoint and/or from the disturbance directly to the control variable*, that is, a coupling from an input signal to the control variable. The control variable adjustment is not error-based. In stead it is based on knowledge about the process in the form of a mathematical model of the process and knowledge about or measurements of the process disturbances.

Perfect feedforward control gives zero control error for *all types of signals* (e.g. a sinusoid and a ramp) in the setpoint and in the disturbance. This sounds good, but feedforward control may be *difficult to implement* since it assumes or is based on a mathematical process model and that all variables of the model at any instant of time must have known values through measurements or in some other way. These requirements are never completely satisfied, and therefore in practice the control error becomes different from zero. We can however assume that the control error becomes smaller with imperfect feedforward control than without feedforward control.

If feedforward control is used, it is typically used together with feedback control. Figure 9.1 shows the structure of a control system with both feedforward and feedback control. The purpose of feedback control is to

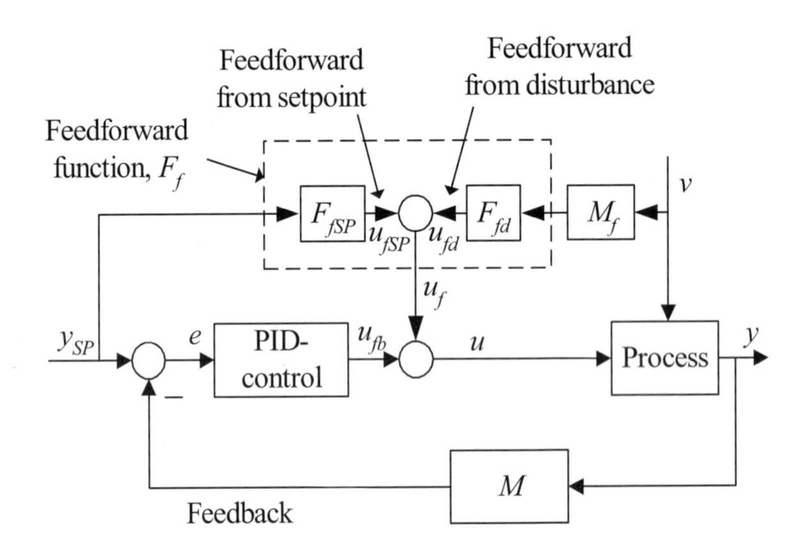

Figure 9.1: Control system with both feedforward and feedback control

reduce the control error due to the inevitable imperfect feedforward control. Practical feedforward control can never be perfect, because of model errors and imprecise measurements.

One interpretation of feedforward control is that it introduces an artificial connection from the disturbance to the process output variable. The purpose of this artificial connection is to counteract the natural connection (from the disturbance to the process output variable).

The feedforward function F_f, which usually consists of a sum of partial functions F_{fSP} and F_{fd} as shown in Figure 9.1, can be developed from a differential equations model or a transfer functions model of the process. Both ways are described in the following sections of this chapter. In both cases the principle is to solve the process model with respect to control variable u with the setpoint y_{SP} substituted for the process output variable y. This control variable, u, is then the additive contribution of the control variable u_f to the total control variable u, cf. Figure 9.1.

The feedforward function F_f can be interpreted as a continuously calculated nominal control signal, that is, that $u_f(t) = u_0(t)$, cf. e.g. Figure 2.11.

Using feedforward together with feedback does not influence the stability of the feedback loop because the feedforward does not introduce new dynamics in the loop.

9.1.2 Designing feedforward control from differential equation models

The feedforward function F_f can be derived quite easily from a differential equations model of the process to be controlled. As mentioned in Section 9.1.1 the principle is to *solve for the control output variable in the process model with the setpoint substituted for the process output variable* (which is the variable to be controlled). The model may be linear or non-linear. (In Section9.1.3 we will derive feedforward functions from transfer function models.)

Example 9.1 ***Feedforward control of a thermal process***

Figure 9.2 shows a heated liquid tank where the temperature T shall be controlled using feedback with PID controller in combination with feedforward control. We assume the following process model, which is based on energy balance:

$$c\rho V\dot{T}(t) = \underbrace{K_h u(t)}_{P} + cw\left[T_{in}(t) - T(t)\right] + U\left[T_e(t) - T(t)\right] \tag{9.1}$$

where T [K] is the temperature of the liquid in the tank, T_{in} [K] is the inlet temperature, T_e [K] is environmental temperature, c [J/(kg K)] is specific heat capacity, w [kg/s] is mass flow (same in as out), V [m^3] is the liquid volume, ρ [kg/m^3] is the liquid density, U [(J/s)/K] is the total heat transfer coefficient, $P = K_h u$ [J/min] is supplied power via heating element where K_h is a parameter (gain) and u [%] is the control signal applied to

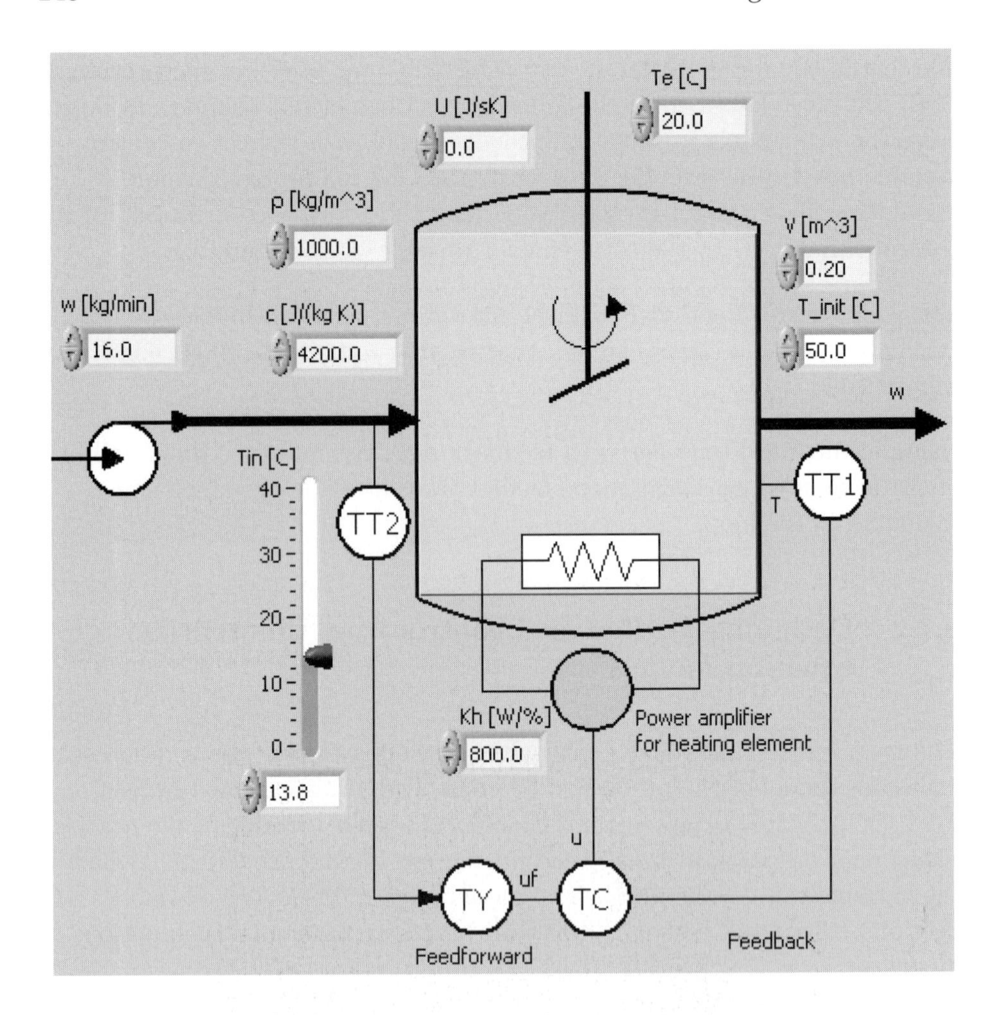

Figure 9.2: Example 9.1: Heated liquid tank where the temperature T shall be controlled with feedforward control in addition to feedback control. TY represents a computing element.

the heating element. $c\rho VT$ is the (temperature dependent) energy of the liquid in the tank. We can consider T_{in} and T_e as disturbances, but the derivation of the feedforward function F_f is not dependent of such a classification. In the following we assume for simplicity that the heat transfer coefficient U is zero so that the heat transport through the walls is zero.

Now, let us derive the feedforward function from the process model (9.1). First, we substitute the temperature T by the temperature setpoint T_{SP}:

$$c\rho V\dot{T}_{SP}(t) = K_h u(t) + cw\left[T_{in}(t) - T_{SP}(t)\right] \tag{9.2}$$

We *solve (9.2) for the control variable* u to get the feedforward control variable u_f:

$$u_f(t) = \frac{1}{K_h}\left\{c\rho V\dot{T}_{SP}(t) - cw\left[T_{in}(t) - T_{SP}(t)\right]\right\} \quad (9.3)$$

$$= \underbrace{\frac{1}{K_h}\left[c\rho V\dot{T}_{SP}(t) + cwT_{SP}(t)\right]}_{u_{fSP}} + \underbrace{\frac{1}{K_h}\left[-cwT_{in}(t)\right]}_{u_{fd}} \quad (9.4)$$

We see that calculation of feedforward control signal u_f requires measurement or knowledge of the following five quantities: c, ρ, V, h, w, K_h and T_{in}, in addition to the setpoint time-derivative, $\dot{T}_{SP}$.

The following cases are simulated:

- **Without feedforward control**, but with feedback control with PI controller with parameters $K_p = 4$ and $T_i = 5$min. T_{SP} is changed as a ramp of slope 0.5K/min. See Figure 9.3. Accurate reading of the steady-state control error shows $e_s = 0.88$K.

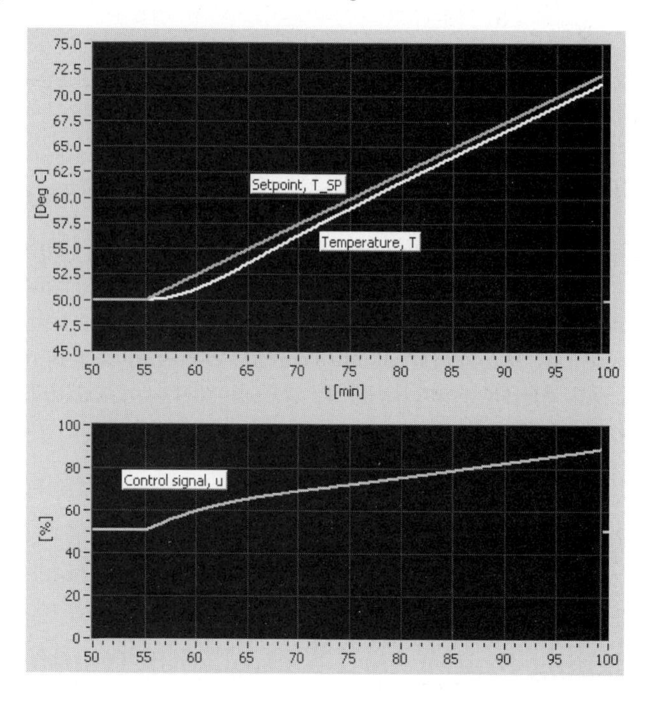

Figure 9.3: Example 9.1: Simulation of temperature control system *without feedforward control*, but with feedback control with PI-controller

- **With feedforward control** according to (9.4) together with feedback control with PI controller. See Figure 9.4. The steady-state

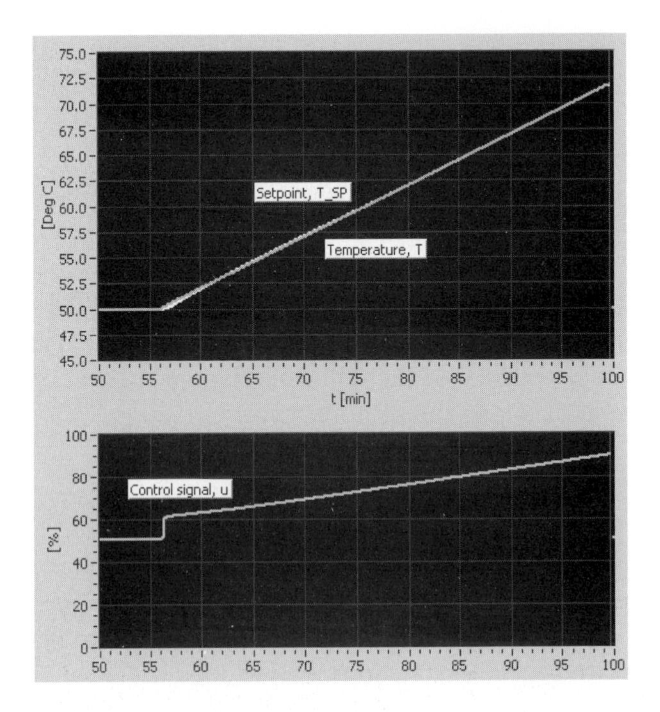

Figure 9.4: Example 9.1: Simulation of temperature control system *with feedforward control* and with feedback control with PI-controller

control error now goes towards zero with increasing time.

- **With feedforward control** according to (9.4) together with feedback control with PI controller. The setpoint is now constant, but there are steps in the disturbance T_{in} (the inlet temperature), see Figure 9.5. We see that the control variable compensates immediately for the variations in the disturbance, which is due to the direct control action inherent in feedforward control. The control error is (in principle) always zero.[1]

[End of Example 9.1]

9.1.3 Designing feedforward control from transfer function models

All *linear* process models can be written on the form

$$y(s) = H_u(s)u(s) + H_v(s)v(s) \tag{9.5}$$

[1] In this simulaton the control error is a little different from zero due to numerical inaccuracies in the way I have implemented the simulator.

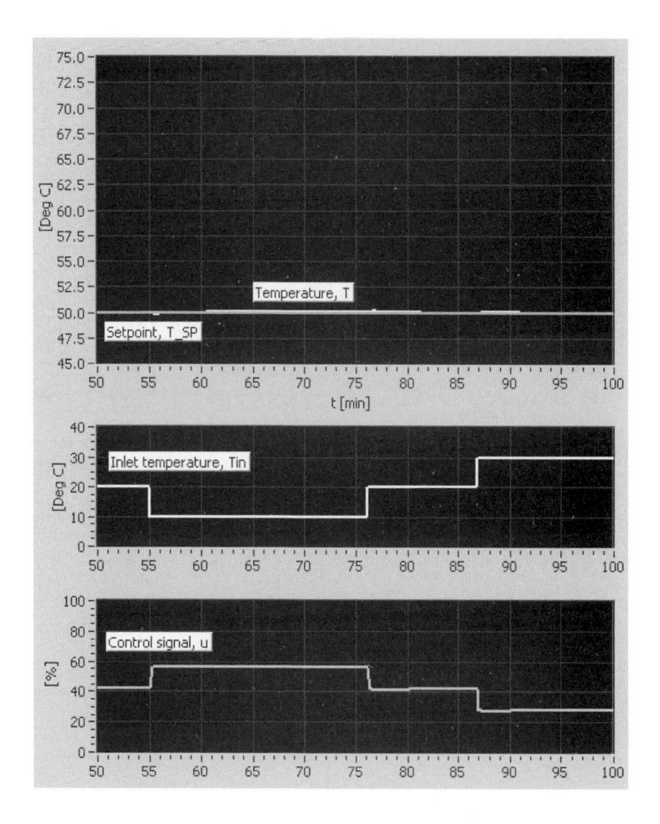

Figure 9.5: Example 9.1: Simulation of temperature control system with feedforward control *together with* feedback control with PI-controller. There is a step change in T_{in}.

where u is the control variable, v is the disturbance and y is the process output variable. We shall find the feedforward function which gives zero control error for any setpoint signal y_{SP} and any disturbance signal v. We start by setting $y_{SP}(s)$ for $y(s)$ in the process model (9.5), and then solve for the control variable $u(s)$. The result is

$$u_f(s) = \underbrace{\frac{1}{H_u(s)}}_{H_{fSP}(s)} y_{SP}(s) + \underbrace{\frac{-H_v(s)}{H_u(s)}}_{H_{fd}(s)} v(s) \quad (9.6)$$

$$= H_{fSP}(s) y_{SP}(s) + H_{fd}(s) v(s) \quad (9.7)$$

where $u_f(s)$ is the control variable in feedforward control. $H_{fSP}(s)$ and $H_{fd}(s)$ are transfer functions which realize feedforward from the setpoint and the disturbance, respectively. ($H_{fd}(s)$ includes the transfer function of the sensor used to measure v.)

Feedforward can be combined with feedback. The purpose of feedback is to reduce the control error due to imperfect feedforward. Figure 9.6 shows a control system with both feedforward and feedback.

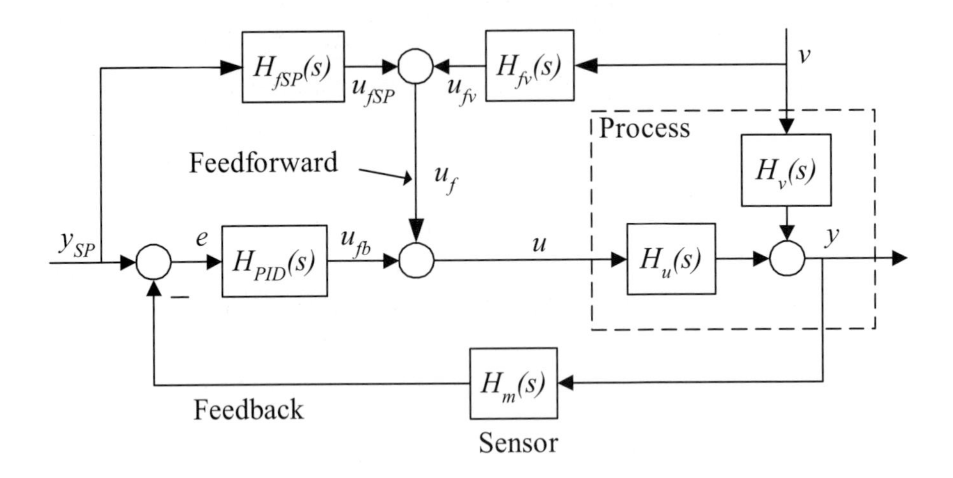

Figure 9.6: Transfer function based block diagram of control system with feedforward and feedback

It is not always possible or desirable to implement the dynamic feedforward functions $H_{fSP}(s)$ and $H_{ff}(s)$ fully. A simplified solution is to implement what we can call the low-frequent dynamic part of the transfer functions. This means that we neglect higher order parts or terms. If we choose to neglect all dynamic terms in $H_{fSP}(s)$ and $H_d(s)$, only the *static* feedforward functions remains:

$$H_{fSP_s} = H_{fSP}(0) = \lim_{s \to 0} H_{fSP}(s) \tag{9.8}$$

and

$$H_{fd_s} = H_{fd}(0) = \lim_{s \to 0} H_{fd}(s) \tag{9.9}$$

Simulations may be used to show if such simplifications give acceptable responses.

Example 9.2 ***Feedforward from disturbance***

Given the following transfer functions process model:

$$y(s) = \underbrace{\frac{K_u}{T_u s + 1}}_{H_u(s)} u(s) + \underbrace{\frac{K_v}{T_v s + 1}}_{H_v(s)} v(s) \tag{9.10}$$

Inserting the above two transfer functions into (9.6) yields the following feedforward transfer function:

$$H_f(s) = \frac{u_{fd}(s)}{v(s)} = \frac{-H_v(s)}{H_u(s)} = -\frac{K_v}{K_u} \cdot \frac{T_u s + 1}{T_v s + 1} \tag{9.11}$$

which is a *lead-lag-function*, which is available as a functional block in most commercial controllers. If $T_u > T_v$ the transfer function has lead-effect because the phase function of the frequency response has positive value which means that the response of the lead-lag function is phase leading (is ahead in phase). If $T_u < T_v$ the transfer function has lag-effect because the phase function of the frequency response has negative value which means that the response of the function is phase lagging (is behind in phase).

[End of Example 9.2]

Example 9.3 ***Feedforward control from the setpoint***

For DC-motors the transfer function from the control variable u to the angular velocity y is approximately

$$\frac{y(s)}{u(s)} = H_u(s) = \frac{K}{(Ts+1)\,s} \tag{9.12}$$

(that is, a first order system with integrator). The feedforward function $H_{fSP}(s)$ in (9.6) becomes

$$H_{fSP}(s) = \frac{u_{fSP}(s)}{(s)} = \frac{1}{H_u(s)} = \frac{(Ts+1)\,s}{K} = \frac{T}{K}s^2 + \frac{1}{K}s \tag{9.13}$$

which in the time-domain corresponds to

$$u_{fSP}(t) = \frac{T}{K}\ddot{y}_{SP}(t) + \frac{1}{K}\dot{y}_{SP}(t) \tag{9.14}$$

To avoid numerical problems of calculating the derivatives in (9.14) the setpoint y_{SP} may be chosen to be sufficiently smooth. For example, setpoint changes could be in the form of parabolic functions of time since this signal has a continuous second order time derivative. Another solution is to use a lowpass filter in the setpoint path, as shown in Figure 7.18.

[End of Example 9.3]

9.2 Cascade control

From earlier chapters we know that a control loop compensates for disturbances so that the control error is small despite the disturbances. If the controller has integral action the steady-state control error is zero. What more can we wish? In some applications it may be desirable if the transient time progression of the error is faster, so that e.g. the IAE index, cf. Section 2.8, is smaller. This can be achieved by *cascade control*, see Figure 9.7.

In a cascade control system there is one or more control loops inside the *primary loop*, and the controllers are in cascade. There is usually one, but

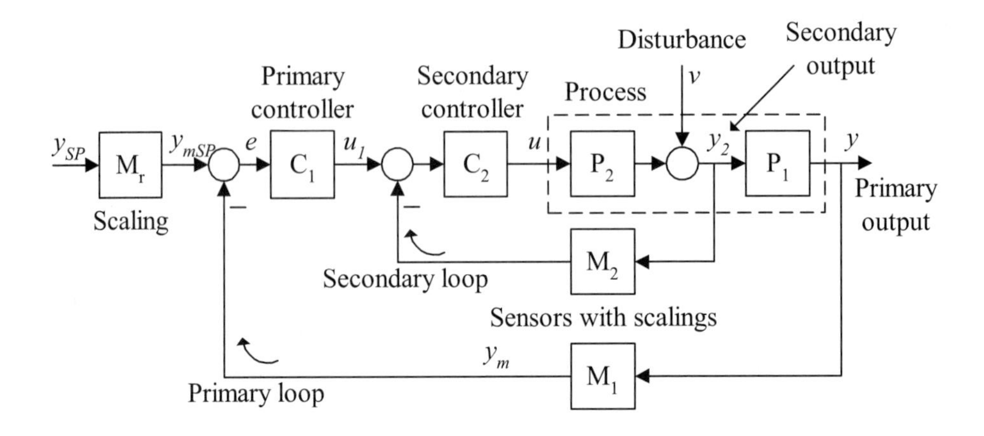

Figure 9.7: Cascade control system

there may be two and even three internal loops inside the primary loop. The (first) loop inside the primary loop is called the *secondary loop*, and the controller in this loop is called the *secondary controller* (or slave controller). The outer loop is called the *primary loop*, and the controller in this loop is called the primary controller (or master-controller). The control signal calculated by the primary controller is the setpoint of the secondary controller.

In most applications the purpose of the secondary loop is to compensate quickly for the disturbance so that its response in the primary output variable of the process is small. For this to happen the secondary loop must register the disturbance. This is done with the sensor M_2 in Figure 9.7.

In addition to getting better disturbance compensation cascade control may give a more *linear relation* between the variables u_1 and y_2, see Figure 9.7 than with usual single loop control. In many applications process part 2 (P_2 in Figure 9.7) is the actuator. In this case the secondary loop can be regarded as a new actuator having better linearity (or proportionality). One example is a control valve where the secondary loop is a flow control loop. With this secondary loop there is a more linear relation between the control signal and the flow than without such a loop. The better linearity may make the tuning of the primary controller (performing e.g. level or temperature control) easier and with more robust stability properties.

The improved control with cascade control can be explained by the increased information about the process – there is at least one more measurement. It is a general principle that the more information you have about the process to be controlled, the better it can be controlled. Note however, that there is still only one control variable to the process, but it is based on two or more measurements.

Since cascade control requires at least two sensors a cascade control system is somewhat more expensive than a single loop control system. Except for cheap control equipment, commercial control equipment are typically prepared for cascade control, so no extra control hardware or software is required.

In which frequency range is the secondary loop effective for compensation for disturbances? This is given by the bandwidth of the secondary loop. A proper bandwidth definition here is the -11 dB-the bandwidth ω_s of the sensitivity function, $S_2(s)$, of the secondary loop, cf. Chapter 6.3.4. $S_2(s)$ is

$$S_2(s) = \frac{1}{1 + L_2(s)} = \frac{1}{1 + H_{c_2}(s)H_{u_2}(s)H_{m_2}(s)} \tag{9.15}$$

where $L_2(s)$ is the loop transfer function of the secondary loop. $H_{c_2}(s)$ is the transfer function of the secondary controller. $H_{u_2}(s)$ is the transfer function from the control variable to the secondary process output variable, y_2. $H_{m_2}(s)$ is the measurement transfer function of the secondary sensor.

As explained above cascade control can give substantial compensation improvement. Cascade control can also give improved tracking of a varying setpoint, but only if the secondary loop has faster dynamics than the process part P_2 itself, cf. Figure 9.7, so that the primary controller "sees" a faster process. If there is a time delay in P_2, the secondary loop will not be faster than P_2 (this is demonstrated in Example 9.4). In most applications improved compensation – not improved tracking – is the main purpose of cascade control.

The secondary controller is typically a P controller or a PI controller. The derivative action is usually not needed to speed up the secondary loop since process part 2 anyway has faster dynamics than process part 1, so the secondary loop becomes fast enough. And in general the noise sensitive derivative term is a drawback. The primary controller is typically a PID controller or a PI controller.

In the secondary controller the P- and the D-term should not have reduced setpoint weights, cf. Section 2.7.1. Why?[2]

How do you *tune* the controllers of a cascade control? You can follow this procedure:

- First the secondary controller is tuned, with the primary controller in manual mode.
- Then the primary controller is tuned, the secondary controller in automatic mode.

Controller tuning can be made using a standard tuning method, e.g. the Ziegler-Nichols' closed loop method, cf. Section 4.4.

Example 9.4 ***Cascade control (simulation)***

In this example the following two control systems are simulated simultaneously (in parallel):

- A cascade control system consisting of two control loops.
- An ordinary single loop control system, which is simulated for comparison.

The process to be controlled is the same in both control systems, and they have the same setpoint, y_{SP}, and the same disturbance, v. The process consists of two partial processes in series, cf. Figure 9.7:

- Process P_1:

$$y(s) = H_{P1}(s)y_2(s) \tag{9.16}$$

  where

$$H_{P1}(s) = \frac{K}{\left(\frac{s}{\omega_0}\right)^2 + 2\zeta\frac{s}{\omega_0} + 1} e^{-\tau s} \tag{9.17}$$

  with

$$K = 1;\ \omega_0 = 0.2\text{rad/s};\ \zeta = 1;\ \tau = 1\text{s} \tag{9.18}$$

[2] Because attenuating or removing the time-varying setpoint (which is equal to the control signal produced by the primary controller) of the secondary loop will reduce the ability of the secondary loop to track these setpoint changes, causing slower tracking of the total control system.

- Process P_2:

$$y_2(s) = H_{P2}(s)u(s) + v(s) \tag{9.19}$$

where

$$H_{P2}(s) = \frac{K}{\left(\frac{s}{\omega_0}\right)^2 + 2\zeta\frac{s}{\omega_0} + 1} e^{-\tau s} \tag{9.20}$$

with

$$K = 1;\ \omega_0 = 2\text{rad/s};\ \zeta = 1;\ \tau = 0.1\text{s} \tag{9.21}$$

Simply stated, process P_2 has ten times quicker dynamics than process P_1 has. The controllers have been tuned according to the Ziegler-Nichols' closed loop method with some fine-tuning to avoid too aggressive control action (increase of T_i from 0.69 to 1). The controller parameter settings are as follows:

- Cascade control system: Primary controller, C_1 (PID):

$$K_p = 2.1;\ T_i = 4.0;\ T_d = 1.0 \tag{9.22}$$

- Cascade control system: Secondary controller, C_2 (PI):

$$K_p = 1.5;\ T_i = 1.0;\ T_d = 0 \tag{9.23}$$

- Single loop control system: Controller, C (PID):

$$K_p = 1.9;\ T_i = 4.0;\ T_d = 1.0 \tag{9.24}$$

Figure 9.8 shows simulated responses with a step in the setpoint. IAE values for the two control systems are shown in Table 9.1. The IAE values

	Cascade control	**Single loop control**
Setpoint step	IAE = 17.18	IAE = 26.85
Disturbance step	IAE = 7.80	IAE = 84.13

Table 9.1: IAE values for cascade control system and for single loop control system

show that the setpoint tracking is better in the cascade control system, but not substantially better.

Figure 9.9 shows simulated responses with a step in the disturbance. The IAE values in Table 9.1 show that the disturbance compensation is much better in the cascade control system.

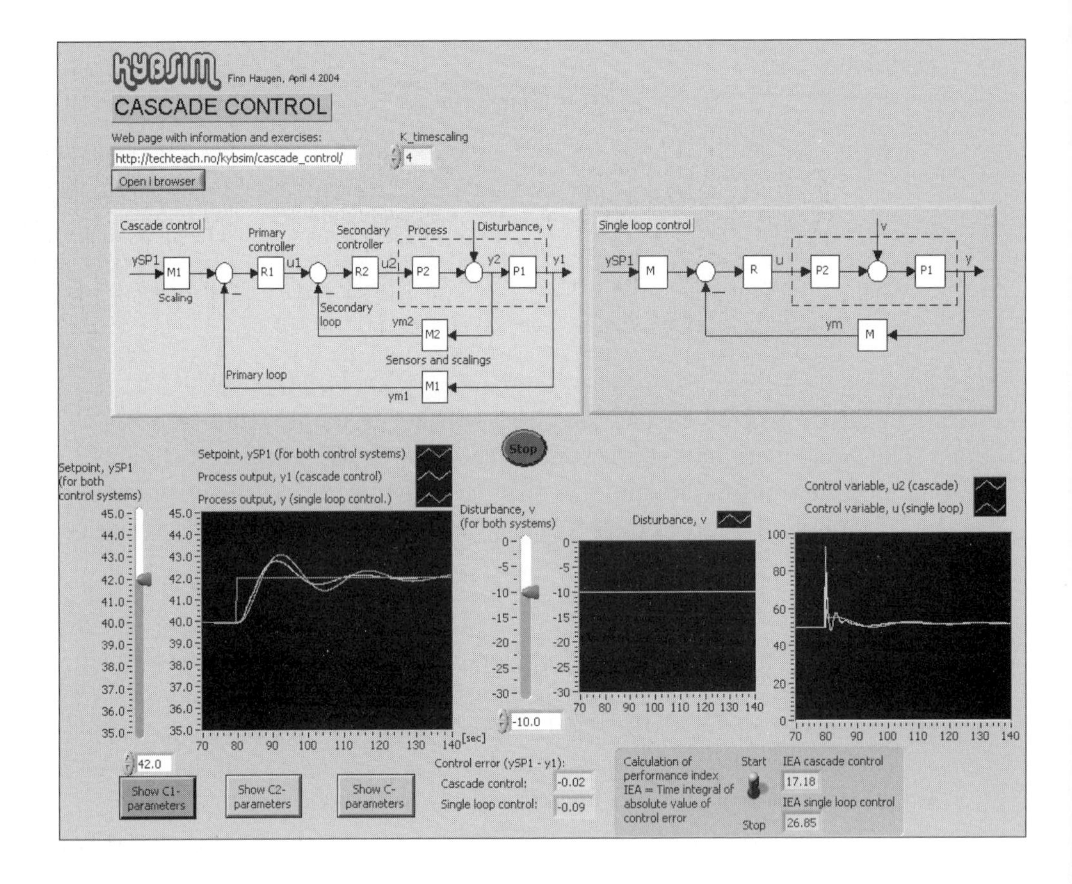

Figure 9.8: Example 9.4: Simulated responses with a step in the setpoint

Figure 9.9 shows that the control variable of the cascade control system works much more aggressively than in the single loop control system, which is due to the relatively quick secondary loop.

[End of Example 9.4]

Cascade control is frequently used in the industry. A few examples are described in the following.

Example 9.5 *Cascade control of the level in wood-chip tank*

Level control of a wood-chip tank has been a frequent example in this book. In the real level control system[3] cascade control is used, although not described in the previous examples. The primary loop performs level control. The secondary loop is a control loop for the mass flow on the

[3] at Södra Cell Tofte in Norway

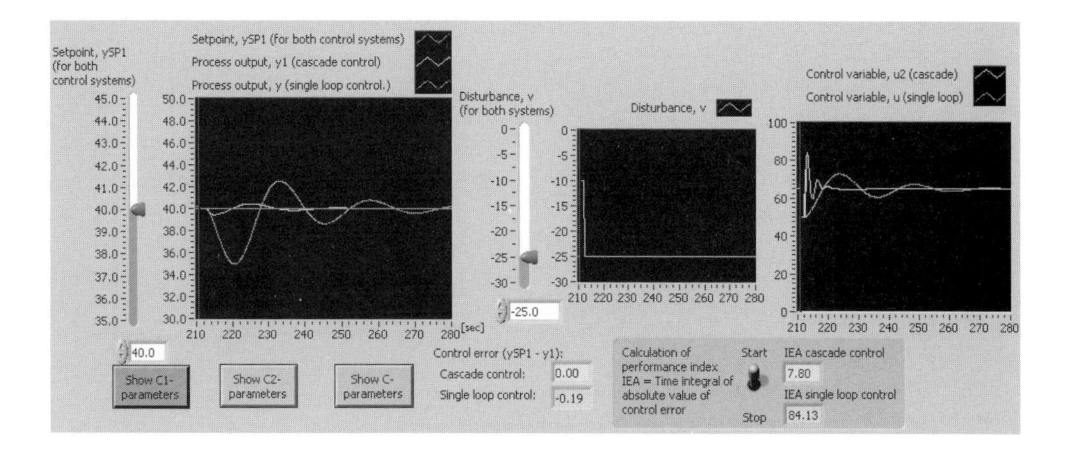

Figure 9.9: Example 9.4: Simulated responses with a step in the disturbance

conveyor belt, see Figure 9.10. The mass flow is measured by a flow sensor (which actually is based on a weight measurement of the belt with chip between two rollers). The purpose of the secondary loop is to give a quick compensation for disturbances in the chip flow due to variations in the chip consistency since the production is switched between spruce, pine and eucalyptus. In addition to this compensation the secondary loop gives a more linear or proportional relation between the control variable u and the mass flow w_s into the conveyor belt (at the flow sensor).

[End of Example 9.5]

Example 9.6 ***Cascade control of a heat exchanger***

Figure 9.11 shows a temperature control system for a heat exchanger. The control variable controls the opening of the hot water valve. The primary loop controls the product temperature. The secondary loop controls the heat flow to compensate for flow variations (disturbances). The valve with flow control system can be regarded a new valve with an approximate proportional relation between the control variable and the heat flow.

[End of Example 9.6]

There are many other examples of cascade control, e.g.:

- **DC-motor:**
  - Primary loop: Speed control based on measurement of the rotational speed using a tachometer as speed sensor.

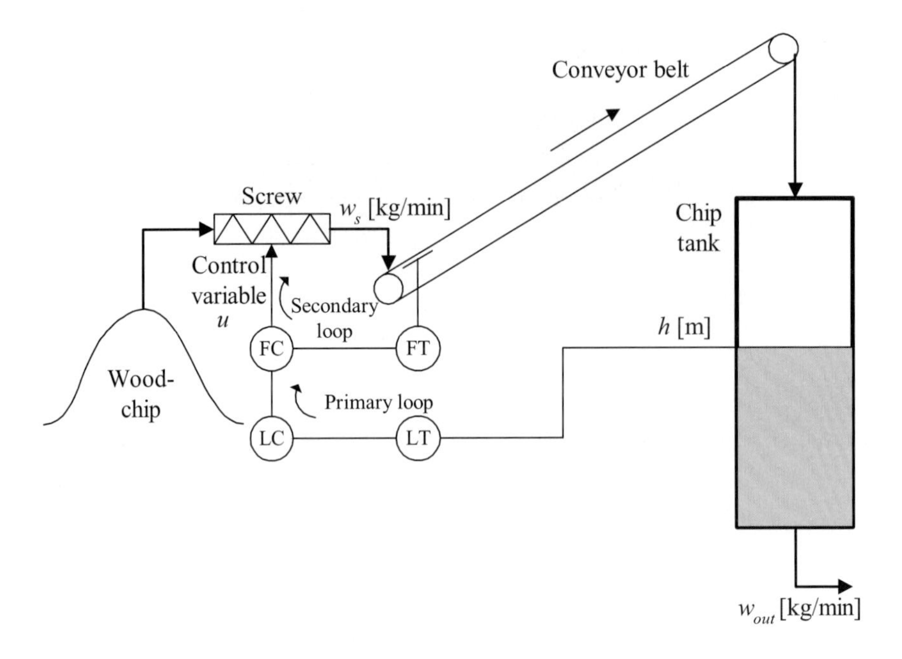

Figure 9.10: Example 9.5: Level control system of a wood-chip tank where the primary loop performs level control, and the secondary loop performs mass flow control. (FT = Flow Transmitter. FC = Flow Controller. LT = Level Transmitter. LC = Level Controller.)

  - Secondary loop: Control of armature current which compensates for nonlinearities of the motor, which in turn may give more linear speed control.

- **Hydraulic motor:**
  - Primary loop: Positional control of the cylinder
  - Secondary loop: Control of the servo valve position (the servo valve controls the direction of oil flow into the cylinder), which results in a more linear valve movement, which in turn gives a more precise control of the cylinder.

- **Control valve:**
  - The primary loop: Flow control of the liquid or the gas through the valve.
  - Secondary loop: Positional control of the valve stem, which gives a proportional valve movement, which in turn may give a more precise flow control. Such an internal positional control system is called positioner.

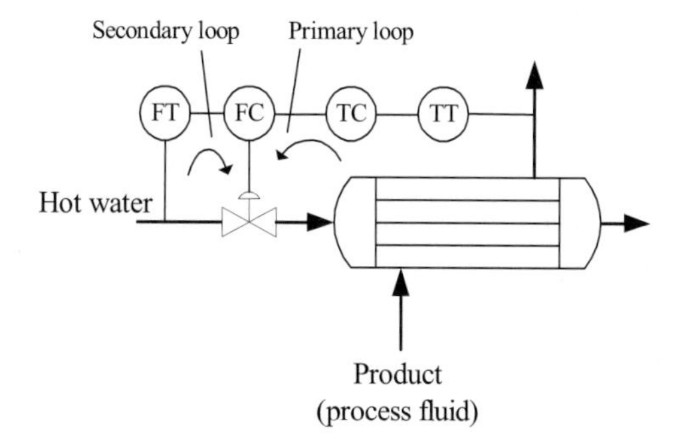

Figure 9.11: Example 9.6: Cascade control of the product temperature of a heat exchanger. (TC = Temperature Controller. TT = Temperature Transmitter. FC = Flow Controller. FT = Flow Transmitter.)

9.3 Ratio control and quality and product flow control

9.3.1 Ratio control

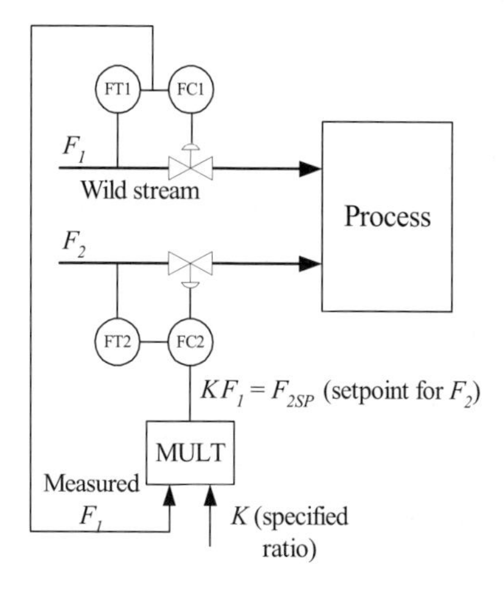

Figure 9.12: Ratio control

The purpose of *ratio control* is to control a mass flow, say F_2, so that the ratio between this flow and another flow, say F_1, is

$$F_2 = KF_1 \tag{9.25}$$

where K is a specified ratio which may have been calculated as an optimal ratio from a process model. One example is the calculation of the ratio between oil inflow and air inflow to a burner to obtain optimal operating condition for the burner. Another example is the nitric acid factory where ammonia and air must be fed to the reactor in a given ratio.

Figure 9.12 shows the structure of ratio control. The setpoint of the flow F_2 is calculated as K times the measured value of F_1, which is denoted the "wild stream". The figure shows a control loop of F_1. The setpoint of F_1 (the setpoint is not shown explicitly in the figure) can be calculated from a specified production rate of the process. The ratio control will then ensure the ratio between the flows as specified.

An alternative way to implement ratio control is to calculate the actual ratio as

$$K_{actual} = \frac{F_2}{F_1} \tag{9.26}$$

Then K_{actual} is used as a measurement signal to a ratio controller with the specified K as the setpoint and F_2 as the control variable, cf. Figure 9.13.

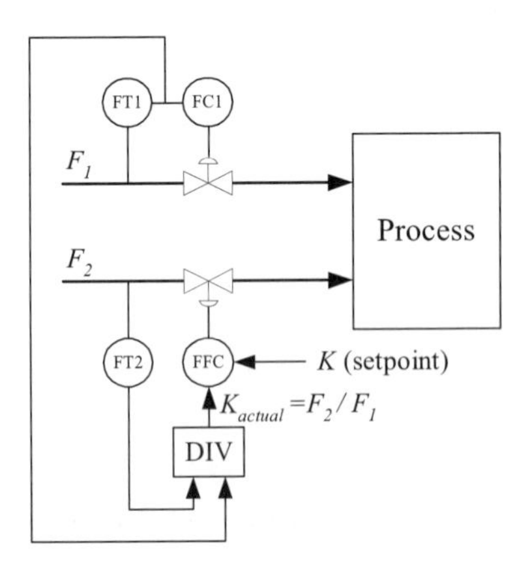

Figure 9.13: An alternative ratio control structure based on measurement of the actual ratio. (FFC = Flow Fraction Controller.)

Although this control structure is logical, it is a drawback that the loop gain, in which K_{actual} is a factor, is a function of the measurements of F_2 and F_1. Hence, this solution is not encouraged[16].

9.3.2 Quality and production rate control

Earlier in this section it was mentioned that the ratio K may origin from an analysis of optimal process operation, say from a specified product quality quantity, say Q_{SP}. Imagine however that there are disturbances so that key components in one of or in both flows F_1 or F_2 vary somewhat. Due to such disturbances it may well happen that the actual product quality is different from Q_{SP}. Such disturbances may also cause the actual product flow to differ from a flow setpoint. These problems can be solved by implementing

- a quality control loop based on feedback from measured quality Q to the ratio parameter K, and
- a product flow control loop based on feedback from measured flow F to one of the feed flows.

Figure 9.14 shows the resulting quality and production rate control system.

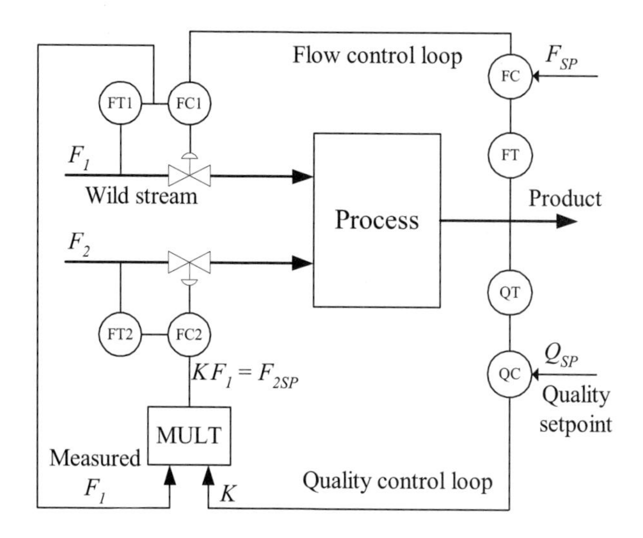

Figure 9.14: Control of quality and product flow. (QT = Quality Transmitter. QC = Quality Controller.)

9.4 Split-range control

In *split-range control* one controller controls two actuators in different ranges of the control signal span, which here is assumed to be 0 – 100%. See Figure 9.15. Figure 9.16 shows an example of split-range temperature control of a thermal process. Two valves are controlled – one for cooling and one for heating, as in a reactor. The temperature controller controls

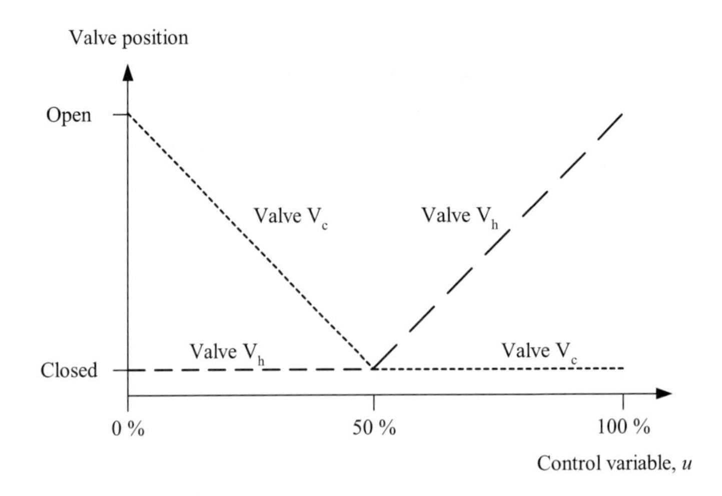

Figure 9.15: Split-range control of two valves

the cold water valve for control signals in the range 0–50%, and it controls the hot water valve for control signals in the range 50–100%, cf. Figure 9.15.

In Figure 9.15 it is indicated that one of the valves are open while the other is active. However in certain applications one valve can still be open while the other is active, see Figure 9.17. One application is pressure control of a process: When the pressure drop compensation is small (as when the process load is small), valve V_1 is *active* and valve V_2 is *closed.* And when the pressure drop compensation is is large (as when the process load is large), valve V_1 is *open* and valve V_2 *is still active.*

9.5 Control of product flow and mass balance in a plant

In the process industry products are created after treatment of the materials in a number of stages in series, which are typically unit processes as blending or heated tanks, buffer tanks, distillation columns, absorbers, reactors etc. The basic control requirements of such a production line are as follows:

- The mass flow of a key component must be controlled, that is, to follow a given production rate or flow setpoint.

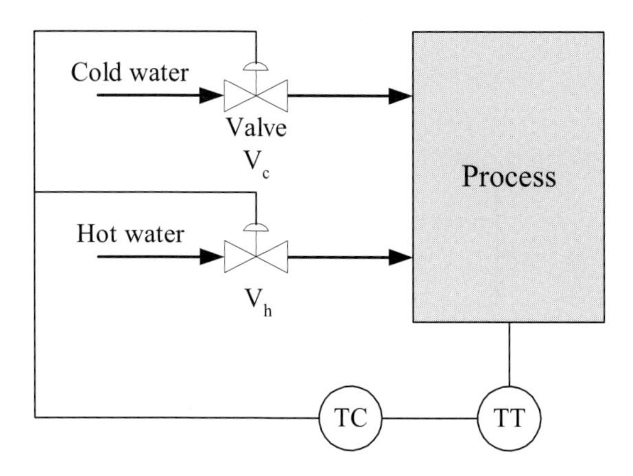

Figure 9.16: Split-range temperature control using two control valves

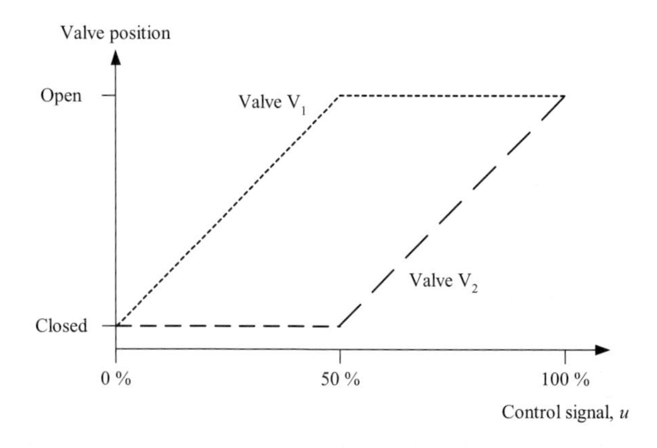

Figure 9.17: In split-range control one valve can be active while another valve is open simultaneuously.

- The mass balance in each process unit (tank etc.) must be maintained – otherwise e.g. the tank may go full or empty.

Figure 9.18 shows the principal control system structure to satisfy these requirements. (It is assumed that the mass is proportional to the level.) The position of the production flow control in the figure is just one example. It may be placed earlier (or later) in the line depending on where the key component(s) are added.

Note that the mass balance of an *upstream* tank (relative to the production flow control) is controlled by manipulating the mass *inflow* to the tank, while the mass balance of a *downstream* tank is controlled by manipulating the mass *outflow* to the tank.

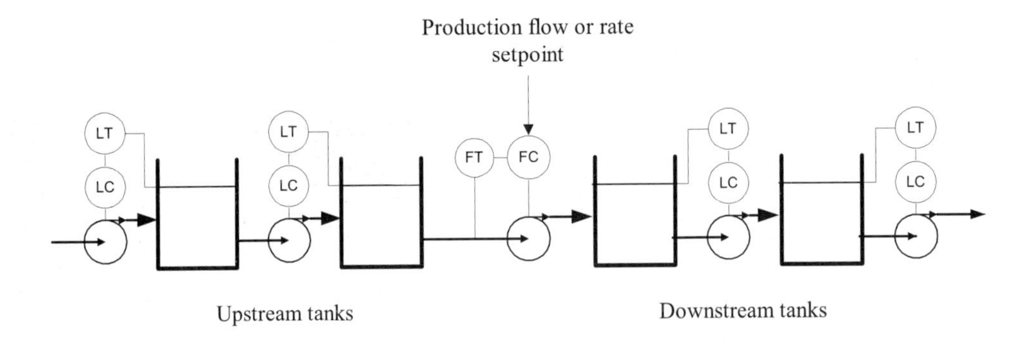

Figure 9.18: Control of a production line to maintain product flow (rate) and mass balances

In Figure 9.18 the mass balances are maintained using level control. If the tanks contains vapours, the mass balances are maintained using *pressure control.* Then pressure sensors (PT = Pressure Transmitter) takes the places of the level sensors (LT = Level Transmitters), and pressure controllers (PC = Pressure Controller) takes the place of level controllers (LC = Level Controller) in Figure 9.18.

Example 9.7 *Control of production line*

Figure 9.19 shows the front panel of a simulator of a general production line. The level controllers are PI controllers which are tuned so that the control loops get proper speed and stability (the parameters may be calculated as explained in Chapter 7.2.2). The production flow F is here controlled using a PI controller. Figure 9.19 shows how the level control loops maintain the mass balances (in steady-state) by compensating for a disturbance which is here caused by a change of the production flow. Note that controller LC2 must have negative gain (i.e. direct action, cf. Section 2.6.8) – why?[4]

[End of Example 9.7]

[4]Because the process has negative gain, as an increase of the control signal gives a reduction of the level/level measurement.

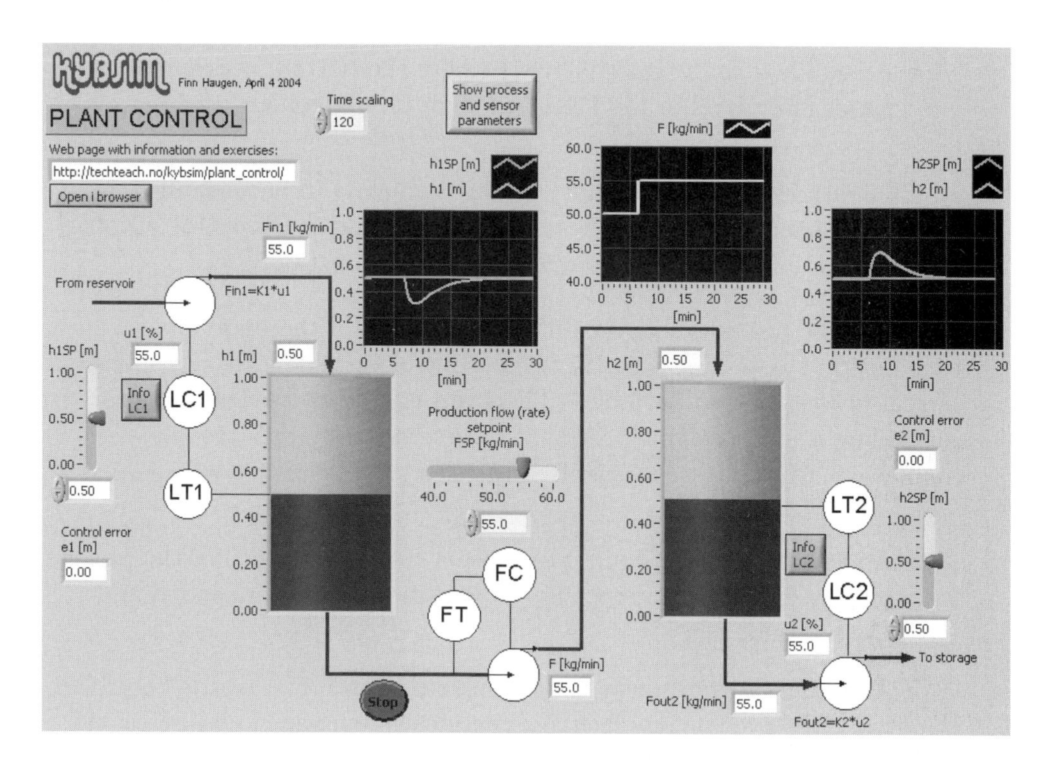

Figure 9.19: Example 9.7: The level control loops maintain the mass balances (in steady-state).

9.6 Multivariable control

9.6.1 Introduction

Multivariable processes has more than one input variables or ore than one output variables. Here are a few examples of multivariable processes:

- A heated liquid tank where both the level and the temperature shall be controlled.
- A distillation column where the top and bottom concentration shall be controlled.
- A robot manipulator where the positions of the manipulators (arms) shall be controlled.
- A chemical reactor where the concentration and the temperature shall be controlled.

- A head box (in a paper factory) where the bottom pressure and the paper mass level in the head box shall be controlled.

To each variable (process output variable) which is to be controlled a setpoint is given. To control these variables a number of control variables are available for manipulation by the controller function.

Multivariable processes can be difficult to control if there are *cross couplings* in the process, that is, if one control variable gives a response in several process output variables. There are mainly two problems of controlling a multivariable process if these cross couplings are not counteracted by the multivariable controller:

- A change in one setpoint will cause a response in each of the process output variables, not only in the output variable corresponding to the setpoint.
- Assuming that ordinary single loop PID control is used, a controller will "observe" a complicated dynamic system which consists of the multivariable process *with* all control loops! This can make it difficult to tune each of the PID controllers, and the stability robustness of the control system may be small.

The following sections describe the most common ways to control multivariable processes.

9.6.2 Single loop control with PID controllers

The simplest yet most common way to control a multivariable process is using *single loop control* with PID controllers. There is one control loop for each process output variable which is to be controlled. The control system structure is shown in Figure 9.20, where subsystems are represented by transfer functions although these subsystems are generally non-linear dynamic systems. Since this process has two control variables and two process output variables, we say that the process is a 2x2 multivariable process.

Pairing of process output variables and control variables

In single loop control of a multivariable process we must determine the pairing of process output variable (its measurement) and control variable

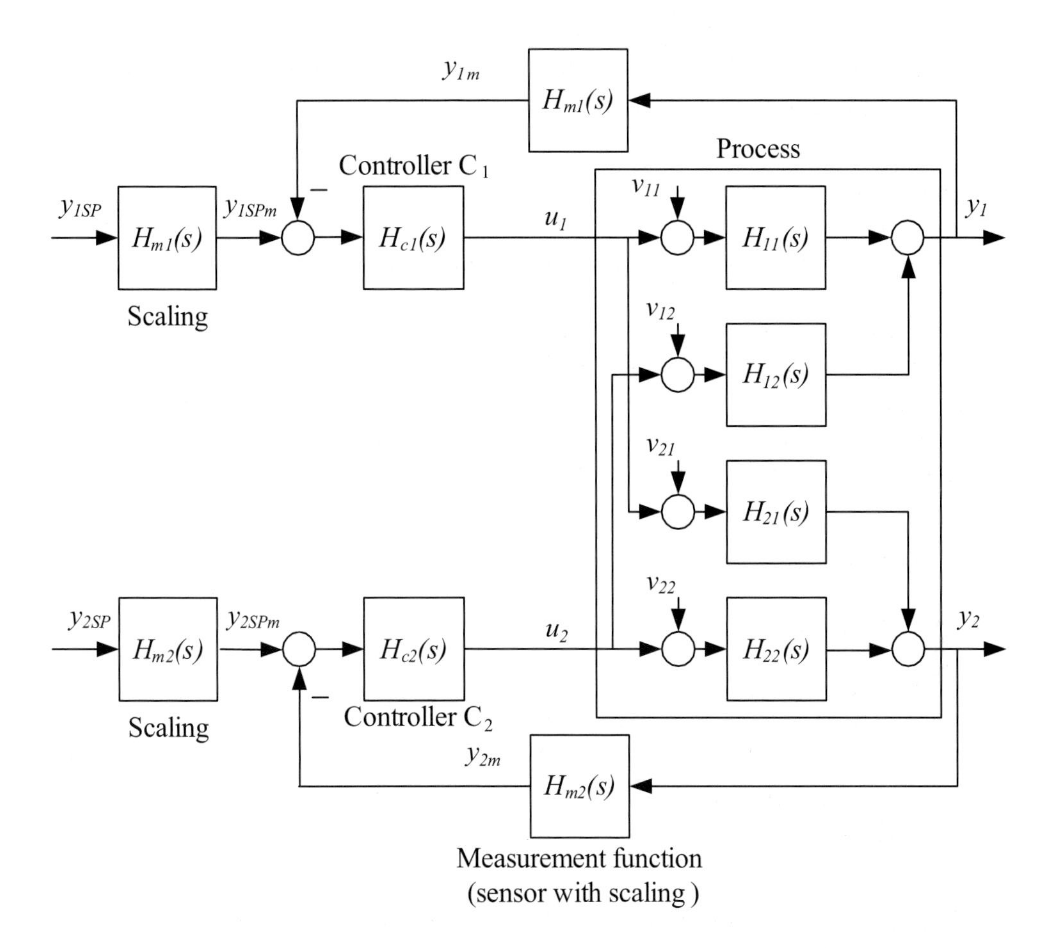

Figure 9.20: Single loop control of a 2x2 multivariable process

(via the PID controller). A natural rule for choosing this pairing is as follows: *The strong process couplings (from control variable to process output variable) should be contained in the control loops.* Following this rule is an effective use of the control variable, and supports stability robustness against variations of the dynamic properties in other parts of the control system. Figure 9.20 shows the correct control system structure if there are strong couplings in $H_{11}(s)$ and in $H_{22}(s)$.

In most cases it is easy to determine the strong pairings. One example is a heated liquid tank where both level and temperature is to be controlled. The two control variables are power supply via a heating element and liquid supply. This process is multivariable with cross couplings since both power supply (control variable 1) and liquid supply (control variable 2) influences both process output variables (level and temperature). (The level is influenced by the power supply through liquid expansion due to temperature increase.) In this case the process output variable/control variable pairing is obvious: Level $\leftrightarrow$ Power and Temperature $\leftrightarrow$ Liquid flow, right?[5]

There are model based methods for analysis of process couplings, as RGA-analysis (Relative Gain Array) and singular value analysis [15].

Controller tuning

In the tuning procedures below you can try the Ziegler-Nichols' closed loop method or the P-I-D tuning method, cf. Chapter 4.

According to [15] a widely used procedure for tuning the PID controllers in single loop multivariable control is as follows:

Procedure 1:

1. Tune the controller in each of the loops in turn with all the other controllers in manual mode.
2. Close all the loops (set all controllers in automatic mode).
3. If there are stability problems, reduce the gain and/or increase the integral time of the controllers in the least important loops.

An alternative procedure [15] for cases where the control of one specific process variable is more important than the control of other variables is as follows:

Procedure 2:

1. Tune the controller of the most important loop. The other controllers are set in manual mode.
2. Tune the other controllers in sequence, with the tuned controllers set in automatic mode.
3. If there are stability problems, reduce the gain and/or increase the integral time of the controllers in the least important loops.

[5] Wrong :-)

Example 9.8 ***Single loop multivariable control***

See Figure 9.20. The process transfer functions are on the form

$$H_{ij}(s) = \frac{y_i(s)}{u_j(s)} = \frac{K_{ij}}{T_{ij}s+1}e^{-\tau_{ij}s} \tag{9.27}$$

with these parameters:

$$K_{11} = 1;\ T_{11} = 1;\ \tau_{11} = 0.5 \tag{9.28}$$

$$K_{12} = 0.5;\ T_{12} = 1;\ \tau_{12} = 0.5 \tag{9.29}$$

$$K_{21} = 0.5;\ T_{21} = 1;\ \tau_{21} = 0.5 \tag{9.30}$$

$$K_{22} = 1;\ T_{22} = 1;\ \tau_{22} = 0.5 \tag{9.31}$$

Thus, there are cross couplings "both ways" in the process since both K_{12} and K_{21} are different from zero.

The measurement transfer functions are $H_{m_1}(s) = 1 = H_{m_2}(s)$. The controllers are PID controllers tuned according to Procedure 1 described above (with the Ziegler-Nichols' closed loop method). The tuning gives

$$K_{p_1} = 2.0;\ T_{i_1} = 0.9;\ T_{d_1} = 0.23 \tag{9.32}$$

$$K_{p_2} = 2.0;\ T_{i_2} = 0.9;\ T_{d_2} = 0.23 \tag{9.33}$$

However, simulations shows that the multivariable control system actually is unstable using the above PID settings. So, re-tuning was necessary. Decreasing the proportional gains from 2.0 to 1.4 was sufficient in this case. The final settings are

$$K_{p_1} = 1.4;\ T_{i_1} = 0.9;\ T_{d_1} = 0.23 \tag{9.34}$$

$$K_{p_2} = 1.4;\ T_{i_2} = 0.9;\ T_{d_2} = 0.23 \tag{9.35}$$

Figure 9.21 shows simulated responses in y_{1_m} and y_{2_m} due to a step in y_{1SP_m}. As expected, the setpoint step gives a cross response in y_{2_m}. The stability of the control system seems to be acceptable.

[End of Example 9.8]

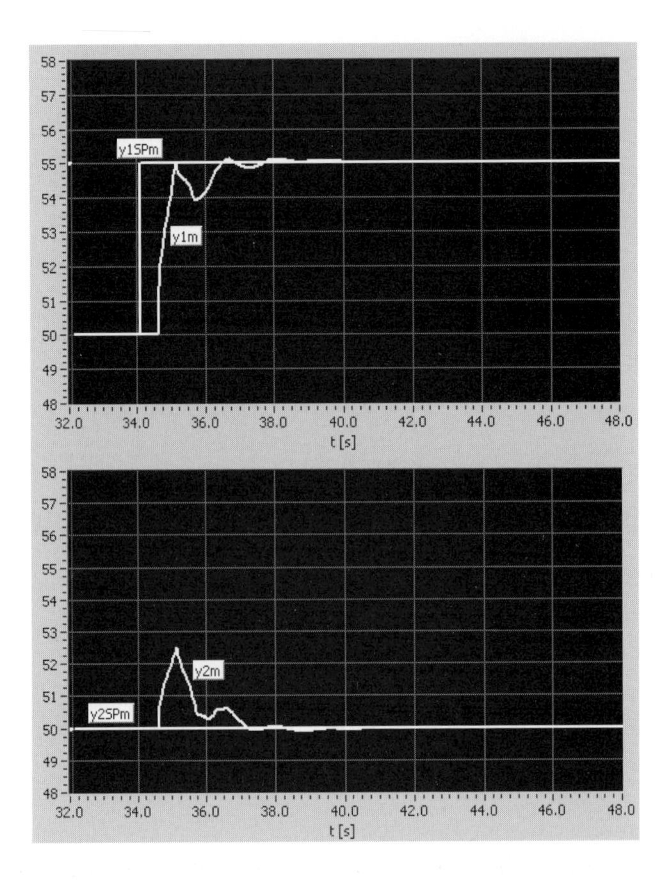

Figure 9.21: Example 9.8: Single loop multivariable control. Simulated responses in y_{1m} and y_{2_m} due to a step in the setpoint $y_{1_{SPm}}$.

9.6.3 Single loop PID control combined with decoupling

If the interaction between the control loops in a multivariable control system is problematic (as if the cross responses are too large or if there are stability problems), you may consider using a *decoupler* together with the PID controllers. With decoupling the controller counteracts the cross couplings in the process so that there are no interaction between the control loops. Thus, the control loops are decoupled, but the process cross couplings are still there, of course.

Several methods of decoupled control exist. Here a method called *linear decoupling* is described. Figure 9.22 shows a block diagram of a multivariable control system with decoupling. The controller consists of two parts:

- A decoupler in series with the process. The purpose of the decoupler

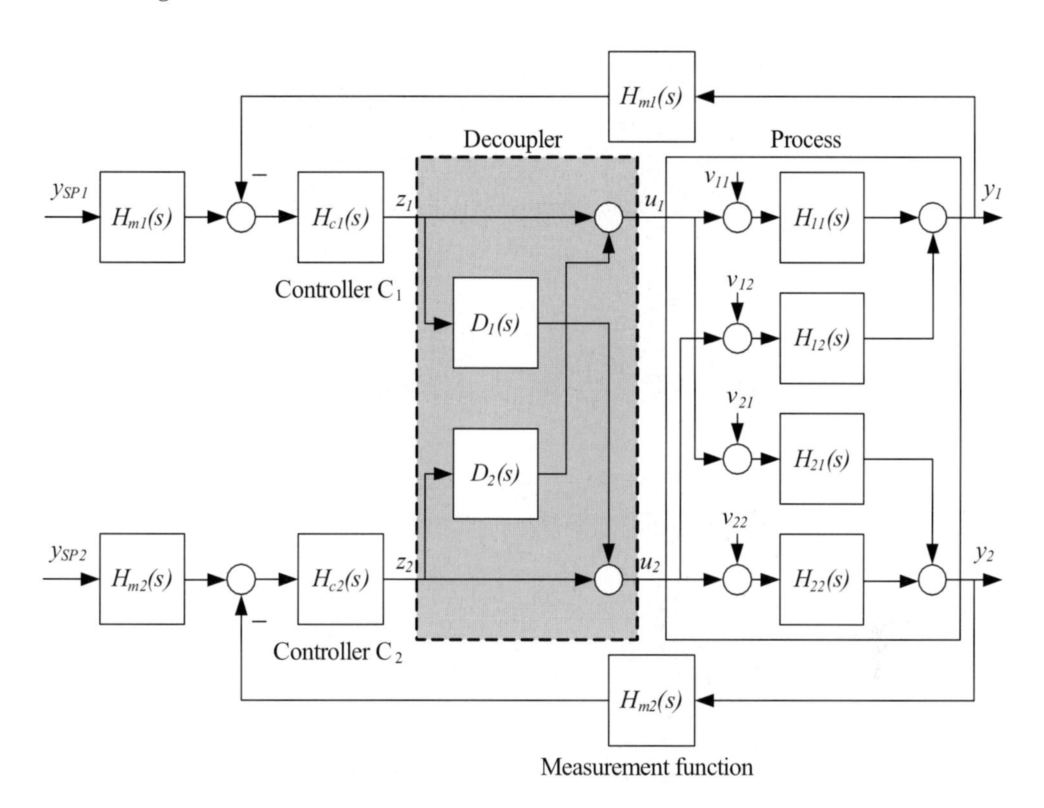

Figure 9.22: Multivariable control system with decoupler and single loop PID-controllers

is to counteract the cross couplings in the process.

- A controller consisting of two independent PID controllers which controls the combination of decoupler and process.

The decoupler transfer functions $D_1(s)$ and $D_2(s)$ must be designed so that the cross couplings via $H_{12}(s)$ and $H_{21}(s)$ in the process are counteracted. If this is achieved, each of the PID controllers sees just one monovariable process, and there will be no interacting control loops. z_1 and z_2 is transformed control variables for the "new" decoupled process. The physical control variables are however still u_1 and u_2.

Let us derive the decoupler transfer functions $D_1(s)$ and $D_2(s)$. We start with $D_1(s)$. It will be derived from the requirement that the net effect that z_1 has on y_1 is zero. Mathematically, cf. the block diagram in Figure 9.22, we require:

$$D_1(s)H_{22}(s)z_1(s) + H_{21}(s)z_1(s) \stackrel{!}{=} 0 \tag{9.36}$$

for all values of z_1. This is satisfied with

$$D_1(s) = -\frac{H_{21}(s)}{H_{22}(s)} \tag{9.37}$$

Similarly, $D_2(s)$ becomes

$$D_2(s) = -\frac{H_{12}(s)}{H_{11}(s)} \tag{9.38}$$

Example 9.9 ***Decoupling***

Assume given the same process model as in Example 9.8. (9.37) and (9.38) becomes

$$D_1(s) = -\frac{H_{21}(s)}{H_{22}(s)} = -\frac{\frac{K_{21}}{T_{21}s+1}e^{-\tau_{21}s}}{\frac{K_{22}}{T_{22}s+1}e^{-\tau_{22}s}} = -\frac{K_{21}}{K_{22}}\frac{T_{22}s+1}{T_{21}s+1}e^{(\tau_{22}-\tau_{21})s} = -0.5 \tag{9.39}$$

$$D_2(s) = -\frac{K_{12}}{K_{11}}\frac{T_{11}s+1}{T_{12}s+1}e^{(\tau_{11}-\tau_{12})s} = -0.5 \tag{9.40}$$

The PID controllers are tuned with the Ziegler-Nichols' closed loop method (with the decoupler in action):

$$K_{p_1} = 2.7;\ T_{i_1} = 0.9;\ T_{d_1} = 0.23 \tag{9.41}$$

$$K_{p_2} = 2.7;\ T_{i_2} = 0.9;\ T_{d_2} = 0.23 \tag{9.42}$$

Figure 9.23 shows simulated responses in y_{1_m} and y_{2_m} due to a step in the setpoint $y_{1_{SP_m}}$. Ideally there is no cross response in y_{2_m}.[6]

[End of Example 9.9]

9.6.4 Model-based predictive control

Model-based predictive control or *MPC* has become an important control method, and it can be regarded as the next most important control method in the industry, next to PID control. Commercial MPC products are available as separate products or as modules included in automation products. MPC can be applied to multivariable and non-linear processes. The controller function is based on a continuous calculation of the optimal

[6] The simulation shows a (very) small response in y_{2_m}, but this is due to imperfect numerical conditions in the simulator.

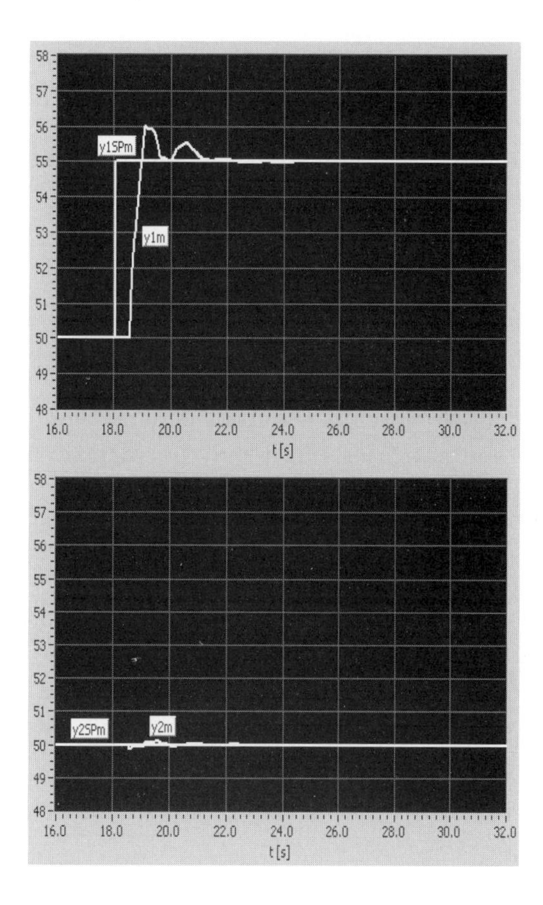

Figure 9.23: Exampe 9.9: Single loop PID-control with decoupling: Simulated responses in y_{1_m} and y_{2_m} due to a step in the setpoint $y_{1_{SPm}}$.

sequence or time-series of the control variable. The optimization criterion which is to be minimized is typically stated as follows:

$$J = \sum_{j=1}^{N} \left\{ [y_{SP}(t_{k+j}) - y(t_{k+j}|t_k)]^2 + \lambda(j)[u(t_{k+j-1})]^2 \right\} \tag{9.43}$$

where y_{SP} is the setpoint, y is the process output variable and u is the control variable. In general these variables can be vectors. λ is a weight function. t_k is the present time. N is the *prediction horizon*, a number of future time steps over which J is defined. This optimization criterion defines the criterion of "good control": The less value of J, the better control.

The optimization criterion must have a *constraint*, which is *the process model* including physical constraints of the control variable and the state variables. The process model form are one of the following (different MPC implementations may assume different model forms):

- Impulse response model (which can be derived from simple experiments on the process)
- Step response model (same comment as above)
- Transfer function model (which is a general linear dynamic model form than the impulse response model and the step response model)
- State-space model (which is the most general model form since it may include nonlinearities and it may be valid over a broad operating range)

Figure 9.24 illustrates how predictive control works.

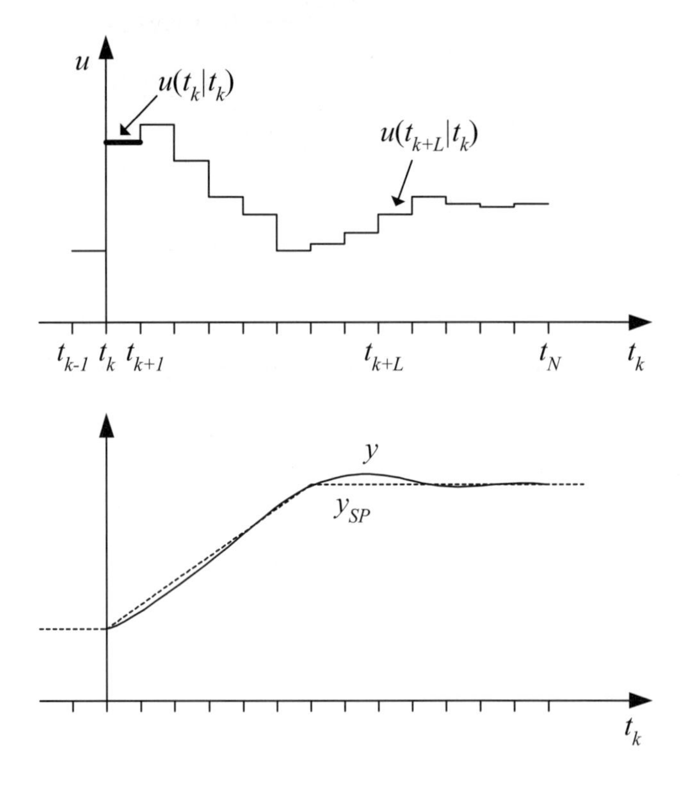

Figure 9.24: How predictive control works

Predictive control is based on the following calculations, which are executed at each time step:

1. The (future) control signal sequence $u(t_{k+j-1}|t_k)$ for $j = 1, \cdots, N$, that is, $u(t_k)$, $u(t_{k+1})$,..., $u(t_{k+N-1})$, is calculated to be the control variable sequence which minimizes J (the sequence is therefore optimal). Terms as $u(t_{k+L}|t_k)$ gives the value of u for time t_{k+L} calculated from data available at time t_k. The same applies to $y(t_{k+L}|t_k)$, which is calculated (predicted) using the process model.

2. Of the optimal control signal sequence, the first term, $u(t_k|t_k)$, is used to actually control the process.

3. At the next time step the points above are repeated.

Above it was assumed that the MPC-controller calculates directly the control variable to the process. However the MPC-controller may also calculate setpoints for *local PID controllers*, see Figure 9.25. This structure

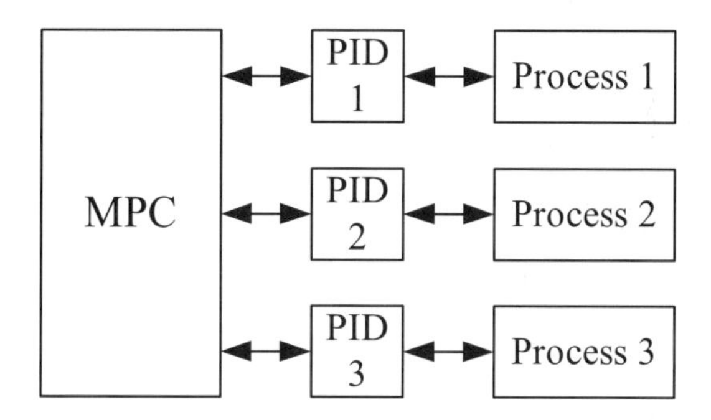

Figure 9.25: The MPC-controller may calculate setpoints for local PID-controllers.

ensures that the process can still be controlled with conventional PID controllers in periods when the MPC-controller is inactive (due to configuration or maintenance). Using PID controllers locally may enhance operation safety.

Appendix A

Codes and symbols used in Process & Instrumentation Diagrams

This appendix gives a brief overview over codes and symbols used in Process & Instrumentation Diagrams – P&IDs. The standards ISA S5.1 (USA) and ISO 3511-1 (International) define the these diagrams. There are also factory internal standards.

A.1 Letter codes

Table A.1 shows the most commonly used letter codes used in Process & Instrumentation Diagrams.

A.2 Instrumentation symbols used in P&IDs

The following figures show common symbols used in Process&Instrumentation Diagrams (P&IDs).

	As first letter	As subsequent letter
A	Alarm	
C		Controller
D	Density. Difference	
F	Flow. Fraction (ratio)	
G	Position	
H	Hand controlled	
I		Indicator
L	Level	
P	Pressure	
Q	Quality	
S	Speed	
T	Temperature	Transmitter (sensor)
V	Viscosity	Valve
Y		Function (e.g. mathematical)
Z		Secure control (e.g. interlock)

Table A.1: Common instrumentation codes used in Process&Instrumentation Diagrams

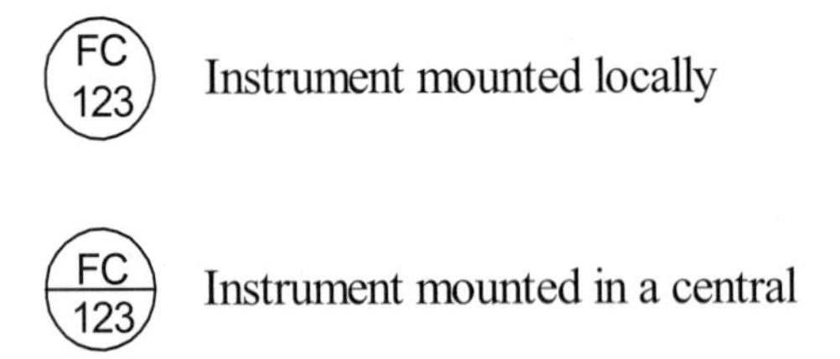

Figure A.1: Main instrument symbols (FC123 is one example of instrument code.) The numbering is based on running numbers, e.g. FC123, TC124 etc., but so that a specific control loop has a fixed number, e.g. FT123, TT124.

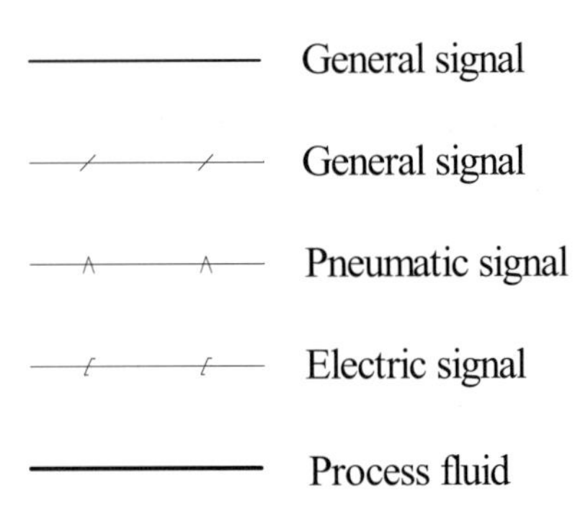

Figure A.2: Symbols of conductors and pipes

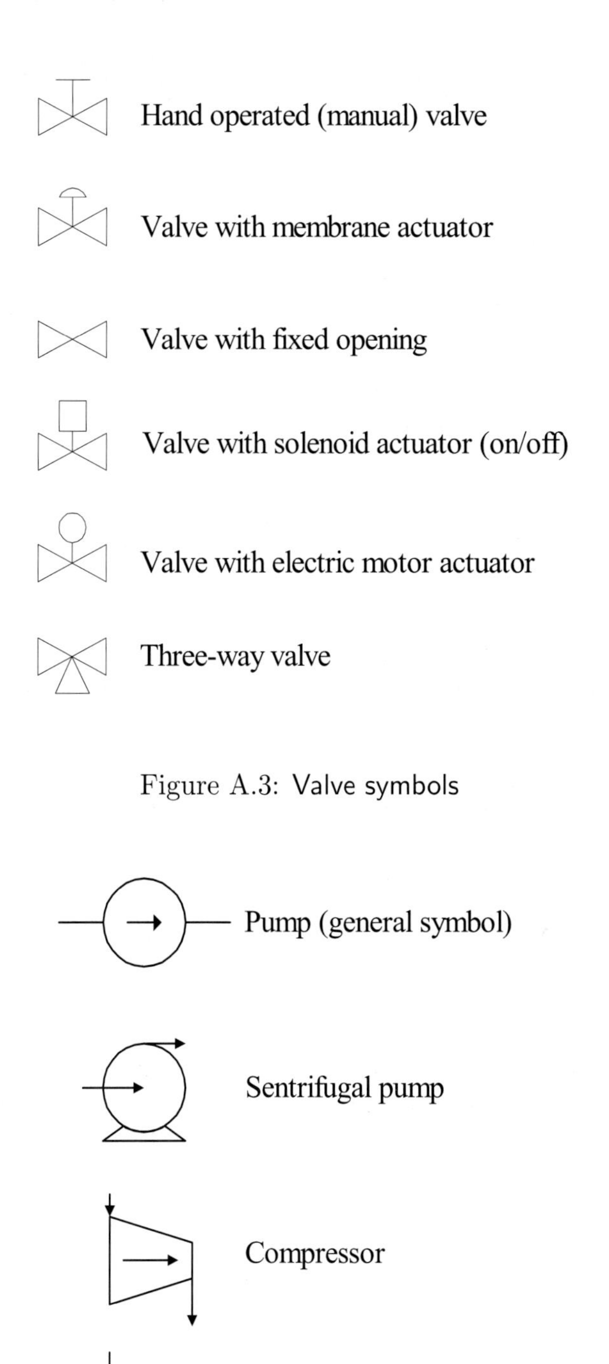

Figure A.3: Valve symbols

Figure A.4: Symbols of pumps, compressors and turbines

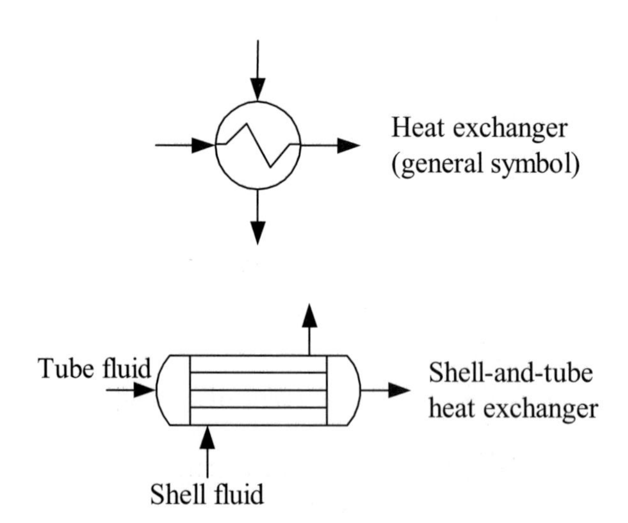

Figure A.5: Symbols of heat exchangers

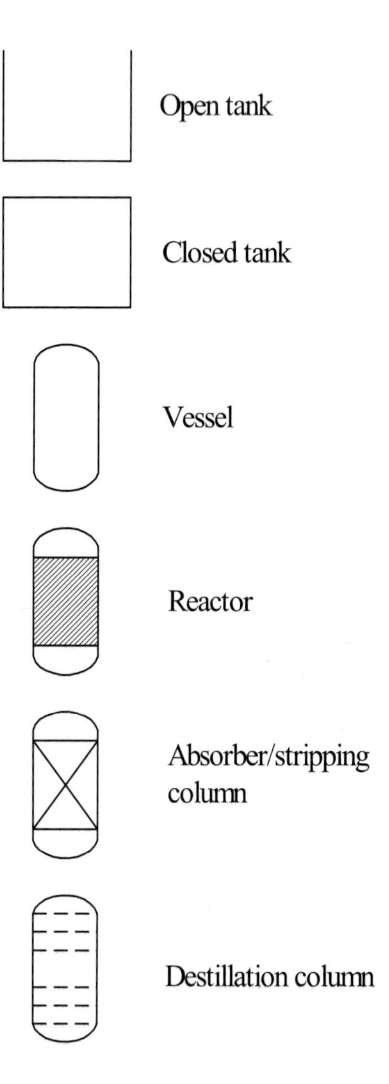

Figure A.6: Symbols of tanks and columns

Appendix B

The Laplace transform

In the ordinary chapters of this book the same symbol (letter) is used for the time-function, say $f(t)$, and the Laplace transform of $f(t)$, that is, $f(s)$. However, in this appendix a unique symbol, namely the corresponding capital letter, is used for the Laplace transform of $f(t)$ that is, $F(s)$, to avoid misunderstandings.

B.1 Properties of the Laplace transform

Below are several important properties of the Laplace transform relevant for this book.

Linear combination:

$$k_1 F_1(s) + k_2 F_2(s) \Longleftrightarrow k_1 f_1(t) + k_2 f_2(t) \tag{B.1}$$

Special case:

$$kF(s) \leftrightarrow kf(t) \tag{B.2}$$

Time delay:

$$F(s)e^{-\tau s} \quad \Longleftrightarrow \quad f(t-\tau) \tag{B.3}$$

Differentiation:

$$s^n F(s) - s^{n-1} f(0) - s^{n-2} \dot{f}(0) - \ldots - \overset{(n-1)}{f}(0) \quad \Longleftrightarrow \quad \overset{(n)}{f}(t) \tag{B.4}$$

Special case: Zero initial conditions:

$$s^n F(s) \quad \Longleftrightarrow \quad \overset{(n)}{f}(t) \tag{B.5}$$

Integration:

$$\frac{1}{s}F(s) \iff \int_0^t f(\tau)d\tau \tag{B.6}$$

Convolution:

$$F_1(s)F_2(s) \iff f_1(t) * f_2(t) = \int_0^t f_1(t-\tau)f_2(\tau)d\tau \tag{B.7}$$

Initial value theorem:

$$\lim_{s\to\infty} sF(s) \iff \lim_{t\to 0} f(t) \tag{B.8}$$

Final Value Theorem:

$$\lim_{s\to 0} sF(s) \iff \lim_{t\to\infty} f(t) \tag{B.9}$$

B.2 Transform pairs

Below are some useful Laplace transform pairs, $F(s) \iff f(t)$. The given time-functions are defined for $t \geq 0$, and it is assumed that $f(t) = 0$ when $t < 0$.

When taking the inverse Laplace transform using the transform pairs listed below, we usually need the linearity rules (B.1) or (B.2) in combination with the transform pair.

Note: You can "remove" factors of type $Ts + 1$ in $F(s)$ by setting $T = 0$. This corresponds to $e^{-t/T} = 0$ in the time-function.

$$F(s) = k \iff f(t) = k\delta(t) \quad \text{(impulse of strength or area } k\text{)} \tag{B.10}$$

$$\frac{k}{s} \iff k \quad \text{(step of amplitude } k\text{)} \tag{B.11}$$

$$\frac{k}{s^2} \iff kt \quad \text{(ramp of slope } k\text{)} \tag{B.12}$$

$$k\frac{n!}{s^{n+1}} \iff kt^n \tag{B.13}$$

$$\frac{k}{Ts+1} \Longleftrightarrow \frac{ke^{-t/T}}{T} \tag{B.14}$$

$$\frac{k}{(Ts+1)^n} \Longleftrightarrow \frac{k}{T^n(n-1)!}t^{n-1}e^{-t/T} \tag{B.15}$$

$$\frac{k}{(Ts+1)s} \Longleftrightarrow k\left(1-e^{-t/T}\right) \tag{B.16}$$

$$\frac{k}{(T_1s+1)(T_2s+1)} \Longleftrightarrow \frac{k}{T_1-T_2}\left(e^{-t/T_1}-e^{-t/T_2}\right) \tag{B.17}$$

$$\frac{k}{(Ts+1)^2s} \Longleftrightarrow k\left[1-\left(1+\frac{t}{T}\right)e^{-t/T}\right] \tag{B.18}$$

$$\frac{k}{(T_1s+1)(T_2s+1)s} \Longleftrightarrow k\left[1+\frac{1}{T_2-T_1}\left(T_1e^{-t/T_1}-T_2e^{-t/T_2}\right)\right] \tag{B.19}$$

$$k\frac{T_1s+1}{(T_2s+1)s} \Longleftrightarrow k\left[1+\left(\frac{T_1}{T_2}-1\right)e^{-t/T_2}\right] \tag{B.20}$$

$$k\left[\frac{s+d}{(s+a)(s+b)(s+c)}\right] \tag{B.21}$$

$$\Longleftrightarrow k\left[\frac{(d-a)e^{-at}}{(b-a)(c-a)}+\frac{(d-b)e^{-bt}}{(c-b)(a-b)}+\frac{(d-c)e^{-ct}}{(a-c)(b-c)}\right]$$

$$\frac{k}{\left(\frac{s}{\omega_0}\right)^2+2\zeta\frac{s}{\omega_0}+1} \Longleftrightarrow k\frac{\omega_0}{\sqrt{1-\zeta^2}}e^{-\zeta\omega_0 t}\sin\left(\sqrt{1-\zeta^2}\omega_0 t\right) \quad (0\leq\zeta<1) \tag{B.22}$$

$$\frac{k}{\left[\left(\frac{s}{\omega_0}\right)^2+2\zeta\frac{s}{\omega_0}+1\right]s} \tag{B.23}$$

$$\Longleftrightarrow k\left[1-\frac{1}{\sqrt{1-\zeta^2}}e^{-\zeta\omega_0 t}\cos\left(\sqrt{1-\zeta^2}\,\omega_0 t-\varphi\right)\right] \quad (0\leq\zeta<1)$$

where

$$\varphi = \arcsin \zeta \tag{B.24}$$

$$\frac{k\omega}{s^2 + \omega^2} \iff k \sin \omega t \tag{B.25}$$

$$\frac{ks}{s^2 + \omega^2} \iff k \cos \omega t \tag{B.26}$$

Bibliography

[1] J. G. Balchen, T. Andresen, B. A. Foss: **Reguleringsteknikk**, Tapir, 2003

[2] G. J. Blickley: **Modern Control Started with Ziegler-Nichols Tuning**, Control Engineering, 2. Oct. 1990

[3] **DS-R Toolbox for MATLAB** (David.Di.Ruscio@hit.no)

[4] G. F. Frankling, J. D. Powell, A. Emami-Naeini: **Feedback Control of Dynamic Systems**, Addison-Wesley, 1994

[5] F. Gerald & P.O. Wheatley: **Applied Numerical Analysis**, 6. Edition. Addison-Wesley, 1999

[6] F. Haugen: **Advanced Control**, Tapir Academic Publisher, 2004

[7] F. Haugen: **Dynamic systems – modelling, analysis and simulation**, Tapir Academic Publisher, 2004

[8] F. Haugen: **Tutorial for Control System Toolbox**, TechTeach, 2004. Web document on http://techteach.no

[9] Th. Kailath: **Linear Systems**, Prentice-Hall, 1980

[10] L. Ljung: **Reglerteori- Moderna analys- och syntesmethods**, Studentlitteratur, 1981

[11] L. Ljung: **System Identification**, Prentice-Hall, 1999

[12] O. Mayr: **Origins of Feedback Control**, The MIT Press, 1971

[13] C. Meyer, D. E. Seborg, R. K. Wood: **A Comparison of the Smith Predictor and Conventional Feedback Control**, Chem. Eng. Sci., Vol . 31, 1976

[14] S. K. Mitra: **Digital Signal Processing**, McGraw-Hill, 2001

[15] D. E . Seborg, Th. F. Edgar, D. A. Mellichamp: **Process Dynamics and Control**, Wiley, 1989

[16] F. G. Shinskey: **Process Control Systems**, McGraw-Hill, 1996

[17] S. Skogestad: **Simple Analytical Rules for Model Reduction and PID Controller Tuning**, J. Process Control, 2002

[18] O. J. Smith: **Closer Control of Loops with Dead Time**, Chem. Eng. Progr. 53

[19] The MathWorks: **System Identification Toolbox** (for MATLAB)

[20] J. G. Ziegler and N. B. Nichols: **Optimum Settings for Automatic Controllers**, Trans. ASME, Vol. 64, 1942, s. 759-768

[21] K. J. Åstrøm and T. Hägglund: **Automatic Tuning of PID Controllers**, Instrument Society of America, 1988

[22] K. J. Åstrøm and B. Wittenmark: **Adaptive Control**, Addison-Wesley, 1989

[23] K. J. Åstrøm and B. Wittenmark: **Computer-controlled Systems**, Prentice-Hall, 1990

[24] K. J. Åstrøm and T. Hägglund: **PID Controllers: Theory, Design, and Tuning**, Instrument Society of America, 1995

[25] K. J. Åstrøm: **Reglerteori**, Almqvist & Wicksell, 1968

Index